AF527809

PHYSIOLOGY OF PLANT CELLS

ADVANCES IN PLANT PHYSIOLOGY SERIES—1

Physiology of Plant Cells

D. K. Arora
Seema Gupta

Anmol Publications Pvt Ltd
NEW DELHI-110002 (INDIA)

Anmol Publications Pvt. Ltd.
4374/4B, Ansari Road, Daryanganj,
New Delhi-110002

Physiology of Plant Cells

First Edition 1996

ISBN 81-7488-235-9

PRINTED IN INDIA

Published by J. L. Kumar for Anmol Publications, Pvt. Ltd. New Delhi and printed at Chetna Offset Press, Delhi.

Preface

'Physiology of Plant Cells' offers a comprehensive account of the funaamental principles of plant physiology. It incorporates and organizes the information concerning Physiology of Plant Cells from relevant authentic sources. It is an humble attempt to present a correct. and precise account of the subject matter of this important branch of Botany which forms an integral part of the studies undertaken by the undergraduate and post-graduate students of this subject. The principles have been outlined in considerable detail alongwith this fundamental facts and theories that explain the life processes of plants. The main object of the present book is to provide a comprehensive and well illustrated account of the prescribed subject matter in a very simple and lucid language and concise manner yet—still maintaining its desired academic standard.

Most of illustrations have been taken from professional papers, standard books and journals. Subject matter has also been obtained from a large number of standard books, reviews and research papers by internationally renowned workers and established authors. That is why there can be no claim of originality except in the manner of treatment.

Though the authors have taken special care to present a current account, yet they are fully aware of their limitations, and the readers may come across the mistakes of various types. For all types of mistakes they extend their due apology.

The authors express their thanks to their friends and colleagues whose continuous inspirations have invited them to bring out this text.

The authors are thankful to Shri J. L. Kumar and other staff members of M/S. Anmol Publications Pvt. Ltd. for bringing out a handsome print of this set in a comparatively short time.

Constructive criticism and suggestion for improvement of the book will be thankfully acknowledged.

Author's

Contents

1

Plant Cell

Discovery of Cells

Leeuwenhoek ground the best lenses of his time, but many other eager explorers were also using microscopes during the latter half of the seventeenth century. It was a time not unlike our own : an age of intense conflicts and of rapid technical improvements that revolutionized man's view of his environment. Some fifteen years before Leeuwenhoek's first letter to the Royal Society, an Italian, Marcello Malpighi, had seen small blood vessels (later named *capillaries*) ; and in 1665 Robert Hooke, a secretary of the Royal Society, found interesting microscopic structure in cork (the bark of a Mediterranean oak) and in the stems of various plants. In the cork he observed neat rows of thick-walled compartments that reminded him of a honeycomb. Because of this resemblance he called the compartments "cells." This discovery turned out to be at least as important to biology as was Leeuwenhoek's.

The story of Hooke's "cells" closely parallels the story of Leeuwenhoek's "animalcules." Many men saw both of them during the following century and a half, but no one really understood their significance (Fig. 1.1).

Hooke himself found that in many living materials (the cork, of course, was dead) the cell was filled with a liquid material. Gradually attention shifted to this liquid content. One of the first to call attention to it was a Frenchman of great genius, Jean Baptiste Pierre Antoine de Monet, Chevalier de Lamarck. In 1809 Lamarck wrote : "Every living body is

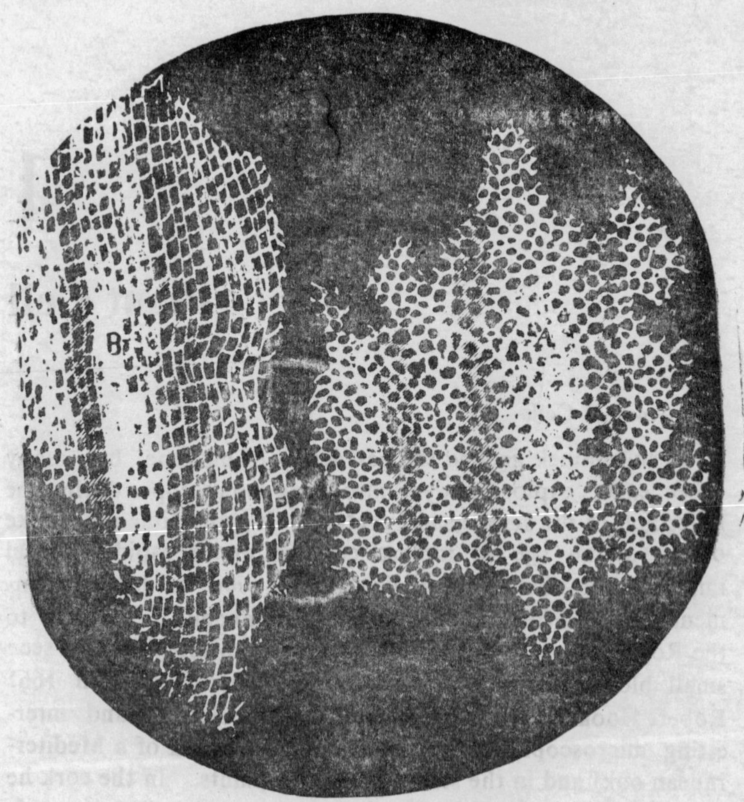

Fig. 1.1. Hooke's drawings of cork, as reproduced in his *Micrographia*, 1665.

essentially a mass of cellular tissue in which more or less complex fluids move more or less rapidly." Some years later Johannes Purkinje gave a name to the "complex fluids—*protoplasm*. And in 1831 Robert Brown reported that cells he had studied contained small bodies, which he called *nuclei* (plural of "nucleus"). So little was the cell understood at the time that the nucleus—now known to be important in many life processes—was mentioned merely in a footnote of Brown's work.

The Cell Theory

A few investigators were beginning to suspect that all large organisms are constructed of microscopic units of living matter.

But there was no science of biology at that time, so botanists had to come to this conclusion from the study of plant materials, and zoologists had to come to the same conclusion from the study of animal materials. In 1824 a Frenchman, Henry Dutrochet, wrote that "the cell is truly the fundamental part of the living organism." But because the boundaries of plant cells are much easier to see than those of animal cells, it was not until 1839 that the idea was fully developed with respect to animals. Two Germans, Matthias Schleiden, a botanist, and Theodor Schwann, a zoologist, did much to popularize the generalization. As Schwann wrote, "We have thrown down a great barrier of separation between the animal and vegetable kingdoms."

Thus, "cell," which once meant an empty space, came to mean a unit of living matter. Leeuwenhoek's "animalcules" were interpreted as the minimum cellular organization; a single cell. All other organisms were simply aggregations (groupings) of cells—very highly organized aggregations, to be sure, but all reducible to cell units. In the development of this generalization—the cell theory—we can see something of the international character of science.

But the cell theory did not lead to a great new era of research immediately. The detailed study of cells had to await another technological improvement—the development of dyes that could make visible their internal structure. This came with the great spurt in chemical knowledge that occurred in the 1850's and 1860's. Soon every life process was being linked to the cell. The cell came to be regarded not only as the unit of structure but the unit of function as well.

Already, in the decade after Schleiden and Schwann announced the cell theory, investigators began to find that cells normally and regularly come into being through the division of parent cells. Soon the idea was established that since the beginning of life there has been no break in the descent of living cells from other cells in the past, and that all of heredity and all of evolution must be embodied in cells.

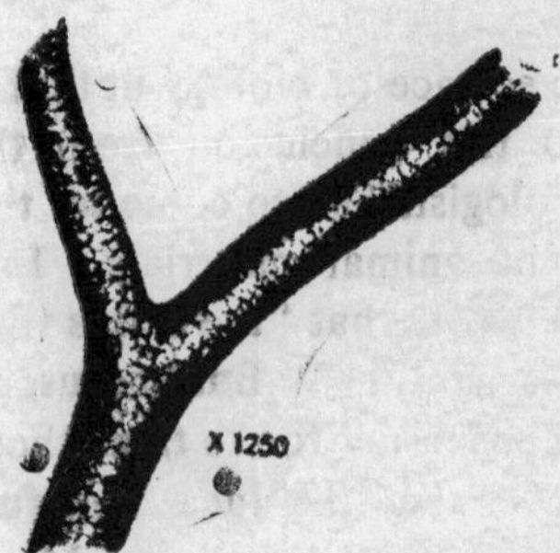

Fig. 1.2. Portion of a hypha of black bread mold (*Rhizopus*).

The cell theory may be summarized as three main ideas : (1) The cell is the unit of structure in living organisms. (2) The cell is the unit of function in living organisms. (3) All cells come from the division of pre-existing cells.

Difficulties in the Theory

Generalizations, theories, and principles are very useful things, but in biology we are on dangerous ground when using the word "all." Today an increasing body of opinion supports the idea that some organisms are *acellular* (without cells). And it is probably impossible to fit the viruses into the cell concept, but we may pass over them as being doubtfully alive. In some fungi, such as those of the genus *Rhizopus* (Fig. 1.2), the hyphae are made up of an undivided mass of protoplasm with many nuclei scattered about in it, whereas cells are usually thought of as containing one nucleus each. The protists of the genus *Paramecium* have a complicated kind of organization that is difficult to associate with the kind of structure usually found in cells ; and they, too, have more than one nucleus. It may be that these and a good many similar organisms once had cellular structure, which they have since lost through the disappearance of the barriers between cells. But the study of cell structure and function will be easier if we simply regard such organisms as exceptions.

Uses of the Term "Cell." The term "cell" has, in its different uses, sometimes included and sometimes excluded the cell wall. Early students of plant anatomy, seeing cells only when the walls were thick and visible at low magnification, called these structures "cells," "pores," "bladders," and described their contents as nutritive juices. The walls were the prominent part of these "receptacles." When the cell content was recognized as the essential part of the cell, attention turned to it as "the cell." Following this recognition of the protoplast as the fundamental unit, the term "cell" was often used to designate the protoplast alone. In the more common sense, however, the term implies protoplast plus wall—these together constituting the obvious unit of structure (Fig. 1.3), a unit for which there exists no term if "cell" is restricted in use to the protoplast alone. The term is further applied even to units of wall alone, as, for example, to the tracheid where the protoplast has disappeared, leaving within the walls a cavity, the *lumen*. The use of the term "cell" to indicate both protoplast and wall is desirable in view of the difficulties of expression otherwise involved and also in view of the intimate relationship between the two. The closeness of this relationship is not yet fully understood. Possibly the wall of a living cell is not merely a nonliving, external secretion of the protoplast, but is itself partly protoplasmic in nature. The use of the term to

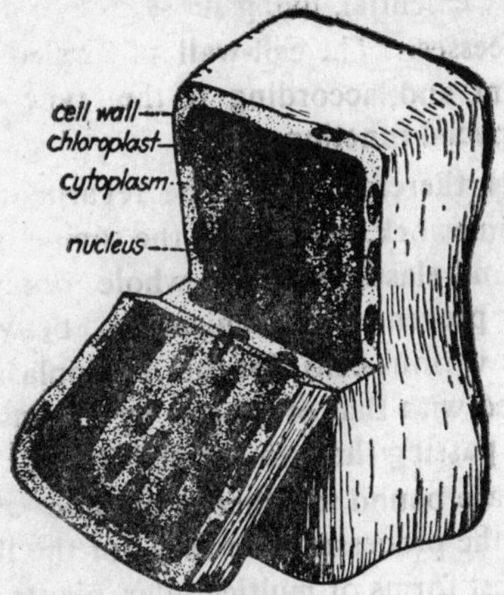

Fig. 1.3. Diagrammatic representation of mature mesophyll cell opened to show central vacuole (filled with cell sap), nucleus and plastids embedded in the peripheral cytoplasm ; wall thin. (*Based on Zea.*)

include both protoplast and wall is, of course, less strict than that underlying the frequently used definition of the cell as "an organized and more or less independent mass of protoplasm constituting the structural unit of an organism." Nevertheless, from the standpoint of anatomy, the term is better used in the looser and more comprehensive sense ; it is so used in this text.

Cellular Complexity of Plants

The cellular structure of plants was first recognised by Robert Hooke and recorded in his *Micrographia*, published in 1667. He observed a resemblance between the microscopic structure of ordinary bottle-cork and the cells of honeycomb, and introduced the term "cell" in this sense.

We know that the substance of the plant-body is not homogeneous, but, in all except the simplest forms, consists of aggregations or unions of microscopic structures, called cells, each *living* cell consisting of a tiny mass of a viscid substance called protoplasm, bounded by a distinct membrane, the cell-wall. These cells can be seen readily by teasing out the substance of a very ripe, mealy apple in water, and examining it under the microscope ; also in thin sections of stems, roots, and other parts of plants could be viewed under the microscope. The protoplasm is the essential, living substance of a cell, and the seat of vital processes. The cell-wall is formed by the activity of the protoplasm, and, according to the type of cell, is built up to perform special functions.

The cell-walls, therefore, are to be regarded as constituting a skeleton or framework, enclosing the living substance, and giving firmness and elasticity to the whole organism. They do not necessarily prevent communication between the living substance of the various cells, for the protoplasm of one cell may be connected with that of others by means of extremely delicate threads passing through the cell-wall. Thus the living cells of a plant are bound together in organic union, and carry on in harmony the processes necessary for the life of the plant.

In the simplest forms of multicellular plants, the organism consists of an aggregation of similar cells, all carrying on very much the same functions ; but in higher forms, correlated with the physiological differentiation which has taken place, there is what is called Histological Differentiation. In other words, many different kinds of cells, more or less definitely arranged in groups, or tissues, can be recognised, the form and structure of the cells in each tissue being related to the functions which they perform. This differentiation becomes more and more marked as we pass from lower to higher types. In the Thallophyta and Bryophyta, all parts of the organism consist of *living* cells, although these cells may present many different forms. For this reason these two divisions are distinguished as "cellular plants."

A considerably higher degree of differentiation is exhibited by the Pteridophyta and Spermatophyta. These plants are rooted in the ground with their shoots exposed to the air, and possess elaborate arrangements for the conduction and distribution of nutritive substances. In addition to typical living cells or varied form, other cells are modified and allow the rapid transmission of fluids from the roots to the leaves or vice versa. Such specialised cells are usually associated in groups forming strands or bundles which supply the whole plant body, and are generally arranged in a definite pattern, according to the plant organ in which they are situated. These strands. concerned with conduction of fluids, are called vascular bundles, and together they form the vascular system of the plant. The Pteridophyta and Spermatophyta are referred to as "vascular plants" because they possess these vascular tissues.

Other cells may be modified in such a way that they give added strength to those organs which require it. Individual cells, or groups of cells forming a tissue, may be wholly given up to this function. Their cell-walls are strengthened, their form altered during development, and they may lose their living protoplasm, in which case they are dead. They are advantageously placed so as, for example, to give support to a stem which not only has to stand erect and bear the weight of its leaves and branches, but has to withstand strains brought to bear on it by external influences such as wind. Such tissues are referred to as mechanical tissues.

As we proceed we shall see that certain tissues may perform more than one function. Thus a tissue may be vascular and also mechanical.

The primitive organism was doubtless unicellular or, in a sense, noncellular. The larger body of more advanced types is multicellular and the cells are limited by walls. The walls perhaps separate regions of activity of individual nuclei but certainly serve also for protection and support to the individual protoplasts and to the entire organism. Apparently, this walled condition is of great importance in large plant bodies both from the standpoint of mechanical stability and of physiological delimitation. Clearly, adaptation to a terrestrial and aerial habitat and the maintenance of a large body under conditions unfavourable to protoplasm have resulted in the specialization of the cell along various lines, especially those involving elaboration of the wall. Hence, among higher plants there are many kinds of cells with much variety in function, structure, arrangement, and with a great deal of complexity of wall structure. Variety, both structural and functional, in kind of cell ; variety in relation of cells to one another ; variety in arrangement of cells and of systems of cells with relation to one another and to the body as a whole—these give great cellular complexity to the more highly organized plants. Indeed, it may be said that in a general way the higher the plant, phylogenetically, the more complex its cellular structure.

Cell Arrangement. Regularity of arrangement of cells is brought about by successive divisions in the same plane. Such arrangement is characteristic of secondary tissues (those formed by a cambium or similar meristem) and is occasional in primary tissues, especially those of pith and cortex. But most primary tissues lack symmetrical arrangement and secondary tissues may lose this during development. Regardless of regularity or irregularity of arrangement and of the method of origin, a group of cells may be compactly arranged forming a close tissue continuous in one or more planes or they may be more or less free from one another, with the resulting formation of *intercellular spaces* between them. Such spaces vary in shape with the shape and arrangement of the surrounding cells, in continuity, in abundance, and in size from microscopically minute openings to very large spaces which render the tissue in which they lie loose, spongy, and light. Such large intercellular spaces are known as *air chambers*, *canals*, *lacunae*, etc. These terms are applied rather loosely—the larger spaces being called chambers and the much elongated ones canals. Spaces formed by the separation of walls followed by the retraction of the separated parts or by spatial movements of the cells are said to be *schizogenous* ; for example, the resin canal of *Pinus* and the spaces of the aerenchyma of *Decodon*. Others, developed by the destruction of cells formerly occupying the position of the cavity are called *lysigenous* ; for example, the oil cavities of citrus fruits. Some cavities are formed by a combination of these two methods and have been called *schizolysigenous*. Some protoxylem lacunae are of the last type. The intercellular spaces of a region may be without uniformity in size and shape or in arrangement of the cells surrounding them ; they may, on the other hand, form definite structural features of a tissue or organ because of the development about them of specialized limiting or supporting "walls," or layers of cells, such as the diaphragms of many aquatic plants. Intercellular spaces commonly form definite systems and constitute, perhaps, functionally conducting or aerating systems. Special types of intercellular-space systems form ducts and canals with cells regularly arranged about them forming a sort of epithelial lining.

Cell Shape. Because protoplasm is of semifluid nature, cells when free and independent tend to be spherical. But where young cells of the same type and essentially the same age lie together and grow in size, mutual contact and compression render them polyhedral with diameters nearly alike. Further growth toward individual or group specialization in relation to specific function leads to great variety in form—ovoid, ellipsoid, cylindrical, tabular, prismatic, lamelliform, fiber-like, and stellate or otherwise lobed or branched. There are, however, two chief types : the *subglobose*, or *polyhedral*, with diameters equal or only slightly different ; and the *elongate*, with one diameter many times that of the others. Transitional types are, of course, numerous.

Undifferentiated or unspecialized cells, in continuous mass, when surrounded and closely compressed by other cells of similar type and size, are on the average 14-sided. They approach in shape the orthic tetrakaidecahedron, a mathematically determined form for bodies of uniform volume and minimal surface that fill space without interstices. This form has 14 surfaces of contact—8 hexagonal and 6 quadrilateral. In plants, uniform cells of this shape are probably never formed. The closest approach is perhaps in parenchyma cells of pith that are arranged in columns. The surface cells of a mass of such cells have an average of 11 faces. Epidermal cells, for example, have about that number of contact faces. When intercellular spaces are present, the number of sides is less—the larger the spaces, the fewer the sides in contact with other cells. Where similar cells of different sizes lie together under these conditions, the larger cells have more than 14, the smaller less than 14 faces. The shape of newly formed cells is dependent upon that of the mother cell and upon the plane and frequency of division.

Cell Size. Cells vary greatly in size ; size, like shape, is in part related to function. Extremely small cells do not occur among the higher plants. Parenchyma cells serving the usual functions with normal protoplasm, have a transverse diameter of 0.01 to 0.1 mm. In pith and fleshy fruits, the diameter of parenchyma cells may reach 1 mm or more, and the cells then become readily visible to the naked eye. Fibers of wood and phloem range in length chiefly from 1 to 3 mm in the angiosperms, and from 2 to 8 mm in the gymnosperms. Cortical, pericyclic. and primary-phloem fibers are often much longer, and—partly because of their considerable length—are of particular economic importance, as, for example, those of flax and hemp. Fibers of excessive length—20 to 550 mm—occur in the Urticaceae and in certain monocotyledons. The largest cells known are latex cells of the type which form branching systems throughout a plant body. However, such cells are perhaps not individuals morphologically, since they are coenocytic and continue to grow almost indefinitely. Latex vessels consist of cells united ontogenetically in series.

Cell Development. All cells are formed from preexisting cells—or from nucleated protoplasmic masses—by division. Cell division is a complex process during which the nucleus and cytoplasm are each divided into two parts, usually equal and alike. In this division, the wall is not directly involved. In the newly formed cells, plasma membranes and intercellular substance are not readily distinguishable, and hence a visible plane of separation is at first lacking or scarcely detectable. But very early, new walls appear along this plane as delicate membranes. Cells, therefore, while still very young, are provided with walls. The small daughter cells soon enlarge to about the size of the mother cell.

Young cells are of more nearly uniform size and shape than mature cells and are simple in structure. Elaborate shape, complex structure, and very large size are not found among young cells. Growth of cells involves increase in size and the development of special shape and structure. Since the wall is present from a very early stage, both protoplast and wall are involved in these changes. As the active part of the cell, the protoplast initiates the changes, the wall being modified with the changing protop'ast. While changes in size and shape are taking place, the physical and chemical nature of the wall is usually such that these accommodations are readily made ; only after cell maturity is attained is a fully developed and unchanging wall present. In the maturing of the protoplast many changes occur. These are, briefly, the reduction of the proportionate size of the nucleus ; the development, usually, of less dense or less richly granular cytoplasm ; the appearance of vacuoles that increase in size, and, in most cases, ultimately fuse to form one large central vacuole which restricts the cytoplasm to a peripheral position adjacent to the wall ; changes in size and type of vacuole, where vacuoles are already present in the mother cell, for example, in cambium initials and their daughter cells ; the increase of plastids in size and number, and the development of special types of plastids. In the maturing of the wall there occurs increase in extent and thickness accompanied by complexity of form and changes in chemical and physical strurture.

Cellular Adjustment during Growth. Growth frequently brings about changes in the relationships of contact and position among neighbouring cells. The changes may involve all surfaces of a cell or be localized. In many cells the rate of growth of the wall is unequal in different areas, and in areas of greater or more rapid growth there are commonly greater changes in intercellular contacts. Development of elaborate or extreme shape and size—especially in such cells as large vessel elements, fibers, and some types of sclereids—brings about great changes. These may involve the deep penetration of cell tips or lobes between adjacent or nearby cells and into intercellular chambers. The steps in these changes are not fully understood. As new contacts are made by a penetrating tip and cell walls are separated, the intercellular cementing substance must be modified or removed. The plasmodesmata of the region are ruptured and the primary pit-fields and even partly developed pit-pairs are divided.

There is lack of agreement as to the manner in which these new wall contacts are made. *Gliding* (or *sliding*) *growth* has commonly been described as the basis of these changes. *Intrusive growth* and *symplastic growth* have recently been proposed as supplanting gliding growth in part or wholly. None of these three methods alone seems to cover the conditions met in all tissue and cell types.

By *gliding growth* is understood the slipping, during growth, of the wall of a cell over the wall of a contiguous cell along the surface of contact so that new areas of contact are made with the same cell and with nearby cells with which no contact previously existed. It has been generally assumed that real movement takes place in this type of growth, that the wall is increasing in area over more than the points of new contact, sometimes throughout, and that a sliding therefore occurs. Examples of apparent gliding growth are to be found (*a*) in the formation of new cambium initials with increase in the girth of the cambium – size, form, and position of the new cells are evidence of growth throughout the cell walls and of slipping and the making of new contacts ; (*b*) in the development from cambium derivatives of mature phloem and xylem elements—

nonmatching pit-fields in the phloem and pits without complementary pits in the xylem are evidence here. The development of the lobes or arms of some types of sclereids is an example of gliding growth of parts of the cell wall.

Gliding growth has also been assumed to cover extreme changes in cell contacts brought about by the forcing, under pressure, of cells into new positions and distorted forms. As an example of this, the distortion in developing secondary xylem of cell arrangement and form about enlarging vessel elements has been cited. In ring-porous woods such as that of *Quercus*, the rapid and very great increase in diameter of a vessel element results in compression, or stretching, even tearing apart, of the smaller cells nearby. These cells find new positions, and their new contacts must often involve large areas of their wall surfaces. A sliding of cells on one another is the obvious method of change. Continuing growth cannot explain so great a change of position. It has been claimed that these changes are the result of cell multiplication among the smaller cells, but no cell division is found at this stage and the addition of new cells in the area would not explain extreme distortion of form.

By *intrusive growth* is understood *differential* expansion of the wall—the increase of cell size locally, with the projection of the newly formed parts of the cell between adjacent cells or their extension into intercellular spaces. The parts of other cells with which there is new contact may or may not also be enlarging. In this type of change it is understood that no real slipping occurs ; the new contacts of a cell are those made by new areas of the wall as they form. The elongation of the apices of fibers, tracheids, and vessel elements has been called intrusive growth, but it is perhaps gliding rather than intrusive growth, or, more likely, it involves changes of both these types. An example of extreme continuation of intrusive growth is found in the latex cells of the milkweeds—Apocynaceae and Asclepiadaceae—whose rapidly growing apices maintain their position in apical meristems throughout the plant.

Most gliding and intrusive growth undoubtedly occurs while the cells are very young, but some apparently takes place after the walls are partly mature. In secondary xylem, walls of contiguous cells, already so mature that secondary walls and the beginnings of pits are present, may be split apart by the developing tips of adjacent partly mature cells. Such pits become functionless and are present in mature cells as *blind pits* (pits without complementary pits in the adjacent cell wall). Blind pits may even lead toward an intercellular space ; the matching pit may by found across the space. (In sections pit-pairs often appear to extend only part way through the wall ; these are usually parts of a pit-pair cut obliquely.) Intrusive growth may bring about a wrinkling of the wall of the intruding cell tip.

By *symplastic growth* is understood growth in size throughout a group of young cells with mutual adjustment in shape by all members of the group. The entire walls are growing. New shapes are formed and new spatial positions taken, but there are no new contacts and no movement either gliding or intrusive. The group of cells grows thoughout with their walls forming a framework that is continuously adjusted under pressure and tension. Differences in rate of division and rate of enlargement play a prominent part in symplastic growth. Growth of the derivatives of the initials in an apical meristem is an example of symplastic growth.

All three types of growth doubtless occur. Symplastic growth is characteristic of groups or masses of cells in early stages of growth. Evidence in support of the occurrence of typical gliding growth in many tissues is strong. Distinction between gliding and intrusive growth is doubtless one of degree rather than kind—the difference depends upon the definition of "tips" of the cell. If only a small part of the cell at the apex is growing and making new contacts with the walls of other cells, growth is intrusive ; if there is growth in regions other than the tips, there must be movement of the gliding type.

PROTOPLASM

Protoplasm is a highly complex mixture of some elements and compounds found in the bodies of living beings. The protoplasm is variously known as the *living matter, living substance* or the *physical basis of life*. It is the basic fundamental substance exhibiting all the vital processes of the cell. The protoplasm was first observed by *Corti* in 1772. In 1835 *Dujardin* a Frenchman described it as a soft and gumy substance and named it as *sarcode i.e. flesh*. *Purkinje*, a Bohemian physiologist (1839), was the first biologist, who gave the name *protoplasm* to this living substance. *Hugo von Mohl*, a German botanist, in 1839, also suggested the name protoplasm for the granular and viscous substance found in plants similar to that found in the animals. The above-mentioned German botanist popularized the word protoplasm as a name given to the living matter found in plants and animals. Protoplasm is the most complex and interesting substance. It is not to be though of a chemical compound but rather as very complex organized system. Protoplasm varies somewhat in its nature from cell to cell and from organism to organism, but basically it must be the same, as evidenced by its common manifestations of metabolism, growth, reproduction and by some other peculiarities.

PHYSICAL NATURE OF PROTOPLASM

Different workers have proposed different theories to explain the physical nature of protoplasm as given under the following heads :

1. *Granular theory*. This theory was propounded by *Altmann* in 1893. According to this theory, protoplasm consists of numerous tiny granules as shown in *Amoeba*, *Henle*, *Maggi*, etc., considered these protoplasmic granules as *plastidules*. *Altmann* recognized them as "elementary organisms", or *bioplasts* (or cytoplasts), Fig. 1.4 C.

2. *Alveolar Theory*. The alveolar nature of protoplasm was suggested by *Butschilii* in 1892. According to him, protoplasm consists of many suspended droplets or alveoli or minute bubbles, resembling the forms of emulsion Fig. 1.4 B.

3. *Fibrillar Theory*. This theory was put forward by *Flemming*. According to him, protoplasm consists of fibres embedded in the inner mass of matrix. The fibrillae are called

mitome or spongioplasm and ground substance is temed paramitome or hyaloplasm, Fig.1.4 D.

4. *Reticulate Theory.* This theory was proposed by *Klein*, *Cornoy* etc., suggesting that the protoplasm consists of a network or reticulum of fibres in the ground substance, Fig. 1. 4 A.

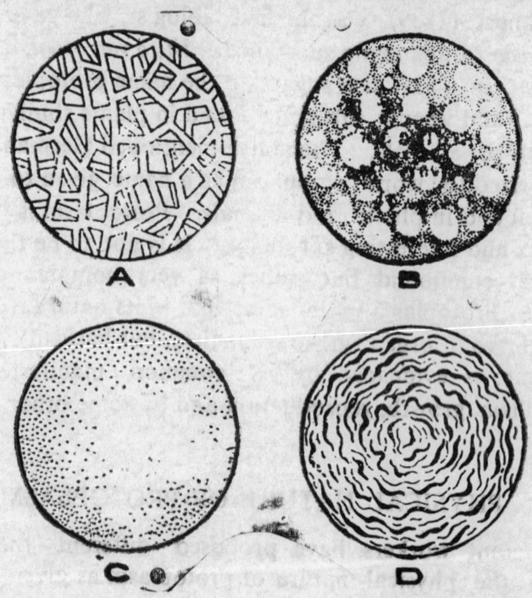

Fig. 1 4. Physical appearance of protoplasm. A. Reticular, B. Alveolar, C. Granular, D. Fibrillar.

5. *Colloidal Theory.* Proposed by *Wilson* in 1925. According to this theory the protoplasm having any of the aforesaid appearance is always a fluid-colloidal system having various chemical inclusions in *gel phase*. A gel is a semi-solid condition which presents jelly-like appearance. Here the molecules are held together by various bonds depends upon their nature and strength. By absorbing water the gel changes to more liquid-like phase. This is known as *sol* and the process is termed as *solation*. The sol can stream easily and by losing water again changes into get-state. Both these states of colloidal matrix are interchangeable according to the various physiological, mechanical and biochemical activities of the cell. This

represents phase reversal in the colloidal system. These sol-gel conditions of colloidal system are the basis for the mechanical behaviour of protoplasm.

PROPERTIES OF PROTOPLASM

1. *Cohesiveness.* The various particles or molecules of protoplasm are adhered with each other by forces, such as Vander Waal's bonds, that hold long chains of molecules together. These Vander Waal's bonds are weak and non-specific forces between non-polar groups of atoms. This property varies with the strength of these forces.

2. *Contractility.* This property is significant in various stomatal operations in plants. The contractility of protoplasm is important for the absorption and removal of water as they generally occur in protoplasm.

3. *Electrical charge.* Protein molecules repel each other because their overall charge is similar. All molecules are either positively charged or negatively charged. However, if the molecules approach one another close enough so that valency forces can act, then they may be attraced to each other.

4. *Precipitation.* Addition to certain amount of electrolyte in a colloidal suspension causes its dispersed particles to colloide, aggregate and finally to precipitate as suspension. For example, the addition of HCl to a colloidal system of arsenic sulphide causes precipitation.

5. *Viscosity.* The viscosity of the ground substance of the cell varies greatly. It may be as low as that of water, or may be very high in the gelating cytoplasm of pseudopodia of *Amoeba.*

6. *Streaming movement or cyclosis.* The protoplasm exhibits various sorts of streaming movements inside the cell. These have been studied in *Euglena, Volvox, Amoeba, Paramecium* etc. The movement involves only the localized portions with no visible changes in the protoplasm. No complete explanation of this has been given as yet. But it is seen that its rate depends upon the rate of cell metabolism. It is due to the fact that the energy is supplied by respiration.

7. *Amoeboid movement.* The amoeboid movement as exhibited by *Amoeba* and other protozoans involves the movement of entire protoplasm of the cell, where the cytoplasm moves as one mass carrying the various inclusions with it. It is due to the continuous change of gel to sol and sol to gel.

8. *Brownian movement.* It is characterized by the zigzag motion of suspended colloidal particles, occurring due to the bombardment of one particle or molecule by other. This type of movement of particles was first of all observed by *Robert Brown* in 1827 in the colloidal solution and hence such movements are known as Brownian movements. The higher the temperature, more rapid the movement and thus viscosity of cell is decreased. This means that high viscosity indicates a more gel-like state of protoplasm and low viscosity, a more sol-like condition.

9. *Tyndall effect.* Colloidal particles of protoplasm have the property of scattering light. When a beam of light is passed through a colloidal solution it becomes visible. This is the Tyndall effect. A colloidal solution of proteins in water shows a typical tyndall cone.

10. *Adsorption.* The tendency of particles, molecules or ions to adhere to the surface of certain solids or liquids is known as adsorption and is exhibited by the particles of colloidal system. The phenomenon helps the matrix to form protein boundaries.

11. *Biological properties.* Protoplasm has all the biological properties of a living organism. It is capable of nutrition, respiration, excretion, metabolism, growth and reproduction. It has the property of irritability, *e.g.*, it responds to stimuli like heat, light and chemicals. It also has the property of conductivity, *i.e.*, of conducting impulses produced by stimuli.

CHEMICAL NATURE OF PROTOPLASM

For the detailed study the chemical components of the cell can be classified as inorganic (mineral salts) and organic (proteins, carbohydrates, nucleic acids, lipids and so forth) substance. Although the most prominent constituent of protoplasm is water—the substance which gives protoplasm its characteristic structure is protein. Lipids are important in all membranes and carbohydrates serve as nutrient stores. The protoplasm of a plant or animal cell contains 75 to 85% water, 10-20% protein, 2-3% lipids, 1% carbohydrates and 1% inorganic material. The following table 1.1 gives approximate figures of the relative amounts of the main inorganic and organic compounds found in active protoplasm.

Table 1.1.

Substance	*Percent*	*Average molecular weight*	*Number of molecules in relation to protein*
Water	85	18	180
Protein	10	36000	1
DNA	0.4	10^6	—
RNA	0.7	4.0×10^4	—
Lipid	2	700	10
Other organic matter	0.4	250	20
Inorganic substances	1.5	55	100

Water. Water is the main component of protoplasm and occurs in large amount, about 85%. It serves as a natural solvent for other materials and also plays a major role in the metabolic activities, since physiologic processes occur exclusively in an aqueous medium. The water molecules also participate in many enzymatic reactions in the cell. No doubt, certain amount of water is formed in the cell as a result of metabolic processes, but this is insufficient to maintain the water balance of the body. Therefore, it must be and is supplied from the outside. Inside the cell, the water exists free, *i.e.* a solvant as well as bound, *i.e.* tied to polar groups of protein molecules by hydrogen bonds.

INORGANIC COMPOUNDS

A number of inorganic salts occur both in free as well as in the ionised state. The elements that occur in quantity in protoplasm are oxygen, carbon, hydrogen, and nitrogen.

Always present but in much smaller amounts are potassium, phosphorus, calcium, sulphur, magnesium and iron. These ten are generally referred to as *essential elements*. Chlorine and sodium are necessary components of most animal protoplasm but are apparently not essential for plant protoplasm. Copper, boron, iodine, magnese, zinc and several other elements which are present in all protoplasm but only in minute quantities, are called *trace elements.* This term must be used with care, because these elements are still important even though they are present only in very small quantities. The essential elements found in human protoplasm are listed below in Table 1.2.

Table 1.2. Showing percentage by weight to the elements in protoplasm.

Elements	*Symbol*	*Per cent*
Oxygen	O	65.04
Carbon	C	18.24
Hydrogen	H	10.05
Nitrogen	N	3.15
Potassium	K	1.60
Phosphorus	P	0.84
Calcium	Ca	0.25
Sulphur	S	0.20
Magnesium	Mg	0.04
Iron	Fe	0.01
Chlorine	Cl	0.27
Sodium	Na	0.26

The elements are almost always present in protoplasm in the form of chemical compounds rather than elements. Many of these compounds are inorganic salts, which are usually in solution. These salts have numerous functions. They serve as sources of elements to be built into other compounds, and some act as buffer in maintaining the proper acidity or alkalinity in the protoplasm. Because salts in solutions are electrical conductors, they also function in some way in connection with the electrical properties of protoplasm, however, this is not well understood at present.

ORGANIC COMPOUNDS

Proteins

Proteins are the framework of protoplasm. About 10% of protoplasm is protein. Proteins are linear polymers of high molecular weight. All proteins contain C, H, O and N, the presence of N distinguishing them from carbohydrates and fats. On an average proteins contain 16% nitrogen. Some proteins also have S in addition, and in a few protein P and other elements may be present. The molecular weight of proteins varies from about 12,000 daltons (bovine insulin) to several million. Cells contain a very large number of proteins. The number may vary from 1000—2000 in simplest bacteria to as many as 100,000 different proteins in human cells.

Protein are found in two form : Simple and conjugate.

The Simple Proteins are compounds, which on hydrolysis, yield exclusively alpha (α) amino-acids. In this group the most important proteins are the albumins (soluble in water and coaguable by heat ; the globulins (insoluble in water, but soluble in dilute salt solutions) ; the protamines, strongly alkaline components of the sperm cell (*e.g.* clupein, salmine, etc.) ; and the some what less basic histones which are found in many cell nuclei.

Conjugate proteins are a combination of a simple protein and another substance called the *prosthetic group*. Unlike the simple protein, the conjugate protein yields on hydrolysis alpha (α) amino-acids and an organic component. Nucleo-proteins are also included in conjugate protein group which play an important role in the cell and whose prosthetic groups are the nucleic acids ; the glyco-protein (mucoproteins) in which protein is combined with a carbohydrate ; the lipo-proteins, which are combinations of proteins with higher fatty acids ; and the chromoproteins, a very widely distributed group, included a series of substances of great biological importance which are characterized by their particular colours. Among the chromoproteins are haemoglobin, in which the globulin is combined with an iron-porphyrin compound, and the haemocyanin which occurs in the blood of various inverte-brates and in which copper is found in a position similar to that of iron in haemoglobin. A series of respiratory enzymes (cytochromes, flavoprotein etc.), belong to the chromoprotein group.

SIGNIFICANCE OF PROTEINS

Proteins are the most significant compounds in the body of living organisms because of the following physiological roles performed by them :

1. *Proteins confer specific individuality to living structures, organisms and the species.* This is known as biochemical individuality—Protein individuality is well illustrated by blood transfusion experiments. The blood of one species of animals cannot be transfused into the circulation of animals of other species. Not only this, the blood proteins of different individuals of the same species are not identical.

2. *Proteins play an important role in life processes.* These can combine with both acid and bases and can neutralize excess of any one of them in the protoplasm. These check the accumulation of H^+ or OH^- ions in the protoplasm.

3. *Proteins have a high chemical potentiality* and, therefore, readily react with many substances, therefore, these act as enzymes and catalyse almost all the life processes of living organism.

4. *Proteins form boundaries of living cells* and their components and thus do not permit an easy access to and from the cell cytoplasm.

5. *Protein differentiation of organs and animals.* How only twenty amino acids lead to the formation of thousands of proteins is a complex but interesting question. For easy understanding of this phenomenon the twenty amino acids are linked to 20 alphabets of English language and the formation of a protein with the construction of a word. For example, an insulin molecule formed of 16 amino acid molecules is similar to a word formed of sixteen letters. In English dictionary 26 alphabets combine to form at least 2,00,000 words. Similarly 20 amino acids can give approximately same number or even more varieties of proteins. Such a wide variety of proteins will naturally bring wide range of differentiation.

AMINO ACIDS

There is a common plan of construction for the thousands of kinds of proteins in living systems. The 20 kinds of naturally-occurring amino acid monomers are strung together in unbranched, linear polymer chains of proteins. These are the 20 amino acids specified in the genetic code that is universal to all organisms. Some other kinds of amino acids are also found in cells, but they are either degradation products or residues that have been modified from one of the 20 commonly occurring amino acids after this latter has been inserted into the polymer chain. Hydroxyproline is a major amino acid constituent of collagen in connective tissue, but proline residues are initially included in the protein and are converted to hydroxyproline after polymerization. Hydroxyproline is not one of the encoded amino acids, but proline is. Many proteins contain fewer than 20 kinds of amino acids. The relative proportions and the absolute numbers of the amino acid repertory vary from one protein to another, as a reflection of the specific in formation in genes, which are the blueprints for protein construction.

CARBOHYDRATES

The carbohydrates (*L.*, *carbo*-carbon or coal, *Gr.*, *hydro*-water) are the compounds of the carbon, hydrogen and oxygen. They form the main source of the energy for all living beings. Only green parts of plants and certain microbes have the power of synthesizing the carbohydrates from the water and CO_2 in the presence of sunlight and chlorophyll by the process of *photosynthesis*. All the animals, non-green parts of the plants (*viz.*, stem, root, etc.), non-green plants (*e.g.*, fungi), bacteria and viruses depend on green parts of plants for the supply of carbohydrates.

Carbohydrates can be classified into :

Monosaccharides. These are simple sugars with an empirical formula $Cn\,(H_2O)n$. These may be of trisoes if they have 3 carbons, tetroses, if they have four carbons, pentsoses type, if they have five carbons, or hexoses, if they have six. Glucose or grape sugar and fructose or fruit sugar are examples of hexoses type. And ribose and deoxyribose (the sugars present in nucleic acids) are examples of pentose. The monosaccharides are readily soluble in water and can pass across the membrane and can also be crystalized. This soluble nature is changed, when these are stored in the animal as well as in plant tissues. In such cases it is usually polymerized to an insoluble form such as glycogen (animal starch) in animals and starch in plants.

D Glucose	*D fructose*	*D Ribose.*
H	CH_2OH	H
C=O	C=O	C=O
HCOH	HOCH	HCOH
HOCH	HCOH	HCOH
HCOH	HCOH	HCOH
CH_2OH	CH_2OH	CH_2OH

Disaccharides. The disaccharides are having the empirical formula $C_{12}H_{22}O_{11}$ and are readily soluble in water and these can also be crystalized and passed through the membrane. The commonest and best known disaccharide is sucrose or table sugars which is stored in such plants as the sugar cane and the sugar beet. On hydrolysis yields monosoaccharides or simple sugars.

Polysaccharides. These are formed by the condensation of many molecules of monosaccharides with a corresponding loss of water molecules. Their empirical formula is $(C_{16}H_{10}O_5)n$. They yield molecules of simple sugars on hydrolysis. Biologically important polysaccharides are the starch, the glycogen, and the cellulose. The former two are the reserve substances in cells of plants and animals respectively where as the cellulose is the characteristic structural element of the plant cells. The starch $(C_6C_{10}O_5)n$ is a mixture of two long polymer molecules, *i.e.* linear amylase and branched amylopectin. It has no structural importance in animals. The glycogen $(C_6H_{12}O_6)n$

is generally called as the starch of animal cells. It is a polymer composed of many molecules of glucose and thus represent an important reserve of energy in the body. Though found in numerous tissues and organs, but contained in liver cells and muscle fibres in greatest proportions. It forms a colloidal solution in the protoplasm. The cellulose is composed of hundred of monosaccharides. It has no structural importance in animals except the tunicates which have a wall of cellulose. It takes part in the formation of cell wall and also a series of other structures which form the supporting skeleton in plants.

Besides cellulose, plant tissues contain several other structural components such as xylem, alginic acids (in algae), pectic acid, etc.

Significance. Thus we can conclude that carbohydrates play three main roles in the cells. These are :

1. They furnish the most readily available fuel.
2. They are important articles of storage.
3. They have a minor role in furnishing part of the structural material of atleast one part of the essential environment of living cells.

LIPIDS

Lipids comprise a rather heterogeneous class of compounds which are sparingly soluble in water, but show considerable solubility in organic solvents like ether, chloroform, benzene, hot alcohol and petroleum ether. Their physical properties indicate the hydrophobic nature of the hydrocarbon structure, although lipids may also contain some hydrophilic groups.

Lipids have several biological functions.

(1) Certain lipids *e.g.* triglycerides are storage compounds for the reserve energy of the body, *e.g.* lipids of the fat body and subcutaneous fat. (2) Lipids are important components of cell membranes. The plasma membrane of eukaryote cells is composed primarily of lipids and proteins, along with carbohydrates in some cases. (3) In nerve fibres the myelin sheath contains lipids which act as electrical insulators. (4) As components of subcutaneous tissue lipids are important for

insulation of heat. (5) Some hormones *e.g.* steroids are lipids. (6) Some lipids are important as vitamins. (7) Lipids also occur as components of some enzyme systems.

Lipids can be classified as follows :

(I) Simple lipids, (II) Compound lipids.

I. Simple lipids

Simple lipids are the esters of the alcohols or the triglycerides containing fatty acids and alcohols.

$$\text{Triglyceride} \xrightarrow[\text{HO}]{\text{Lipase}} \text{Glycerol}+3\ \text{Fatty acids.}$$

The constituent fatty acids may be *saturated, e.g.*, palmatic stearic acids or *unsaturated, e.g.*, oleic, linoleic, arachedonic and clupanadonic acids. The lipids of the matrix of animal cells contain fatty acids, *viz.*, palamatic, stearic palmitoleic, oleic, linoleic and inolenic acids. The simple lipids of the matrix are as follows :

1. *Natural fats.* The natural fats are the naturally occurring fats which occur in animal and plant cells as stored food substances.

2. *Waxes.* Waxes are esters of higher fatty acid with alcohols other than glycerols. In the human body the commonest waxes are the cholesterol esters. They are most abundant in the blood, suprarenal glands, the gonads and the sebaceous glands of the skin. The natural waxes are solid at ordinary temperature and are not so readialy hydrolysed as fats. Three common waxes are as follow :

(*a*) Beeswax (*b*) Lanoline (*c*) Spermaceti.

II. Compound lipids

These are the lipids which on hydrolysis yield other compounds, in addition to the alcohols and acids. They are of following types :

1. *Phospholipids.* Phospholipids are soluble in both water and fat solvents, and, therefore serve an important role in binding both types of compounds together. The contain in addition to fatty acids and glycerol ; phosphoric acid and a nitrogenous base.

Three types of phospholipids have been indentified :

(*a*) *Lecithin.* When placed in water, it swells up and readily forms an emulsion. Due to presence of acid and basic groups, it combines both with bases and acids. Thus these are important as a structural material in the cell membrane since by its hydrophilic group it maintains the continuity between the aqueous outside and the aqueous inside of the cell; yet fat soluble material dissolves in it and enters the cell because of its hydrophobic group.

(*b*) *Cephalin.* It is similar to lecithin. Cephalin and lecithin occur together in the tissue. *Howell* claims that the substance thrombokinase which initiates blood clotting is identical with cephalin. It is found specially in nerve tissues, egg yolk, etc.

(*c*) *Sphingomyelin.* These are derivatives of sphingosinol. It is also found in nerve tissues, egg yolk etc.

(*d*) *Plasmalogen.* It is another type of phospholipid present in the brain, liver and muscle tissue.

2. *Glycolipids.* These are compounds of carbohydrates, fatty acids and sphingosinol which contain nitrogen, but no phosphoric group. They are also called as glycosphingosides. They are specially abundant in the brain.

3. *Gangliosides.* Found in nerve cells, spleen and red blood cells.

4. *Lipoprotein.* These are the complexes of lipids and proteins. Lipoprotein is present in cell membrane, egg yolk etc.

5. *Sulpholipids.* They are sulphuric esters of sphingosine cerebronic acid and galactose. Sulpholipids are observed in white matter of brain and to a lesser extent in liver, kidney, salivary gland, testes, etc.

III. Steroids

Steroids are not true fats. These are characterized by OH group thus are called sterols and are of highly importance for the body. Male and female sex hormones, adrenocortical hormones, vitamin D, cholesterol are major steroids. They are found in cell membranes and other cell structure containing lipids.

IV. Carotenoids (Lipochromes)

These are red or orange cell pigments, soluble in organic solvents, but insoluble in water. This group includes the *Carotenes* in carrot and grass, *Xanthophyll* in leaves, vitamin A, egg yolk pigment. These are synthesized by the mitochondria of the cell.

NUCLEOTIDES AND NUCLEIC ACID

Nucleotides are involved in at least two major cellular functions : (1) they are monomeric units from which DNA and RNA polymers are constructed, and (2) they act as agents in certain energy-transferring reactions during metabolism. A mononucleotide is made up of one nitrogen-containing organic base, one pentose sugar, and one phosphate residue derived from phosphoric acid. When there is no phosphate group, the sugar-base combination is called a *nucleoside* (Table 1.3). For this reason, *nucleotides* (phosphate-sugar-base) are also called nucleoside phosphates. Nucleoside mono-, di-, and tri-phosphates contain one, two, or three phosphate groups respectively.

The nitrogenous bases commonly found in nucleic acids and their nucleotide building blocks are derivatives of purine and

Table 1.3. Showing nomenclature of nucleic acids and their constituent units.

Base	*Nucleoside*	*Nucleotide*	*Nucleic Acid*
Purines :			
Adenine (A)	Adenosine	Adenylic acid	RNA
	Deoxyadenosine	Deoxyadenylic acid	DNA
Guanine(G)	Guanosine	Guanylic acid	RNA
	Deoxyguanosine	Deoxyadenylic acid	DNA
Pyrimidines :			
Cytosine(C)	Cytidine	Cytidylic acid	RNA
	Deoxycytidine	Deoxyadenylic acid	DNA
Thymine(T)	Thymidine	Thymidylic acid	DNA
Uracil(V)	Uridine	Uridylic acid	RNA

pyrimidine. The commonly occuring purines *adenine* and *guanine* are found in both DNA and RNA, as in the pyrimidine compound *cytosine*, the second kind of pyrimidine in DNA is *thymine*, while its demethylated form, *uracil*, occurs in RNA.

Since each kind of nucleic acid contains one unique pyrimidine, it is convenient to study synthesis and activity of DNA or RNA using isotopically-labelled precursors containing one or the other of these bases. Usually the nucleosides uridine or thymidine, or their nucleotide forms, are added to the biological system under study.

The only difference in the pentose sugars of nucleotides is the presence of a hydroxyl group at carbon atom 2 of D-ribose in RNA, but a hydrogen at carbon-2 of 2-deoxy-D-ribose in DNA monomers and polymers. This seemingly simple difference is partly responsible for profound differences in stabilities, pairing potential, and functions of DNA and RNA.

Polynucleotides of both DNA and RNA varieties are built from mononucleotides that are linked covalently via phosphodiester bridges between the 3′ position of one unit and the 5′ position of the next. Since there is no restriction on the vertical sequence of adjacent mononucleotides in either DNA or RNA, a considerable variety of molecules is possible even though only 4 kinds of nucleotide monomers (one for each of the four kinds of bases in the combination with sugar and phosphate) are used in polymer construction. The theoretical variety is calculated as 4^n, where 4 is the number of different kinds of nucleotides, and n is the number of monomers in the polymer. For a molecule made of only 75 monomeric units, as in some of the smallest RNAs, there may be 4^{75} different arrangements of the constituent units. Each arrangement theoretically constitutes a molecule of different specificity. Where the average gene may include about 500 nucleotides in a DNA sequence, 4^{500} different sequences are theoretically possible and, therefore, that many different and specific genes. Despite the apparently meager number of monomer types an astronomically high number of possible genes can be constructed.

Such variety can easily account for all past and present life forms.

The DNA Double Helix. DNA molecules usually have regular helical configurations because most DNA molecules consist of two *complementary polynucleotide strands.* The two strands are held together by *hydrogen bonds* between complementary pairs of purines and pyrimidines. Adenine always binds with thymine while guanine always binds with cytosine. This repeated hydrogen bonding within the double helix structure and bonding between virtually all the surface atoms in the sugar and phosphate groups with water molecules serve to stabilize the structure (Fig. 1.5).

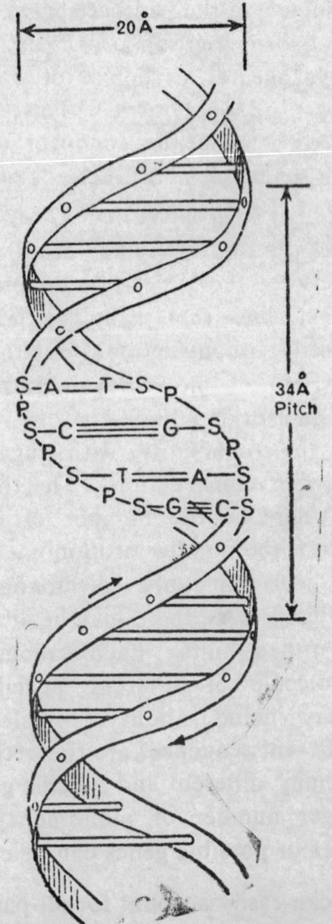

Fig. 1.5. DNA and its paired bases.

Since the purine-pyrimidine pairs are found in the center of the molecule, their flat surfaces can stack on top of each other and thereby limit their contact with water. In double-helical molecules, a regular structure is possible because the complementary base pairs are exactly the same size. Single poly-uncleotide chains could not have a regular backbone structure because pyrimidines are smaller than purines, which would cause the angle of helical rotation to vary with the sequence of bases.

DNA double helix molecules are very stable at physiological temperatures because : (1) disruption of the double helix breaks hydrogen bounds and brings hydrophobic purines and pyrimidines into contact with water, which is energetically unsatisfactory ; and (2) there are many weak bonds within the DNA molecule, arranged so that most of them cannot break without many others breaking at the same time. Even though some hydrogen bonds may be broken by thermal motion, hydrogen bonds in the rest of the molecule remain intact and the molecule does not fall apart. In fact, when held together by more than ten nucleotide pairs the double helices are quite stable at room weak bonds is stability of molecular shape, in proteins as well as in nucleic acids. At abnormally high termperatures there is more frequent breakage of weak bond, which becomes less stable as temperatures rise above physiological levels. Once a significant number of weak bonds have been broken, a protein or nucleic acid molecule usually loses its original form and changes to an inctive or denatured form.

The Young Cell

Cells arise from pre-existing cells by a process of cell-division. All plants begin life as a single cell. They may remain unicellular, or become multicellular as the result of repeated cell-division. As we pass from the lower to the higher plants, along with the increasing size and complexity of the plant-body, we find that cell-division becomes localised. All living cells are potential dividing cells, but cells which have become specialised for various functions do not usually divide unless stimulated to do so, as by wounding. In the higher plants dividing-cells are found, particularly at the tips of roots and stems, which are consequently termed the growing points. In members of the Bryophyta and Pteridophyta a single apical cell of definite form may repeatedly divide. In the Spermatophyta two or more layers of cells may be concerned. Such cells are said to be meristematic, and the growing points primary meristems. Secondary meristems may arise later in living cells which have originated from the primary meristems.

The growth of the multicellular plant-body depends, in the first place, on cell-division, then on the enlargement of the cells thus produced, and their differentiation and adoption of various functions.

In order to examine the structure of a meristematic cell it is necessary to cut thin sections, *e.g.*, of a root-tip, and examine them under the high magnification of a compound microscope. Transverse and longitudinal sections are required to give a complete picture of the individual cell. In longitudinal section it may appear more or less rectangular, and in transverse section it may be rectangular or several-sided Fig. 1.6. In the solid, therefore, it is shaped something like a

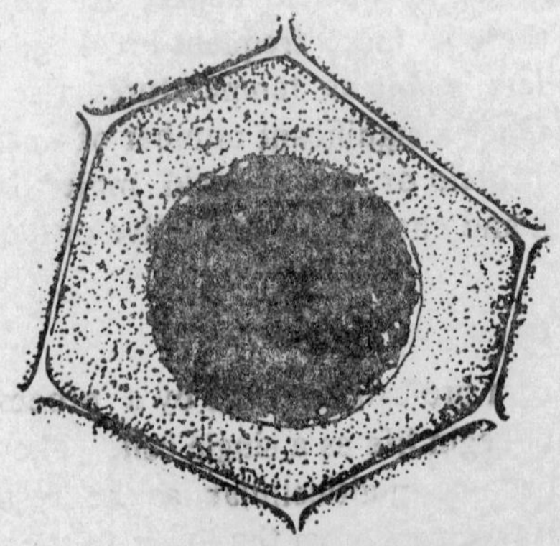

Fig. 1.6. Meristematic cell, resting condition.

prism or truncated pyramid with four or more sides, the short axis parallel to the axis of the root.

The meristematic cell is bounded by a thin cell-wall and contains the protoplast. The latter consists of a relatively large central nucleus and the surrounding cytoplasm. Within the cytoplasm are suspended specialised protoplasmic bodies called plastids or chromatophores, and non-living inclusions in the form of minute granules and droplets.

The Protoplasmic Substance

Observation of the living cell indicates clearly that the *protoplasm* is not homogeneous, but its physical nature is still rather a matter for conjecture. There is little doubt, however, that it is a colloidal system, usually with many of the properties of a liquid, being in fact a liquid (sol) in which are dispersed either liquid or solid particles which are either aggregates of small molecules, or are very large melecules. The particles dispersed in the continuous liquid phase of the protoplasm are extremely minute, and often only just visible with the most powerful microscope. Many of the properties of the protoplasm are the outcome of its colloidal nature.

Chemically, the dry substance of the protoplasm appears to consist mainly of proteins with appreciable amounts of fats and fatty substances and of carbohydrates, and smaller amounts of other organic compounds and of mineral matter. Analysis of the protoplasm always, of course, includes any substances present in it at the time, that is, both living and non-living protoplasmic inclusions. But it must be remembered that methods of chemical analysis bring about the death of the protoplasm, and so the analysis is really of *dead*, and not living protoplasm.

The analyses do, however, show the presence of substances (fat-like compounds and proteins) which it is known are capable of forming the disperse phase in a colloidal system. The evidence of chemical analysis and of observation on the physical properties of the protoplasm suggest that in the protoplasm we have a continuous phase of water containing proteins and fatty substances, and dispersed in this are granules (solid or liquid) of protein, either alone or united with other substances, the whole being of the consistency of a slightly viscous liquid.

The surface layer of the protoplasm differs from the inner part, and in a system such as that outlined above we might expect to find in the outermost layers a concentration of fatty substance. Any such surface layer is certainly of a complex nature and of vital importance, as its properties will govern the entry of substances into, and their exit, from the protoplasm.

The Nucleus

The *nucleus* of the meristematic cell may be observed in the resting condition or in any of the various stages of nuclear division. Recent work on nuclear structure appears to show that two types of resting nucleus occur in different plants. One type is known as the "solid" nucleus, the other as the "pro-chromosomal" or vesicular type. It must be borne in mind, however, that detailed study of nuclear structure is only possible, so far, as the result of first "fixing," then sectioning the meristematic tissue. The sections have to be "stained" with certain dyes. Technique, as well as treatment, plays an important role in this process : choice of fixative and of stain, amount of staining and, above all, resolving power of the microscope at high magnifications are all fundamental to successful interpretation of what is observed. Moreover, certain observed phenomena may be the result of treatment, and recorded observations must make all due allowance for these.

The prochromosomal type of nucleus (Fig. 1.7A) consists of a vesicle of sap around the periphery of which is the whole of the stainable material of the nucleus, the chromatin, included in relatively small bodies called prochromosomes. One or more nucleoli are also present.

The "solid" type of resting nucleus (Fig. 1.7B) is devoid of free sap and, after fixation, has the appearance of a uniform meshwork of chromatin filling the whole nucleus except for the space occupied by one or more nucleoli. The latter, at

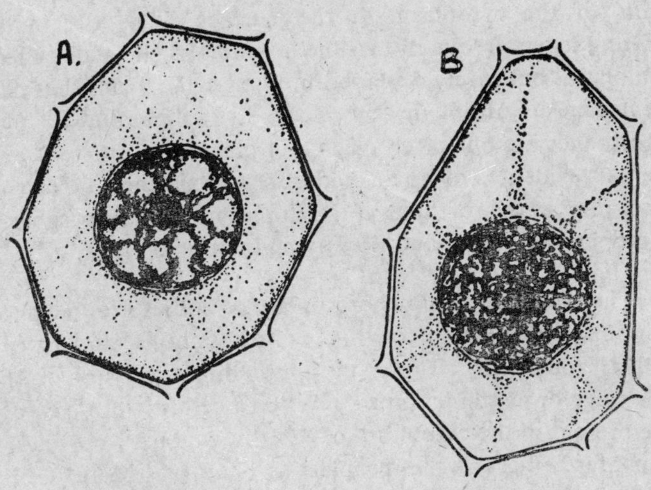

Fig. 1.7. Two types of resting nucleus.
A. Prochromosomal. B. Solid.

least in the early stages, are fluid. This type of nucleus is found in members of the Liliaceae, in *Vicia* and *Osmunda* (the royal-fern). The chromatin meshwork is frequently referred to as a "reticulum," composed of "karyotin."

The nucleus plays an important part in governing the metabolic processes which go on within the cell, resulting in cell-division, growth and differentiation. The nucleus also contains the hereditary characters which are passed from parent to offspring.

The Cytoplasm

In mature living cells the *cytoplasm* (that part of the protoplasm distinct from the nucleus) forms a lining to the cell-wall. Internally the cytoplasm is in contact with the contents of the vacuole, and externally with the cell-wall and the liquid with which it is saturated. The cytoplasm is not homogeneous. On physical grounds we should expect the surface layers of the cytoplasm to differ in composition and structure from the innermost parts, a knowledge of the laws of surface tension leading us to expect an accumulation of the fat-like components of the cytoplasm at the surface. The differentiated limiting layers frequently visible in animal cells and called the ectoplasm, but rarely observed in plant cells, form the plasma membranes. Although the specially differentiated plasma membranes are but rarely visible in plant cells, there is experimental evidence of their widespread existence. The innermost part of the cytoplasm (endoplasm) contains specialised protoplasmic bodies and non-living inclusions.

The most conspicuous protoplasmic bodies are the plastids. In young meristematic cells they are minute, and have been called proplastids. They multiply by direct division, either at this stage or at cell maturity. When a cell divides, proplastids are present in the cytoplasm of each daughter cell, and as the daughter cell grows the proplastids mature. Mature plastids are classified somewhat arbitrarily according to their colour into two chief types ; chromoplasts, which are coloured, and leucoplasts which are colourless.

Ribosomes

These are tiny, darkly staining particles present either attached to ER or freely distributed in cytoplasm. Ribosomes were first discovered in 1941 by Claude, who called them *microsomes*, Palade (1956) named them ribosomes. Usually the membrane bound ribosomes (membrane piece+ribosomes) are called the microsomes. Ribosomes are very significant as the sites of protein synthesis. They are called engines used by cell in protein synthesis. In cells which are highly active in protein synthesis, simple ribosomes get transformed into clusters called *polyribosomes*.

The number and concentration of ribosomes are related to the metabolic state of the cell.

Measurement of the sedimentation coefficient by ultracentifugation has revealed the presence of two sizes in ribosomes. In bacteria they have a coefficient of 70s whereas in eukaryotic cells they have about 80s.

Ribosomes are generally spheroidal and have a size of 250Å×150Å. Each spheroid has two sub-units, a larger (140-160Å) and a smaller (50-60Å) one. The smaller one is attached in the form of a cap to the larger sub-unit.

Chemical analysis of ribosomes reveals the presence of RNA and proteins.

Lysosomes

Discovered in 1955 by de Duve, these are the cytoplasmic organelles containing enzymes that bring about *lysis* (break-down). These are more abundant in animal cells that in plant cells. Popularly these are called '*suicide bags*', since they bring about the breakdown of the cell. In shape, lysomes are highly variable. In size, they measure from 0.25-0.8μ Lysosomes are enclosed in a single layered membrane which is made up of lipoproteins. Inside the membrane is present a homogenous dense granular matrix. The matrix is composed of the lytic enzymes, *e.g.*, acid phosphatase, ribonuclease, β galactosidase etc.

Various theories have been put forward to account for the origin of lysosomes. The most important theories are :

(1) They develop from extracellular pinocytic vesicles ;

(2) They develop as cytoplasmi inclusions containing digested material.

Recent evidences seem to point out that lysosomes may be derived from the golgi complex.

Centrosome (centriole)

Centrosomes are usually present close to the nucleus and are two in number. They are intimately associated with the cell division. Their presence in the plant cell is rather doubtful.

Centrosome is made up of a rounded cylinder of about 3000-5000Å in length and 1500-1800Å in diameter. Each centrosome consists of a stack of nine fibres arranged around a central axis. Each fibre is comprised of three secondary fibres. All these fibres are interconnected and are enclosed in a thick membrane.

Plastids

The *plastids* are differentiated bits of protoplasm—"organs," or areas, of metabolic activity associated with particular functions. They have a limiting, apparently semipermeable membrane and complex internal structure. They are often coloured and usually conspicuous. In size they are small, and generally many occur in a cell (Fig. 1.8). They are variable in shape, but rounded types are most common. Spherical, ovoid, diskoid, granular, rod-like plastids all occur frequently. Large plastids of peculiar shape are present in many of the algae and rarely elsewhere. They may be found in all living cells of a plant and probably are present in every cell in its early stages of development. Later they become restricted to certain cells and are abundant only in those which have specialized functions, such as photosynthesis, storage, and colour manifestation.

Origin. Plastids are present in large numbers in young meristematic cells where they are minute, the smallest being at the limit of microscopic visibility. At this stage, when they are known as *proplastids,* they are rounded bodies that do not resemble plastids. As the cell grows, proplastids multiply freely and mature plastids gradually develop. Increase in number by division continues at all stages but less frequently in mature plastids. Probably plastids arise only from preexisting plastids.

Types of Plastids. Several fairly distinct types of plastids occur, but all are of similar fundamental nature. They fall into two-chief classes : coloured plastids, *chromoplasts*, and colourless plastids, *leucoplasts*. Green, chlorophyll-bearing chromoplasts are known as *chloroplasts* and are commonly set apart as a third major type because of their important function in food manufacture. This leaves the term "chromoplast" to cover all coloured plastids except chloroplasts, and this is its general use.

Chloroplasts. The chloroplasts of the higher plants are mostly uniform in size and chiefly flattened-ellipsoid or disk-like in shape. They are numerous—from a few to very many in a cell—and small, averaging 5 micra in diameter. They multiply by constriction, and may at other times change shape, appearing as though semiliquid. The chlorophyll is wholly or largely restricted to numerous small granules known as *grana*. Apparently these deeply coloured bodies are commonly so arranged and so closely packed that the plastid appears structurally homogeneous. Sometimes an obscure layering can be seen. Chloroplasts of shade leaves are somewhat larger than those of leaves in full light on the same plant and their chlorophyll content is possibly greater per unit volume.

Chromoplasts. Nongreen chromoplasts range in colour through yellow, orange, and yellow-red. The colour is given chiefly by xanthophyll, carotin, and carotinoids. Chromoplasts show great variety in shape but are chiefly irregular; granular, angular, acicular, and forked types occur (Fig. 1.8). The irregular and sharp-pointed shapes are believed to be caused in part by the presence of the coloured substances, especially carotin and carotinoids, in crystalline form, as in the root of *Daucus* (Fig. 1.8A). The functions of these plastids are obscure. They are associated with colour in flowers and fruits but occur also in other regions such as roots. Chromoplasts commonly represent transformed chloroplasts, but may orm directly from small leucoplasts.

Leucoplasts. The term "leucoplasts" covers various types of colourless plastids. The early stages of all plastids are culled leucoplasts but such young plastids are best called *proplastids*, and only mature colourless plastids should be called leucoplasts. Like chromoplasts, leucoplasts vary in shape; extreme forms are rod-like. They change shape readily and are probably always highly plastic. They are concerned with food storage and doubtless have other unknown functions. The type of leucoplast associated with starch-grain formation in storage regions is known as *amyloplast*. Leucoplasts related to the formation and storage of oils and fatty substances are known as *elaioplasts*. These are at least in part amyloplasts.

that vary in function at certain times. The leucoplasts of hairs and of other epidermal cells are probably degenerate or dormant plastids of other types.

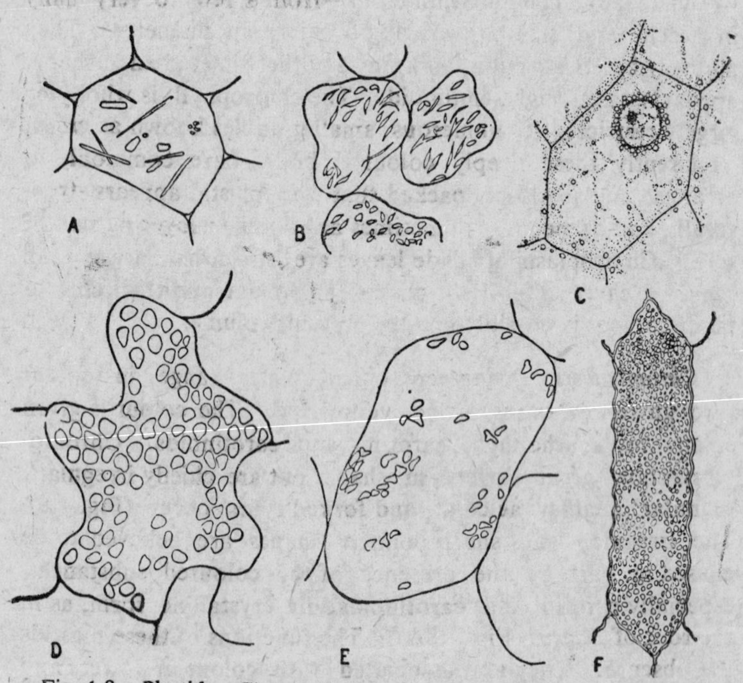

Fig. 1 8. Plastids. Chromoplasts : *A*, in cortex cell of root of *Daucus* ; *B*, in pulp cells of fruit of *Arisaema* ; *D*, in cell of petal of *Forsythia* ; *E*, in pulp cells of fruit of *Lycopersicon* ; *F*, in cell of corolla of *Taraxacum*. Leucoplasts ; *C*, in young endosperm cells of *Zea*.

That all plastids are alike in nature is clear from the readiness with which one type is transformed into another. For example, the chloroplasts of young fruits and of developing petals may become the chromoplasts of the ripe fruit and of the mature flower respectively ; the leucoplasts of a potato tuber become chloroplasts on exposure to light.

Occurrence of Plastid Types

Chloroplasts may occur in any part of a plant that is exposed to the light ; they also occur in some tissues that are apparently without light, such as the wood of many Rosaceae and Ericaceae, and in embryos and endosperm, as in the seeds of some citrus fruits. Red and yellow chromoplasts likewise may occur in any organ of the plant, and their presence is not related to the presence of light. They are chiefly to be found, of course, in flowers and fruits. Leucoplasts occur mostly in parts not exposed to light. Further statements concerning the distribution and the function of plastids are to be found in the discussion of various tissues, tissue systems, and organs, such as collenchyma, the cortex, the leaf, and the petal.

Many protoplasmic bodies smaller than plastids occur in cytoplasm ; these are known as *chondriosomes* and *mitochondria.* Some of those known as chondriosomes are doubtless proplastids. The nature and distribution of these smaller cytoplasmic structures lie outside the field of the present treatment.

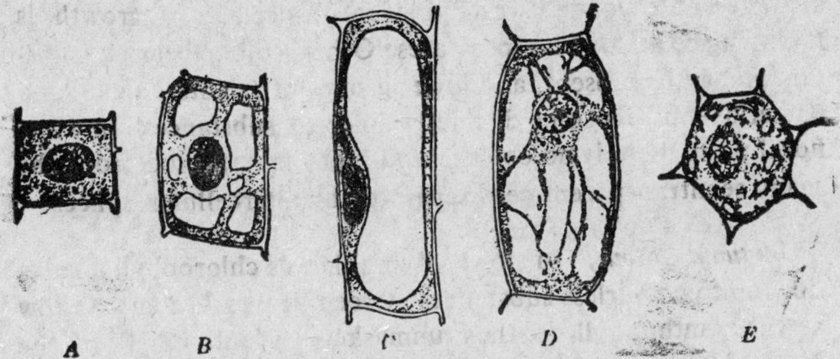

Fig 1.9. *A—C*, diagram of a plant cell in three successive stages of development : the vacuoles increase in volume and fuse and the cytoplasm becomes limited to the parietal region. *D.* cell of stamen hair of *Tradescantia*, indicating direction of streaming movements in the cytoplasmic strands. *E.* parenchyma cell from cortex of *Polygonella*, showing nucleus, plastids, and scanty cytoplasm.

Colour in Cells. Substances that give colour to cells are found chiefly in the plastids and the cell sap. The cell wall, cytoplasm, and nucleus are commonly nearly colourless. The green colouring substance of chloroplasts is *chlorophyll*, in two forms known as *a* and *b*. With the chlorophyll, but masked by it, are yellow pigments, *carotinoids.* Included in the carotinoids are carotins, carotenes, and xanthophylls. These substances give the yellow colour of "golden" varieties of plants (where the chlorophyll content is low) and of leaves variegated with yellow. Leaves variegated with white have little or no colour in the plastids of the pale areas. Abundant carotinoids (sometimes in crystalline form) in chromoplasts give yellow, orange, and yellow-red colours to flowers and fruits. Another group of pigments, the *flavones*, are water-soluble and colour the cell sap. In some genera, for example, *Verbascum*, they give yellow colour to corollas. The *anthocyanins*, oxidation products of flavones, also water-soluble, colour cell sap and give red, purple, and blue colour to many flowers and fruits. A reddish, anthocyanin-coloured cell sap, together with the green colour of the chloroplasts, gives the bronze or purple colour to the leaves of purple-leaved forms of many plants. Anthocyanins are also present in young twigs and leaves, especially if growth is taking place at low temperatures. Other colour-bearing substances may be present, and lower groups of plants—algae and bacteria—possess some distinctly different substances. In white floral parts there is no colour, and light is reflected from the many semitransparent cells separated by intercellular spaces.

Autumn Colour. In slowly dying leaves chlorophyll breaks down into colourless substances. Green leaves become yellow as the xanthophyll is thus unmasked. Xanthophyll in the disorganizing chloroplasts in this way gives much of the yellow colour seen in autumn foliage. Reds and purples are cell-sap colours, resulting from oxidation of flavones, and are especially brilliant when formed in the presence of sugars and abundant sunshine. These cell-sap colours, in combination with some persisting chlorophyll, xanthophyll, browning cell walls, tannins, and less common colour-giving substances, give great variety in autumn colour. The changes from green foliage are not caused by frost, although frost may hasten the process. Many plants whose leaves colour brilliantly in cool temperate climates show equally bright colour when grown in frostless climates. Dying branches of trees often show full "autumn" colour in midsummer.

Vacuoles and Cell Sap. The cytoplasm of mature and of many young cells contains one or more cavities known as *vacuoles.* Conspicuous cavities usually develop as the cell matures, beginning as one or several small cavities and enlarging and fusing until in the mature cell there is usually one large central vacuole which restricts the cytoplasm to a thin layer lining the wall (Fig. 1.9). The nucleus, plastids, and most inclusions are contained in this peripheral layer. Frequently strands of the cytoplasm extend from the outer layer as irregularly anastomosing branches through the central vacuole ; under these conditions the nucleus may occupy a median position in the cell. The vacuoles and their liquid content, the *cell sap*, constitute the cell *vacuome.*

The cell sap is a watery, nonprotoplasmic liquid consisting of water with various substances in solution—inorganic salts, carbohydrates, proteins, amides, alkaloids, pigments, etc. These substances may be mineral nutrients, elaborated food, waste products, or substances of unknown relation to metabolism. In the physiology of the cell the significance of these substances dissolved in the cell sap is not well understood.

We have noticed already that the cytoplasm of meristematic cells contains non-living inclusions. Some of these are in the form of minute droplets. Each droplet is a small vacuole filled with cell-sap. These small vacuoles enlarge and finally coalesce, so that ultimately one large vacuole is formed. The enlargement of the vacuoles is accompanied by the entry of water into the cell. Therefore it will be evident that with the entry of water into the cell, the size of the cell will increase, and, in fact, the cell attains a volume many times greater than the volume of the original meristematic cell. During this process of cell enlargement by vacuolation the nucleus and cytoplasm do not greatly increase in size, and in the fully vacuolated cell the cytoplasm forms a parietal layer which lines the cell-wall internally, and a number of strands which extend from this parietal layer across the central vacuole. The nucleus may be suspended in the centre of the vacuole by these cytoplasmic strands, or it may be embedded in the parietal layer of cytoplasm. This layer of cytoplasm is sometimes referred to as the primordial utricle, and besides the nucleus it contains plastids, and often various non-living inclusions as well. When cell enlargement is proceeding, the cell-wall is tightly stretched and becomes increased in area, whilst its thickness is maintained or even increased by the deposition of cellulose (generally), formed by the activity of the living protoplasm.

The cell-sap which fills the vacuole consists of an aqueous solution of various inorganic and organic compounds. Inorganic ions are always present, especially those resulting from the dissociation of the nitrates, sulphates and phosphates of certain alkaline and alkaline-earth metals, and carbon dioxide in solution is also present.

A number of sugars have been isolated from the cell-sap of various plants. Glucose (dextrose) and fructose (laevulose), monosaccharides with the empirical formula $C_6H_{12}O_6$, are especially common, small amounts of glucose being found in most leaves which are exposed to light, and larger amounts in many fruits. These sugars readily reduce Fehling's solution and ammoniacal silver nitrate, and they share this reducing property with a disaccharide, maltose ($C_{12}H_{22}O_{11}$), which is sometimes present in the sap of leaf cells. Sucrose, or cane sugar ($C_{12}H_{22}O_{11}$), is a non-reducing disaccharide widely distributed in plants, being abundant notably in the root of the sugar-beet and the stem of the sugar-cane. These form the commercial sources of "cane sugar."

A complex, soluble polysaccharide, inulin, occurs in the cell-sap of the tubers of dehlia and Jerusalem artichoke and in the root of the dandelion. Although soluble in water inulin is insoluble in alcohol, and hence when inulin-containing tissues are preserved in alcohol, the inulin is precipitated in the cells in the form of sphaerites (sphere crystals).

Other carbohydrates of less common occurrence in the cell-sap include dextrin, mannitol and various pentosans, whilst soluble pectin, which is believed to be a compound of certain sugars with other substances, sometimes occurs in appreciable amount.

Glycosides are also of common occurrence in the cell-sap. These may be regarded as compounds of glucose, generally with aromatic substances, so that when the glycoside is decomposed, and the aromatic substance set free, an aromatic smell is discernible. An example is amygdalin in the bitter almond. This, when decomposed, gives rise to glucose, benzaldehyde, and hydrocyanic acid, and it is to the benzaldehyde particularly that the characteristic odour and taste of bitter almonds is due. Indeed, artificial almond essence consists mainly or wholly of synthetic benzaldehyde. These glycosides are decomposed by enzymes, and generally the appropriate decomposing enzyme is present in the same tissue, but not necessarily in the same cell, as the glycoside. Thus, in bitter almonds the enzyme emulsin splits up the amygdalin. The glucose set free by the decomposition of the glycoside can, in all probability, be utilised by the plant in its metabolism.

Mucilaginous compounds occur in cell-sap, too, and it is to these compounds that the slimy character of many cell-saps is due. Mucilages are found abundantly in the cell-sap of many bulbs, *e.g.* onion, bluebell, and in the leaves of many succulent plants, and, as we shall see later, are also formed from cell-walls. Mucilages swell or dissolve in water, but are insoluble in alcohol. They are polysaccharide in nature and, when hydrolysed, may yield sugars which reduce Fehling's solution. Reagents in common use for testing for mucilages are solution of ruthenium red, alkaline solution of corallin, dilute solution of methylene blue and chlor-zinc iodide.

Tannins may be present in cell-sap. Treated with ferric salts they give either a blue-black or green colour. They are complex substances, and are classified according to their behaviour when treated with boiling dilute sulphuric acid. They are soluble in water and alcohol and are best examined after extracting, but if it is desired to examine them *in situ*, the plant tissue containing them should be sectioned dry and mounted direct in the reagent, *e.g.* dilute solution of ferric chloride.

Anthocyanin pigments are frequently present in cell-sap. These pigments as a rule give blue or red (or intermediate) colours to the cell-sap, and are the cause of many floral colorations ; whilst a similar type of compound is responsible for the red coloration of the root of the garden beet.

The cell-sap is often acid in reaction, either because of the presence of free organic acids, or their acid salts. Malic, citric and tartaric acids are all of common occurrence, and the acidity of many fruits is due to the presence of these acids in the cell-sap. Oxalic acid, too, is of general occurrence, but it is usually present, not as the free acid, but in the form of its insoluble calcium salt, and as such is usually regarded as a waste product of cell metabolism, and of no further use to the plant, whereas the other organic acids mentioned may be used in respiration or in other ways.

Calcium oxalate occurs in a crystalline form, and the crystals may be of various types, the commonest belonging to the monosymmetric (monoclinic) system. These crystals are widely distributed in plant tissues in the form of:

(1) Single crystals, *e.g.* square prisms, double pyramids, rhombohedra, etc. Sometimes these are so large as to fill the cell, distorting the cell-wall to their shape. The cells containing these crystals may be scattered throughout a tissue, as in Quillaia bark or Quassia wood, or they may be arranged as a crystal sheath around bundles of sclerenchymatous fibres as in Cascara bark and Liquorice root, or vuscular bundles as in Senna leaf. Included under single crystals are needle-shaped (acicular) ones named raphides, usually associated together in bundles in enlarged parenchymatous cells containing mucilage. Such bundles of raphides are particularly met with in Monocotyledons.

(2) Druses, cluster or rosette-crystals, which are spheroidal groups of tetragonal crystals, frequently built up around an

organic core. Large druses are present in *Rheum*, and minute ones, associated with aleurone grains, in the endosperm cells of fruits of Umbelliferae.

(3) Crystal-sand, masses of micro-sphenoidal crystals packed into a cell. Examples are seen in leaves and roots of *Atropa belladonna.*

Crystals of calcium oxalate dissolve in any of the mineral acids without effervescence ; they are insoluble in dilute acetic acid, even on warming. Dilute hydrochloric and acetic acids are therefore employed in microchemical tests for calcium oxalate. By this means it is distinguished from calcium carbonate, which dissolves in dilute acetic acid.

Calcium carbonate occurs only infrequently in plants ; it is best known in those curious structures called cystoliths, which consist of a peg-like ingrowth of the cell-wall into the cavity of epidermal cells of *Ficus* leaf. In and around this protrusion calcium carbonate is deposited. Cystoliths are found mainly in the members of the families Moraceae and Acanthaceae, and may be stalked or sessile.

Soluble nitrogenous compounds may be present in the cell-sap. We have mentioned already that inorganic nitrates may occur. Ammonium compounds may be present too, and also relatively simple organic compounds of the nature of amino acids and amides, and some of these may represent steps in protein synthesis.

Besides the proteins which form a part of the living protoplasm, non-living proteins may occur either dissolved in the cell-sap or in the form of crystal-like bodies, crystalloids. In many seeds, especially those containing oil, the vacuoles contain large amounts of dissolved proteins, and as the seed dries out and ripens these proteins may be transformed into *aleurone grains.* Each grain has a membrane, and this encloses a mass of protein in which bodies are embedded. One of these, the larger, is the crystalloid, which consists of protein. The other globoid bodies, which are smaller than the crystalloid, consist of protein combined with phosphates. Treatment with solvents resolves the aleurone grain into its components. Cold water dissolves away the ground matter of the grain, leaving the crystalloid and globoids inside the membrane. Dilute caustic potash slowly dissolves the membrane and the crystalloid, leaving the globoids unaffected.

Fats and *oils* are frequently present in plant cells, especially in certain seeds. They occur as droplets or globules, and may be present either in the cell-sap or protoplasm, or both. Fats and oils are compounds of glycerol and various fatty acids, mainly oleic, palmitic and stearic acids, and they often form an important kind of food reserve. Similar chemically, fats and oils differ only in their consistency, fats being solid at ordinary temperatures, and oils liquid. Fats and oils found in plants are generally liquid, but there are a few exceptions. Cocoa butter, from cocoa "bean," and the fat from the seeds of the nutmeg both have high melting points and are solid at normal temperatures, but in the plant they are probably present, not as solid fat particles, but in the form of an emulsion. Microscopically, fat globules may be distinguished by their reactions with certain stains. They stain red with tincture of alkanna, Soudan III, and Scharlach R.

We have by no means exhausted the possible constituents of the cell-sap, but have dealt briefly with the most important ones. Other substances sometimes present include various organic basic substances such as the purine bases and the alkaloids. The functions of these in the plant are not clear, but the alkaloids at least are of considerable therapeutic value.

Resins are found also, but are of limited distribution, being confined to special cell or ducts in certain plants, and odorous ethereal or essential oils are produced by some plants. These essential oils are usually mixtures of complex organic compounds such as alcohols, esters, ketones, aldehydes and hydrocarbons, not all of which are sweet smelling. Essential oil of lemon, for instance, consists mainly of inert hydrocarbons, whilst the smell and taste are due chiefly to a small amount (about 5%) of the aldehyde citral which it contains.

Resin also is not itself a simple chemical substance, but a mixture of substances insoluble in water and containing resin acids, resin esters and resenes, all of which are complex substances of high molecular weight. The relative proportions of these compounds vary in different resins. Colophony contains a high proportion of resin acid (abietic acid). Myrrh also is rich in resin acids. Benzoin, Storax, Balsams of Peru and Tolu, and Asafoetida consist mainly of resin esters and free aromatic acids. Sandarac and Mastic contain a considerable proportion of resenes.

Thickening and Alteration of the Cell-Wall

We have already seen that in the young cell the cell-wall is thin and composed mainly of cellulose deposited on a middle lamella, probably consisting of calcium pectate. As the cell increases in size, usually by vacuolation, fresh cellulose, built up by the activity of the living protoplasm, may be incorporated in the cell-wall in two ways : ***intussusception***, where the new cellulose particles are deposited between those of the existing cell-wall so that the latter grows in area ; and *apposition*, in which the cellulose particles are laid down on the surface of the original wall, by which means the latter is thickened. It is probable that both methods may go on simultaneously, the former predominating during active enlargement of the cell, the latter when the cell-wall is increasing in thickness.

The thickened cell-wall will have the same composition as the young cell-wall, and will possess the same or very similar physical and chemical properties, but often, besides being thickened, the cell-wall becomes altered by impregnation with substances other than cellulose. It may be cutinised, lignified become mucilaginous, or be impregnated with mineral matter such as silica.

Hemicelluloses

On the original cellulose cell-wall we may get deposited hemicelluloses. Whereas cellulose is built up from glucose molecules, hemicellulose (sometimes called reserve cellulose) is built up from glucose and the molecules of other sugars such as fructose, galactose and mannose. Hemicelluloses are deposited on the cell-walls especially in some seeds, *e.g.*, nux vomica, lupin, date and coffee, where they represent reserve food-materials which will be used when the seed germinates. During the growth of such cell-walls the cytoplasm maintains intimate contact with the surface of the cell-wall, and also through its thickness, by means of delicate threads, which form protoplasmic connexions between contiguous cells. These *plasmodesma* can be demonstrated, *e.g.*, in the collenchyma of cherry-laurel leaves, in the thick-walled cells of the endosperm of nux vomica, where they are more or less evenly distributed, and of date-stones, where they are limited to pits.

Cuticularisation

Cuticularisation is due to the formation of a waxy substance called *cutin*, which forms a continuous layer, called the *cuticle*, over the external walls of epidermal cells. The cuticle is probably secreted through the epidermal cell-wall by the activity of the living protoplasm. It is almost impermeable to water, and hence helps to protect the cells below it from excessive loss of water. It also gives firmness to the cell-walls. The cuticle varies in thickness, being thicker in plants which grow in dry, sunny situations ; its surface may be smooth, giving the glossy appearance to leaves, as in many evergreens, and to fruits, such as apples and tomatoes, or it may be slightly ridged, appearing finely striated under the microscope, as in leaves of *Atropa belladonna*.

Cutin may also impregnate the cell-wall, in which case it may affect the outer walls only, but sometimes also the lateral walls, of the epidermis.

Wax

Covering the cuticle there may be a layer of wax, giving a matt or glaucous surface, *e.g.* to leaves, and forming the "bloom" of fruits such as grapes and plums, Occasionally the wax layer is quite thick, as on the leaves of the wax palm (*Copernicia*), and can be utilised commercially as vegetable wax.

Suberisation

This is due to the formation of a fat-like substance called *suberin*. Cork cells have suberised walls. It is the middle layer of the cell-wall which is suberised, and thus rendered almost impermeable to water.

Cutin and suberin are stained yellow by iodine solution, and yellow or brown by Schulze's solution. They are not acted on by sulphuric acid.

Lignification

Lignification is due to the deposition in the cell-wall of a number of substances collectively known as *lignin*, which become intimately mixed with both the middle lamella and cellulose layers of the cell-wall. The mode of origin of the lignin is obscure, but lignification of the cell-wall is usually accompanied, or followed, by the death of the protoplasm in the cell. Hence cells with lignified walls are generally dead. Lignification confers increased strength and rigidity on the cell-wall, and at the same time reduces its elasticity. It leads to profound changes in the chemical reactions of the wall which, if impregnated with lignin, is stained a bright yellow with aniline sulphate or chloride, yellow or brown with iodine, yellow with Schulze's solution, and bright red with phloroglucin and hydrochloric acid.

Delignification can be accomplished by prolonged boiling of lignified tissues, *e.g.*, wood, in solutions of calcium bisulphite or caustic soda under pressure, or by alternate treatment with chlorine and sodium hydroxide. The cellulose wall remains. Cellulose wadding is delignified gymnospermous wood.

Mucilaginous Cell-Walls

When dry, these are hard and horny, but when moistened they absorb water, soften and swell up. The cell-walls of seaweeds are generally mucilaginous, hence their slimy character. Mucilaginous cell-walls are found in the seed coats of many seeds, and in the specialised mucilage cells found in certain plants such as slippery elm bark (*Ulmus fulva*), and marshmallow root (*Althaea officinalis*).

Chemically the mucilages are regarded as substances formed by the union of many and diverse sugars (*e.g.*, glucose, arabinose, xylose, etc.), together with complex acidic substances. Similar to mucilages in composition are *gums*, which often result from an alteration of the cell-wall, either naturally or as the result of injury. The process is known as gummosis, and results in the complete loss of cellular structure. Tragacanth exudes from the stems of *Astragalus gummifer* when they are incised, and Gum Arabic from species of *Acacia*, especially induced by wounding. Both plants are members of the family Leguminosae.

Mineral Substances

Mineral substances are sometimes deposited in cell-walls. High percentages of ***silica*** are found impregnating the walls of the epidermal cells of grasses, sedges, horsetails (*Equisetum*). etc., sometimes so completely that, if the tissue is burned, a complete siliceous skeleton of the cells is left behind. This impregnation of the cell-wall may help to confer rigidity on the tissues and possibly afford protection against small animals. Certain small marine plants (diatoms) have a siliceous wall. A number of marine Algae, and a few freshwater plants have a cell-wall which becomes encrusted with *calcium carbonate*, whilst isolated crystals of calcium oxalate, and, less frequently, of calcium carbonate, may occur in the cell-walls of certain land plants.

Ergastic Substances. Many kinds of solid particles, organic and inorganic, as well as such substances as oils, gums, resins, etc., are frequently present within the protoplast, either in the cytoplasm or in the vacuole. These represent, like the dissolved substances, food products, such as starch and aleurone grains ; waste products, such as crystals ; and other substances of doubtful or unknown function, such as rubber, mucilages, tannins, latex, alkaloids. Gums, resins, tannins, etc., are often present in the lumina of nonliving cells ; for example, in the heartwood of *Sequoia* and mahogany (*Swietenia*), and in cork cells of many trees. Some of these substances, for example, starch, occur in most plants ; others are characteristic of certain large or small groups of plants, and are lacking in others.

Crystals. Crystals of various chemical nature occur freely in plant cells—of these, salts of calcium, chiefly calcium oxalate, constitute the majority. Crystals of other salts of calcium and of various other inorganic substances, such as silica and gypsum, occur less frequently; crystals of many organic substances, such as carotin, berberin, and saponin, are frequent. All parts of the plant may contain crystals, though these structures are more abundant in certain regions, as in the pith, cortex, and phloem, than in others. There are many forms of crystals (Fig. 1.10). Of these, solitary, rhombohedral crystals, sheaf-like bundles of long acicular crystals known as *raphides* (Fig. 1.10*E*), and clustered crystals in globose masses called *druses* (Fig. 1.10 *A,B,D,F*) are most common. The individual crystals of druses may be raphides. Solitary raphides, small prismatic crystals, and minute crystals called *crystal sand* are other frequent types. *Cystoliths*, often considered crystalline bodies, consist in large part of cell-wall substance.

One form of crystal may occur in a given cell—the usual condition where crystals are clustered—or two or more types may be found in the same protoplast. Many inorganic crystals appear to involve organic matter in the course of their formation. The larger crystals of xylem often show this condition (Fig. 1.10 *G,H,K,L*), and druses frequently contain a prominent organic center (Fig. 1.10*F*).

The large crystals of somewhat various shapes—prismatic, rhomboid, etc.—occur chiefly in fibers of xylem and of phloem and in parenchyma cells associated with fibers. Raphides occur in thin-walled, mucilage containing parenchyma-cells of soft tissues, such as storage parenchyma of underground parts, fruit pulp, and the tissues of aquatic plants generally. They are present in monocotyledons chiefly. Druses are characteristic of parenchyma cells of cortex and pith, especially of stems and petioles; they also are abundant in phloem.

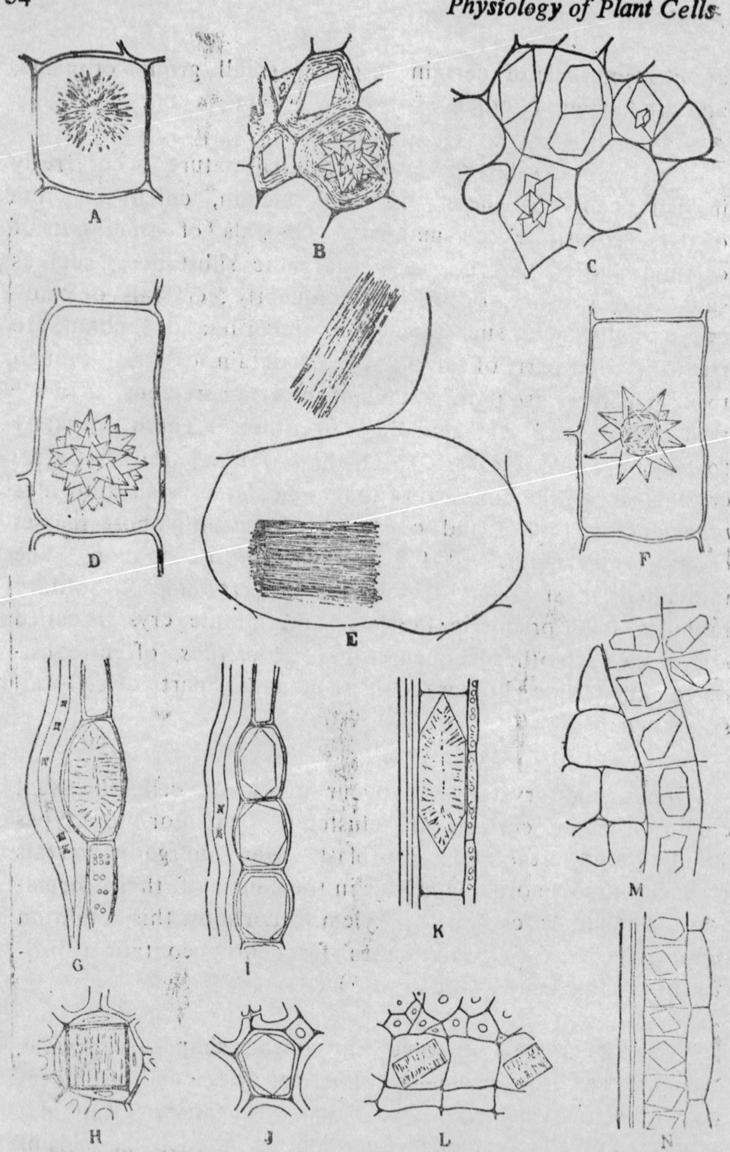

Fig. 1.10. Crystals. *A*, druse in cortical cell of stem of *Viburnum Lentago* ; *B*, druse and rhombohedral crystals in stone cells of nutshels of *Carya glabra* ; *C*, solitary and grouped crystals in pith cells of *Populus grandidentata* ; *D*, druse in cortical cells of *Carica Papaya* ; *E*, "bundeles" of raphides in pulp cells of fruit of *Smilacina racemosa* ; *F*, drus, with organic center, in phloem parenchyma cell of *Juglans nigra* ; *G*, *H* longitudinal and transverse sections of crystal in wood parenchyma of *Carya Pecan* ; *I*, *J*, longitudinal and transverse sections of crystals in wood parenchyma of *Juglans nigra* ; *K*, *L*, longitudinal and transverse sections of crystals in phloem parenchyma of *Tilia americana* ; *M*, various forms of crystals in phloem parenchyma of *Malus pumila* ; *N*, rhombohedral crystals in phloem parenchyma of *Salix nigra*.

Crystals lie in the protoplast—usually in the vacuoles—or in the lumen when the protoplast has disappeared, as in fibers. Occasionally, they are partly or wholly embedded in the cell wall ; more frequently they are suspended in the lumen by projections of the walls. These projections are rod-like ; or they are sac-like, covering the crystal and holding it in a central position. Large crystals may fill the lumen and then determine the interior contour of the wall (Fig. 1.10*B*) ; in elongate cells the lumen may be filled at a given level only (Fig. 1.10*K*).

Cells may be given over entirely to crystal storage, and the protoplasts become much reduced or disappear, but typical active cells may also contain abundant crystals. The early stages of crystals and found in very young cells ; apical meristems often contain young druses.

In large part, inorganic crystals are probably waste products, the result of metabolic processes. Their development in tissues which soon cease to be functional, such as pith, cortex, and secondary phloem, is suggestive of this.

Starch. Food materials are present either as transitory material or in the more or less permanent form of storage particles. Newly formed food may be in solution or exist as solid particles. Starch grains are the most common kind of solid food material found. They are of numorous types and vary in size and in form over a considerable range (Fig. 1.11). In shape they are mostly rounded or oval. Crowding results in the formation of angular shapes, and symmetrical, polyhedral grains are characteristic of some plants. Other plants possess compound grains which closely resemble simple grains, but show their compound nature when broken up into their constituent parts which are minute simple grains. All grains in a cell may be simple or compound, or both types may occur in the same cell (Fig. 1.11*D*). Structurally the grain is made up of radially arranged tapering needles of amylose. The needles lie with their tips at the center, their bases fused to form a border. Both the central area, the *hilum*, and the border are highly refractive. Concentric layering about the

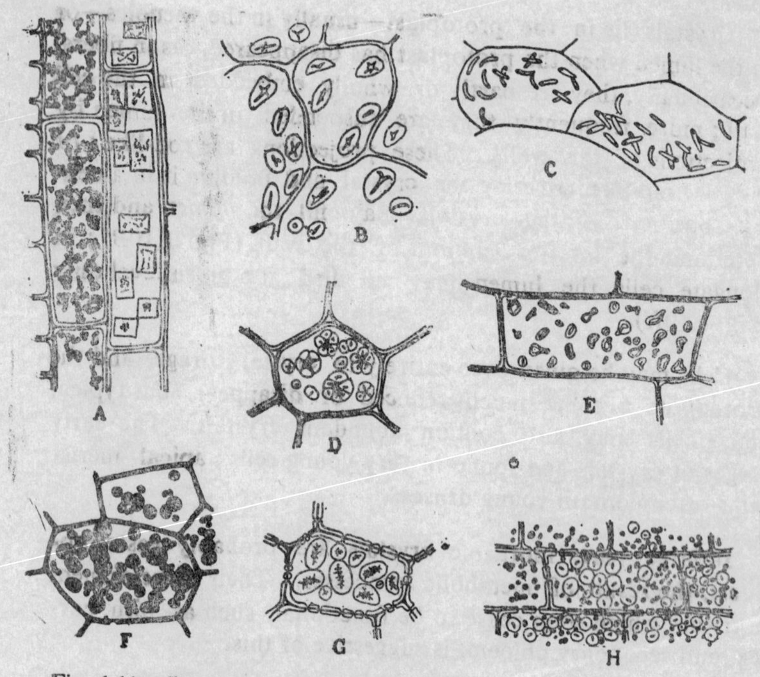

Fig. 1.11. Starch grains and tannin. Tannin : *A*, in phloem parenchyma of *Pinus* (also crystals) ; *F*, in pith cells of *Fragaria* ; *H*, in ray cells of wood of *Malus pumila* (also starch grains). Starch grains ; *B*, in pith cells of *Alsophila* ; *C*, in outer pericarp of *Musa* ; *D*, in cotyledon of *Pisum* ; *E*, in ray cell of phloem of *Ailanthus* ; *G*, in cotyledon of *Phaseolus*.

hilum is commonly seen, although this may be obscure without special treatment. The layering may become strongly excentric in the larger grains of some plants, for example, the potato. The layered appearance is probably the result of variation in the amount of water in the amylose ; and this variation is due to alternations in growing conditions—in light intensity, temperature, and humidity. Grains formed under constant growth conditions are not lamellated. The hilum is rounded or angular, sometimes lobed, forked, or stellate, and is highly refractive to light. By the presence of hila and by staining properties, starch grains may be distinguished from

...bodies, and from other solid

...tarch grains are first formed within chloroplasts and are ...ten found in that position ; leucoplasts in storage cells build up starch grains within themselves from translocated food brought from green cells. Starch grains are thus formed both primarily and secondarily within plastids. It is not known with certainty whether, when a grain is mature, it is freed from the plastid or is still surrounded by a very delicate layer of plastid substance.

Inulin. This is a carbohydrate, soluble in water but insoluble in alcohol. It is generally found in the tuberous roots of the members of the family *Asteraceae* such as *Dahlia, Helianthus, Cichorium* etc. In these members inulin forms fan shaped crystals (Fig. 1.12). The inulin of *Dahlia* tubers can be precipitated by keeping them in alcohol.

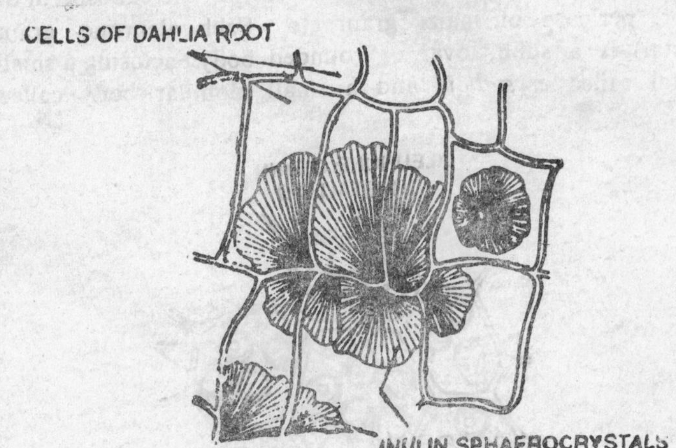

Fig. 1.12. Sphaerocrystals of inulin in the roots of Dahlia.

Hemicellulose. In many seeds hemicellulose is stored as reserve food, *e.g., palms.*

Cellulose. This is mostly present in the cell wall. It is made up of a long chain of glucose residues.

Sugars. These are simple ... number of carbon atoms in the molecule they a... into Trioses (3C), Tetroses (3C), Pentoses (5C), Hexo... etc. Sugars are mostly present in fruits (grape), stems (su... cane), roots (beet root) etc.

Nitrogenous Products. These are reserve food materials containing nitrogen. There are two types of nitrogenous food materials *viz.*, proteins and amino compounds. ***Proteins*** are very complex macromolecules having carbon, hydrogen, oxygen and nitrogen. Sometimes, in addition to these, sulphur and phosphorous are also present.

Proteins are soluble in strong acids and alkalis. Some of them are water soluble also. Chemically proteins are composed of monomers called *amino acids*—which are considered the building blocks of proteins.

Proteins are stored in seeds or grains in the form of ***aleurone crystals*** (Fig. 1.13). These are present in the endosperm of castor, pericarp of maize grain etc. Each aleurone grain (crystal) is a solid, oval or rounded body enclosing a small crystal called *crystalloid* and a small globular body called

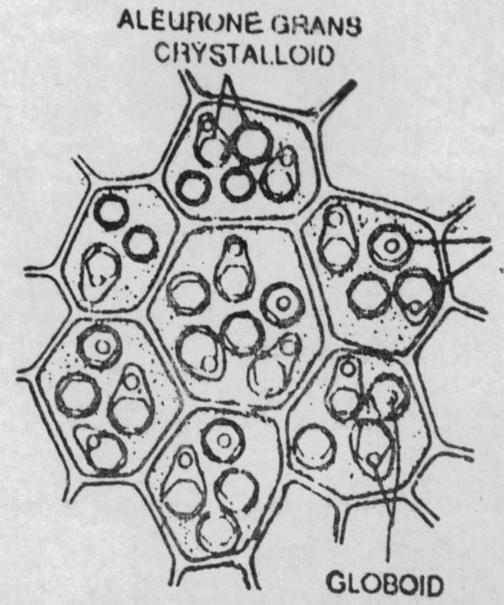

Fig. 1.13. Aleurone grains in the endosperm of Ricinus communis (Castor).

globoid. The crystalloid is proteinaceous and occupies the major portion of the grain, while the globoid consists of phosphates of calcium and magnesium and occupies the rest of the grain.

Amino Compounds occur in the form of amino-acids and amines in the cell sap. These are more frequently present in the growing parts rather than in storage tissues.

Fats and Fatty oils. Chemically fats are esters of fatty acids and glycerol. They occur in the form of oil globules in cytoplasm. They are present mainly in seeds. Chemically fats and fatty oils are composed of carbon, hydrogen and oxygen. They are insoluble in water but soluble in organic solvents. Fats store comparatively more energy than carbohydrates. Fats are solids at room temperature whereas oils are liquids at room temperature.

Secretory Products. In addition to the food materials the cytoplasm also secretes a number of chemical compounds which do not help either in nutrition or in growth.

Secretory products are of three types *viz.*, *Enzymes*, *Pigments* and *Nectar*. Enzymes are complex proteins and participate in various types of biological reactions. They are generally called organic catalysts, because they are not used up in a reaction, but enhance its rate. *Pigments* are colouring substances and are present in plastids or in the cell sap. These are usually present in leaves, and in the petals of flowers. Anthocyanins are the principal pigments which impart different colours to the petals. Chlorophyll, xanthophyll and carotenoids are pigments present in photosynthetic organs. *Nectar* are sweet substances sacreted by special glands called nectaries which are present in flowers.

Waste Products. During the metabolic activity cytoplasm secretes various types of waste products such as *tannins*, *mineral crystals*, *latex*, and *gums*. These are present either in cytoplasm or in the cell sap.

A. Tannins. They are non-nitrogenous substances found in the cell sap. Chemically they are derivatives of phenols and related glucosides. They are found in cell walls and also in the heart wood and the bark. Tannins are also found in leaves and unripe fruits. When the fruits ripen they are converted into glucose and other substances. Generally monocotyledons have poorly developed tannins. Tannins may be present in individual cells or in *tannin sacs*.

B. Alkaloids. These are non-nitrogenous products found in various part of the plant. Some of the common alkaloids are *Nicotine* (Tobacco), *Caffeine* (coffee), *Quinine* (Cinchona), *Ephedrine* (Ephedra), *Theaine* (Tea) etc.

C. Mineral crystals. Various types of mineral crystals are found in plant cells. They may be present either in cell walls or in a cavity in the cytoplasm. Majority of the crystals contain calcium carbonate or calcium oxalate. The following are some of the types of mineral crystals.

1. *Cystoliths.* These are crystals of calcium carbonate, generally found in the leaves of various flowering plants. The members of the family *Moraceae*, *Urticaceae* etc., have cystoliths. Each cystolith has a peg like projection of the cell wall (Fig. 1.14) into the cell cavity. Cystoliths consist of many small crystals grouped together like a bunch of grapes.

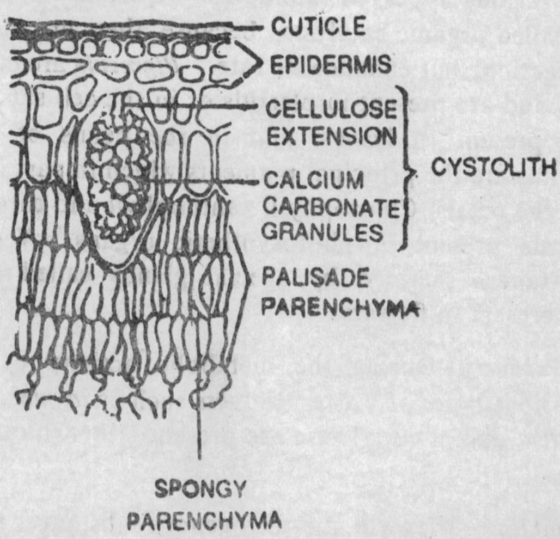

Fig. 1.14. Calcium carbonate cystolith in the leaf of Ficus elastica.

2. *Calcium oxalate crystals*. These are found in the vacuoles and are of the following types.

(*a*) *Raphides*. These are long needle like crystals arranged parallely in the form of stacks or bundles (Fig. 1.15). These are found in sacs and when the sacs are ruptured come out with a jerk. Rephides are commonly seen in the leaves of *Colocasria*, *Pistia* etc.

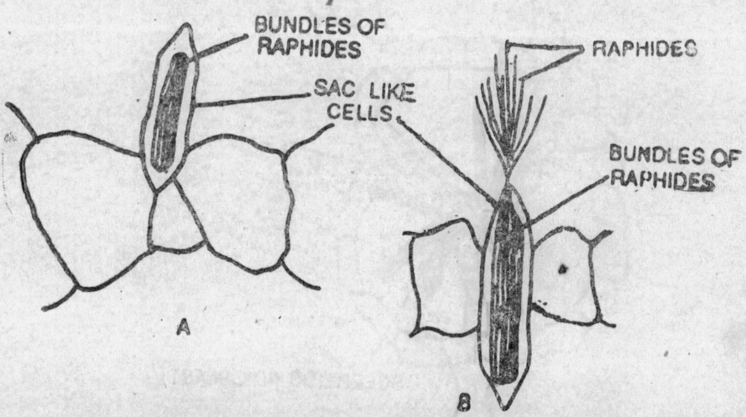

Fig. 1.15. Raphides-A, raphides in colocasia, B. ejection of a raphide.

(*b*) *Druses, Rosette or cluster crystals*. These are compound crystals arranged in the form of a rosette (Fig. 1.16).

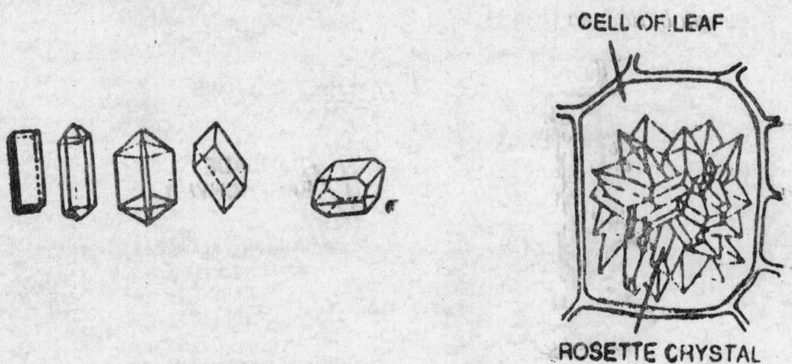

Fig. 1.16. Calcium oxalate crystals. Prismatic crystals (left) and Rosette crystal (right).

These are found in the leaves of *Eucalyptus*, *Ixora* etc. Crystals found in globose masses are called druses. Shining crystals are called prismatic crystals (Fig. 1.16).

(*c*) *Idioblasts.* These are stellate crystals and are usually found in water plants (*Limnanthemum*, *Nymphea* etc.) offering mechanical support. (Fig. 1.17).

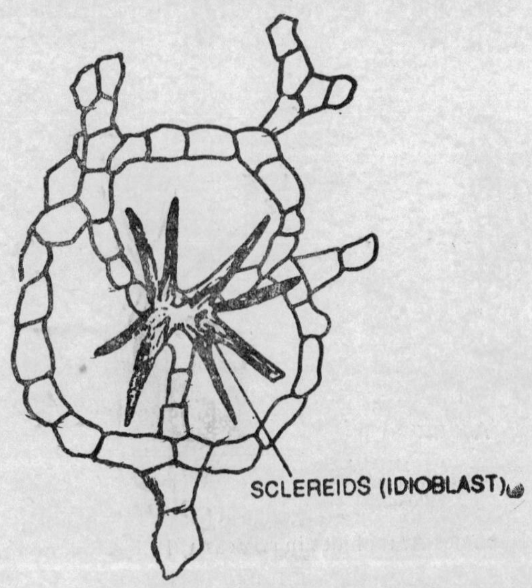

Fig. 1.17. Idioblast in a leaf cell.

(*d*) *Sand crystals.* These are found in masses packed into a cell. Such crystals are seen in the roots and leaves of *Atropa belladona* (Fig. 1.18).

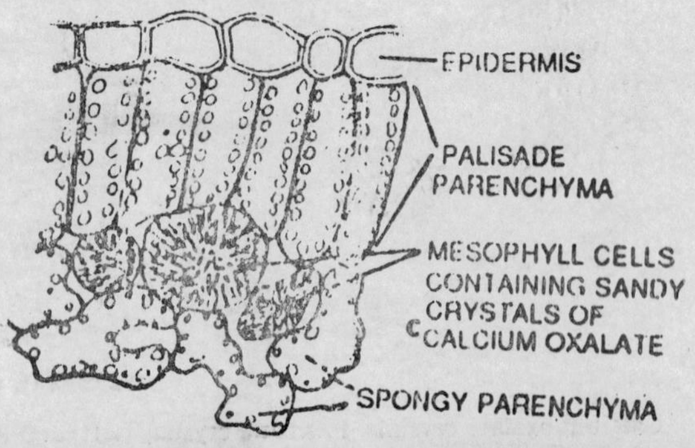

Fig. 1.18. Sand crystals of Atropa belladona.

(*e*) *Latex.* In some plants, a milky or watery juice comes out of latex vessels or cells. Latex contain starch grains proteins, oils, tannins, gums, resins, alkaloids, salts, enzymes and many poisonous substances. Laticiterous cells are seen in *Carica, Euphorbia, Ficus, Hevea, Calatropis* etc. In some cases the dried and processed latex is the source of natural rubber.

(*f*) *Essential oils.* These are volatile oils and are secreted by oil glands. Oil glands are found in almost all parts of the plants, specially in leaves (Fig. 1.19) and in the skin of fruits (*citrus* fruits). Essential oils are also found in the petals of Rose, Jasmine etc.

(*g*) *Gums and resins.* These are produced in Gymnosperms (*Pinus*) and Angiosperms (*Acacia, Feronia* etc.) Canada balsam a widely used resin in microscopic studies is produced from *abies balsamia.*

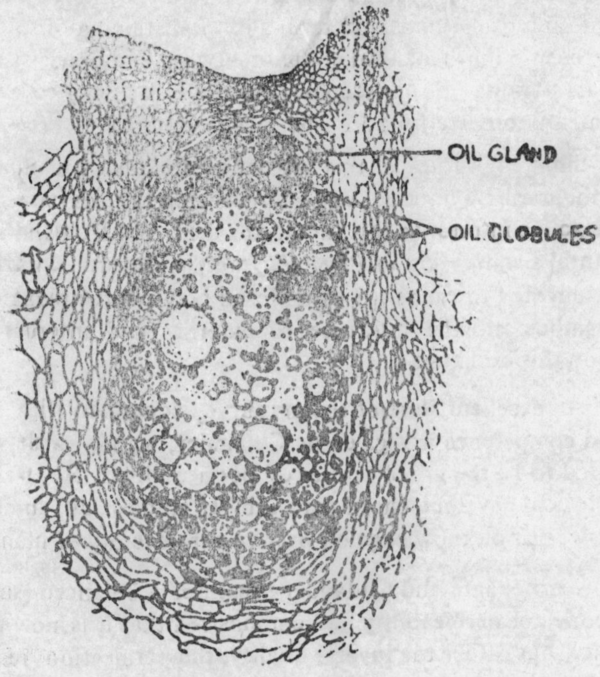

Fig. 1.19. Long section of an oil gland from Cirtus aurantiaca.

PERMEABILITY OF THE PLANT CELL

The term permeability is used in the physiological literature to describe two different but related concepts : (*a*) capacity of cells or tissue to take up or exchange material with their environment (*b*) permeability of membranes to gases, liquids dissolved substances.

This review is almost exclusively limited to the discussion of permeability in the latter sense, and of such membranes as the plasmalemma, tonoplast, and protoplasmic layer of a vacuolated cell.

No section on vital staining is included here, since main interest in this method is upon the adsorption and accumulation of the dye. No new data on passive permeability for vital strains have been collected. However, advanced instrumentation may help to obtain such information by this very useful method in protoplasmic research. A most comprehensive presentation of methods and results of vital staining of plant cells has been published. An earlier summary emphasizes the use of vital staining as a cytochemical tool in living cells. Other contributions are found in the recent volumes of *Protoplasma*.

Since the survey of this subject by Collander, a number of monographs, symposia, and summarizing articles have reflected the active interest in this important field. In Ruhland's monumental *Encydopedia of Plant Physiology*, several contributions are devoted to exchange of material between the plant cell and its milieu and the influence of external and internal factors upon this exchange.

An excellent state-of-the-art survey composed with mastery and competence was given by Collander, who is widely considered to be the grand master of permeability. He traced the historical development of permeability studies and summarized the wealth of knowledge on passive permeability of plant cells.

Wartiovaara and Collander presented a balanced survey on theories of permeability and emphasized that it is not permeability, but rather the inverse quality, the permeation resistance, which is of primary biological importance for isolating the cell selectively from its environment. Stein surveyed various problems of permeability and membrane structure, with emphasis on the molecular aspects and primary reference to characteristics of membranes in animal and human tissues. Monographs devoted to recent research on membrane structure and membrane components and the membrane in general were not yet available before the completion of this manuscript.

Schoffeniels described the permeability characteristics of animal and human tissues and cells primarily for substances occurring in their natural environment and also discussed the electrical properties and structures of these membranes.

Symposia, conferences, or special issues were devoted to particular aspects of passive permeability; to the plasma membrane with an excellent contribution on the history of the plasma membrane concept; to biophysics and physiology of biological transport and cell membranes; to colloid chemistry, ultrastructure of membranes, and active uptake; and to membrane structure and function. Biophysics of permeability and transport and distribution of material in cells of higher plants were discussed in meetings scheduled for 1968. Some data on permeability of algae were compiled by Stadelmann.

Sorption theory. A theory that the interior of the protoplasm alone determines the exchange of material with cell surroundings and that in plant cells the tonoplast alone acts as semipermeable membrane was advanced by Troshin. Similarly, Ling in his "association-induction hypothesis" denied any control function of the cell membrane for material exchange. However, he later assumed that K-ion exchange was controlled by the outer cell surface. Both authors were mainly concerned with animal cells and apparently were unaware of the wealth of evidence supporting the membrane theory which was obtained with plant cells. Also some important cellular functions are in disagreement with the sorption theory. However, it is possible that, depending upon the substance entering the cell, membrane permeability or sorption capacity of protoplasm combined control uptake.

Active uptake of nonelectrolytes. Active ("nonosmotic") uptake of nonelectrolytes which are used in experiments on passive permeability (*e.g.*, urea, malonamide, methylurea) and water into plant cells, especially during plasmolysis, was again inferred by Follmann and Follmann-Schrag from differences in plasmolysis and cell content concentration on diatoms and *Cladophora* with and without treatment with metabolic inhibitors. Hofler and Url found similar experimental results working with sodium azide and dinitrophenol applied to cells of *Blechnum spicant* and *Gentiana sturmiana*. These authors and Hofler discussed in detail the data from which the hypothesis of nonosmotic uptake in plasmolysis experiments was developed. Hofler and Url suggested that recovery in hypertonic sugar solutions or irregularities of the protoplast volume in stepwise increased sugar concentration can be interpreted without difficulty by changes of passive permeability and exosmosis. Effects of sodium azide on plasmolysis and deplasmolysis of *Rhoeo* epidermal cells are best explained by changes of passive permeability. Urea uptake by petiole segments of *Nymphaea alba* and of the peduncle of *Taraxacum officinale* was tested and found to occur only as long as concentration equilibrium with the external solution was reached. Accumulation never was recorded. In all these experiments, evidence for support of the hypothesis of active uptake of these nonelectrolytes was lacking. Therefore, Bogen's criticism of earlier results seemed to be unfounded.

Permeability formulas. Kusel presented a formula for an absolute permeability constant for solutes for filamentous algae like *Oscillatoria*, where contraction and expansion of the whole cell thread without plasmolysis was observed after perfusion of the preparation with a solution of the permeator. The formula gives a good approximation for the permeability constant. However, the factor 2.5 introduced at the end of the derivation must be omitted, since the permeability constant as derived up to this point is already reduced to unit concentration difference.

Much effort has been devoted to application of irreversible thermodynamics to the passive permeation process. The classical kinetic approach resulted in a permeability constant from a modified Fick's law of diffusion, referring to the relationship of a specific substance (permeator) to a given membrane. The irreversible thermodynamics of permeation starts from more general premises which allow for inclusion of the important case of simultaneous passage of solvent and solute through a membrane under pressure and/or concentration difference. Three phenomenological coefficients result :

(*a*) The coefficient of the volume flow of water through a membrane under a pressure gradient, when no permeator is present, L_p. Dimension : $length^2 \times mass^{-1} \times time$; unit : $cm \times sec^{-1} \times atm^{-1}$.

(*b*) The solute permeability coefficient at zero volume flow, ω, which is a magnitude expressing the permeation of a solute. Dimension : $length^{-1} \times mass^{-1} \times time \times mole$; unit : $cm \times sec^{-1} \times atm^{-1} \times mole \times cm^{-3}$.

(*c*) The reflection coefficient σ, a measure of the degree of semipermeability of a membrane. Sigma indicates a ratio and is dimensionless.

These three coefficients are related to the usual permeability constants for water K_w and solute K_s : when K_w is expressed in a unit (*e.g.*, cm/sec) with the same dimension as L_p, it is $K_w = L_p$. The permeability constant for solutes is $K_s = \omega.R.T.$ This formula is strictly correct only when no simultaneous water flow through the membrane occurs, *i.e.*, when K_s is measured by chemical or isotopic analysis. For osmotic and plasmolytic measurements the formula $K_{so} = \omega.R.T/\sigma^2$ is derived, where K_{so} is the permeability constant determined by osmotic or plasmolytic measurements. The importance of σ in this formula, however, may be questioned since in experiments on *Chara australis* and *Nitella translucens* K_s and K_{so} values were found to be the same.

The phenomenological coefficients are considered to contribute little information about the kinetics of permeation or the structure of the membrane. Also the derivation of a set of three implicit functions which relate phenomenological coefficients to frictional coefficients does not significantly advance the understanding of the membrane phenomena involved. The concept of the reflection coefficient seems to be an important contribution to the theory of membrane permeability, since it is a new parameter for quantitative measurement of a membrane quality.

Some investigators have calculated a number of mathematical formulas and models based on the phenomenological approach. These theoretical results frequently exceed any possibility of experimental proof and applicability. They, therefore, often contribute very little to the understanding of permeability problems. It seems to me much more urgent to expand accurate experimental data concerning permeability for an increased number of cell types, substances, and conditions.

Mathematical treatments were applied by Hall to analyze the permeation process through a spherical membrane of a haemolyzing erythrocyte. A formula was established for the absolute water permeability constant for cylindrical cells when deplasmolyzing in a nonpermeating and by pertonic solution.

Advancements in methods. Substances with high specific activity of radioactive tracers are now available. Use of these substances will expand the application of compartmental analysis for ions or nonelectrolytes, where it is now possible to determine the permeability of the tonoplast membrane and plasmalemma separately. The application of the plasmometric method for permeability measurements and the possible errors involved are discussed by Hofler. Microphotographic time-lapse techniques have provided detailed analyses of the time-course of deplasmolysis in plasmometric experiment. The numerical value of the degree of plasmolysis [as introduced by Hofler] for irregularly shaped cells or cells with concave plasmolysis forms can be determined by using an integrating eyepiece. This procedure may be considered an important improvement since it overcomes the limits of the plasmometric method and makes irregularly shaped cells accessible for quantitative evaluation of osmotic and permeability measurements.

The time course of changes in volume or weight of tissue disks (*e.g.*, parenchyma cells of the hypocotyl tuber of *Beta vulgaris*), when transferred into a solution with different osmotic pressure, was found not to reflect cell permeability but rather extracellular (or extraprotoplast) diffusion of the solute, so that this method is not available for determinating cellular permeability constants.

The reliability of plasmolytic water permeability determinations is occasionally questioned on the grounds of the uptake of saccharose and mannitol in leaf disks or entire plants after they are exposed for several hours or longer to a solution of these chemicals. However, these authors overlook that it is routine in plasmolytic work to check for any sugar permeability or change in degree of plasmolysis after plasmolytic equilibrium is reached. A great number of observations on plasmolysis with sugars and mannitol accumulated over the last decades. The results prove that with very few exceptions (*e.g.*, some diatoms) these substances do not (or only negligibly) penetrate into the vacuole during the time of a plasmolytic experiment for determination of the water permeability.

NONELECTROLYTES

The relation between partition coefficient and permeability constant was further investigated for special classes of molecules. Substances with similar partition coefficients permeate more slowly when molecules are higher branched than when they are not. Relatively large molecules (polypropylene glycols with mean mol wt of 425 and 1025) do not permeate at expected from their lipid solubility. Their size is a more important factor than their partition coefficients for permeation.

Permeability constants of 2.06 to 4.82 μ/sec for rapidly permeating substances, such as alcohols, were found with *Chara* and *Nitella*. Radioactive tracer and washing techniques were used. The results were comparable to those found microchemically by Collander. The authors considered the effect of the "unstirred layer" which gives appreciable corrections for the permeability constant of rapidly permeating substances.

Permeation rates of urea and formamide, expressed as $\triangle G$ (change of the degree of plasmolysis in cylindrical cells per time unit), of different cell types of six species of liverworts and five species of higher plants were determined to compare the degree of "porosity" of plasma membranes. A variation of the $\triangle G$ ratio for formamide/urea from 7.7 to 57.3 for liverworts and of 2.0 to 6.3 for higher plants was found. The ratio of the distribution coefficients for formamide/urea is about 5.0. The $\triangle G$ values for formamide are among the few accurate data about its permeation into plant cells. Unfortunately these results are of limited value only, since urea was chosen as reference substance. An appreciable portion of urea is generally thought to penetrate through the "pores". Also, absolute permeability constants were not calculated, this omission rendering invalid any comparisons with results of other investigators.

Oscillatoria margaritifera and *O. lloydiana* have permeability series with the same sequence as that for *Chara* and *Majanthemum*, but the permeability values are about 100 times higher. The water permeability seems to be extremely high also. Permeation of substances of small molecular size is very little enhanced, thus a low "pore" permeability of the plasma membranes of these organisms in indicated.

Permeability series were recently established for *Blechnum spicant*, *Gentiana germanica*, and *G. ciliata*.

Relative values of saccharose permeability of some diatoms were measurad by Ubeleis. An exception to the generally high sugar permeability of diatom cells is *Caloneis obtusa*, which showed no appreciable permeability for glucose, while the other nonelectrolytes used in permeability experiments permeated at about the rate found for other plant cells.

Protoplasmic permeability for methyl and ethyl alcohol has been reinvestigated by using a perfusion chamber in an improved plasmolytic technique. Longitudinal cuttings of the stem base from eight different species of flowering plants and two species of algae were used. The cuttings were plasmolyzed in specific concentrations of glucose, and transferred to a mixture which contains glucose in a concentration equimolar to that used for plasmolysis plus 5 to 20 per cent alcohol (total concentration). The protoplasts underwent a temporary (of a few minutes duration) length change prior to the equilibrium size. When alcohol permeated faster into the vacuole than water could leave it, the protoplast expanded. If alcohol permeated more slowly than water could leave, the protoplast contrated. Examples of each of these types and of intermediates were found in the experimental material. However, the situation appeared to be more complex since the protoplast size for most cell types tested at equilibrium is smaller in the mixture, and yet regains its original equilibrium size when transferred back into pure glucose solution. There also are changes from contraction type to expansion type, depending upon the concentration of alcohol applied, which supposedly are not caused by pathological alterations. For all mixture concentrations the subepidermal cells of *Impatiens sultani* exhibit the concentration type, while *Cladophora utriculosa* and *C. fracta* belong to the expansion type.

Values for the reflection coefficient [first mentioned by De Vries as "isotonic coefficient"], most of them calculated for artificial and animal cell membranes, were recently compiled by Stein. For methanol, ethanol, and DHO with *Nitella* and *Chara* cells, σ is 0.25, 0.29, and 0.C033 to 0.0061, the last value depending upon the DHO concentration applied. The value for σ for urea and *Allium cepa* epidermis cells determined from plasmometric evaluations was 0.87 to 0.93. However, it seems that in most of this experimental work the differences in water activity in equimolar concentrations of different substances were not taken into account.

A list of relative and absolute permeability constants for various substances with lower plants was compiled by Stadelmann.

IONS AND ACIDS

The complex problem of active ion uptake in plant cells and its involvement with metabolism has been treated thoroughly in recent monographs.

Passive permeability of the plasmalemma and the entire protoplasm layer for ions has been reviewed briefly by Sutcliffe. Different mechanisms such as diffusion, ion exchange, adsorption, and mass flow may be involved in passive ion permeation. In addition, permeabilities of the plasma membranes (plasmalemma and tonoplast) for anions and cations and their species may vary considerably.

The giant Characean cells are especially suited for studies of passive permeation of ions at the cellular level with microchemical and tracer techniques. The ion concentration in the cell wall, the protoplasm layer, and the central vacuole can be determined. It is also possible, under certain assumptions, to calculate fluxes and permeability constants for an individual ion species (Table 1.4).

In *Vallisneria* leaves the plasmalemma shows no passive permeability for the Cl ion (leakage) during the process of active Cl uptake. However, by internal and external factors (antagonistic effects of other ions in the external solution), the structure of the plasma membrane can be changed so that leakage and passive uptake occurs. Under optimum conditions no passive permeability for ions can be demonstrated for the cell membranes of *Vallisneria* leaves.

Fatty acids with straight carbon chain permeate into *Saccharomyces cerevisiae* cells corresponding to their relative ether solubility. Acids possessing branched carbon chains, however, have a smaller permeability constant than expected from ether solubility.

WATER

Absolute permeability constants for water been recalculated for a variety of plant cells from relative water permeability constants derived earlier, mostly with the osmotic method. To increase accuracy the recalculations should use differences of water activities rather than concentration differences.

Table 1.4. Passive Ion Fluxes (pmoles.cm^{-2} sec^{-1}) and Permeability Constants (nm/sec)[a] under Steady State Conditions[b]

Species	*Fluxes*				*Permeability constants*				*Reference*
	Plasmalemma		*Tonoplast*		*Plasmalemma*		*Tonoplast*		
	Na	K	Na	K	Na	K	Na	K	
Chara globularis	0.7[d]	2.8[d]	—	—	0.1[b]	3[c]	—	—	(41, p. 516)
	6.2[e]	6.7[e]							
Nitellopsis obtusa	8[f]	4[f]	0.4[d]	0.25[f]	0.5[c]	10[c]	0.64[e]	11[d]	(99, pp. 342, 343, 345,
			0.6[e]					0.025[e]	350, 351)
Nitella axillaris	—	0.47[f]	—	0.72[f]	—	13[c]	—	—	(33, p. 1118)
Nitella translucens	0.55[d]	0.3[d]	—	—	0.9[d]	5[d]	—	—	(98, pp. 875, 877)
	0.1[e]	0.8[e]							
Chara australis	—	—	—	—	10[c]	100[c]	—	—	(66, p. 42)

[a]1 nm (= nanometer) = 10^{-9} m = 10^{-7} cm.
[b]After (66, p. 43); with additions.
[c]From (66, p. 43); influx or efflux not specified.
[d]Influx
[e]Efflux.
[f]Influx and Efflux.

One previously overlooked cause of the discrepancy between absolute water permeability constants obtained from osmotic experiments and those based on diffusion experiments may be "back diffusion." This occurs in such experiments in which permeation resistance is doubled because volume of the vacuole did not change. Hence, for each molecule of heavy water penetrating into the vacuole a molecule of H_2O has to leave the vacuole. When considering "back diffusion" most of the values of the absolute water permeability constant are between 1.0 and 15 μ/sec. This agrees well with a list of water permeability values for animal cells, which lie within the same limits, in most cases.

Artificial bimolecular phospholipid membranes have a water permeability constant of about 2.8 to 4.4 μ/sec in diffusion experiments. Constants calculated from osmotic experiments for the same kind of membrane are 17.3 to 104 μ/sec. However, this range of latter values might be too low because of experimental conditions under which they were obtained. It is still difficult to explain the discrepancies between the permeability constants obtained by these two methods.

Dainty and Hope and Dainty and Ginzburg found extremely high values for water permeability (up to 142 μ/sec) in transcellular osmosis experiments on *Nitella translucens*. They assumed that diffusion experiments gave values that were too low because cell diameter is rate limiting. Initial stages of transcellular osmosis are complicated by the additional action of cell wall shrinkage. The permeability constants derived are concentration dependent. Further evaluation of transcellular osmosis experiments might be necessary to explain existing discrepancies.

The half-time value for water withdrawal in secondary roots of *Vicia faba* in hypertonic solutions (80 to 100 sec) was almost the same as for exchange of DHO (80 sec). Hence the same mechanism apparently determined the rate in both cases. The H_2O/DHO exchange is only about 12 times slower than for free diffusion and the endodermis does not increase diffusion resistance for water. When preplasmolyzed roots are tested for their H_2O/DHO exchange, at first a rapid exchange occurs with water in the "Vorraum" (the space between the plasmolyzed protoplast and the cell wall) and in cell walls. The half-time value for exchange of water with the protoplast is higher than in nonplasmolyzed tissue, probably because the volume of the vacuole and the total protoplasm surface is smaller.

Comparison of the diffusion rate of DHO and H_2O^{18} indicates that undissociated water molecules permeate into the cells. For *Tolypella nidifica* the half-time value for the entrance of H_2O_2 is about 50 sec. Experiments with heavy water show that an equilibrium of the DHO concentration is reached between the cell water and the milieu when measured after 30 min.

The accuracy of the low values of the water permeability constant (0.3 μ/sec) reported by Myers for unplasmolyzed parenchyma cells of the hypocotyl tuber of *Beta vulgaris* has been questioned. Occasionally investigators have overlooked these later references and have referred to Myer's data as providing proof of an effect of plasmolysis on water permeability.

CHANGES IN PERMEABILITY

Interpretation of quantitatively determined permeability changes may contribute significantly to further understanding of membrane structure and function if a sufficiently large body of coherent data can be analyzed. At present such information is very scarce. A few examples of recently reported permeability changes caused by a variety of factors will be mentioned here and will hopefully stimulate more detailed studies in this area.

Permeability changes may be caused by endogenous conditions within cells or by natural or artificial changes in their environment. Reversibility of changes indicates the harmlessness of the causal factor. An increase in permeability often reflects pathological or premortal conditions. Little is known about the site of action of any factor, as to whether it influences the sieve effect of membranes or changes the capacity of membranes to act as a lipid solvent. The latter factor seems to be more easily restored after cells have been damaged, while the principle of the sieve effect is assumed to be more sensitive, more unstable, and less easily restored when changed by external factors.

A change in permeability of cell membranes does not just increase with strength of the causing factor. Occasionally, application of the same factor at a low intensity results in a permeability decrease, which changes to permeability increase with increasing strength of the causing factor. This reversal of the effect is one of the causes for contradictory reports on the action of external factors on permeability. A number of explanations for the observed permeability changes exist but none is based on adequate experimental evidence.

Reversal of permeability changes could be explained by postulating micelle leaflet transitions in the lipid membrane. Any altering factor would first cause the transition micelles into the more stable leaflet structure (this would decrease the water permeability by diminishing the number of "pores" between the micelles). When the intensity of the altering factor increases further, breakdown of the lipid membrane structure begins, with a subsequent increase of water permeability.

Internal factors. Failure to consider changes in these factors may account for apparent discrepancies in results obtained by different investigators. The effect of various cations (*e.g.*, calcium ions and potassium ions) on water permeability was different at different seasons of the year. This can be explained on the basis of their action on the membrane coacervates and seems to indicate seasonal changes in water permeability.

External physical factors. Alpha-irradiation with about 75 krad decreased urea permeability of onion epidermal cells, whereas dosages greater than about 1 Mrad increased permeability for urea. A 50 per cent increase in water permeability was found after about 2 to 5 krad betaradiation. Only 250 to 500 and from this source were needed to cause a decrease in water permeability when tested about 20 to 72 hr after exposure.

External chemical factors. Most of the permeability changes reported concern the action of solutes on permeability. While inorganic substances (inorganic ions) were intensively investigated earlier, present interest concerns organic substances.

Solute permeability. Urea permeability of the tonoplast of one end of a *Nitella translucens* cell was reduced by a factor of 3 after the same end of the cell was placed in a molal solution of saccharose.

Humic acid with pH 7.0 increased urea permeability of epidermal cells of *Gentiana rochelii* and subepidermal cells of the stem base of *Majanthemum bifolium* with a maximum at 0.1 m*M* humic acid.

Water permeability. Fusaric acid increased water permeability of subepidermal cells of the stem base of *Majanthemum bifolium* and other test plants from concentrations of 1 n*M* to 1 μM. At higher concentrations the water permeability decreased again and at 100 m*M* the value of the control was reached.

Concentrations of 10 n*M* to 10 μM of humic acid increased water permeability of cells of the outer epidermic of *Allium cepa* but higher concentrations had no effect.

Indolylacetic acid buffered to pH 7.5 caused a decrease of water permeability of *Majanthemum* cells in concentrations down to 100 μM, but 0.1 to 10 μM caused an increase, while 10 n*M* produced no effect. Further work on permeability changes induced by indolylacetic acid was reviewed by Pohl.

Gibberellic acid at concentrations of 100 μM caused a maximum increase of water permeability in cells of the outer epidermis of *Allium cepa.* The effectiveness decreased with lower concentrations and 10 n*M* caused almost no change. Concentrations of 1 m*M* decreased the water permeability and still higher concentrations were toxic under the experimental conditions applied.

2,4-Dinitrophenol and sodium-iodine acetate decreased water permeability of epidermal cells of the lower leaf side of *Rhoeo discolor* at 0.1 m*M* and 10 m*M*, respectively. Higher concentrations were toxic while threshold concentrations for any effect on water permeability were found to be 0.1 μM and 0.1 mM, respectively.

This criterion was used by Van Overbeek and Blondeau to determine toxicity of organic acids. These substances were found to be increasingly toxic (*i.e.*, effective in altering protoplasmic permeability) with increasing concentrations of the undissociated molecule (*i.e.*, the lipid-soluble portion). Acids of lower molecular weight were more effective than those of high ones.

Similar experiments in which isolated sections of *Zea mays* coleoptiles were treated with hydrocarbons of different chain length show that the smallest hydrocarbon molecules were the most effective in producing permeability changes. Aromatics were more effective than paraffines and olefines. Van Overbeek and Blondeau suggested that incorporation of the toxic molecules in the plasmalemma is the cause for the permeability breakdown. However, the passage of these toxic molecules inside the plasma and their action on centers important for maintenance of the membrane should also be considered.

Toxic increase of water permeability was found when 1 mM decenyl succinic acid was applied for 30 min at pH 4 to cells of the inner epidermis of *Allium cepa*. Here too the undissociated molecule is damaging (while the anion is not harmful) but decreases the water permeability when the substance is applied above about 1 m$M\times$min. Under these conditions decenyl succinic acid seems also to enter the mesoplasm, since occasional cap plasmolysis was observed.

To distinguish effects of certain chemical treatments on the hydraulic permeability constant from those on the reflection coefficient, Glinka and Reinhold introduced criteria based on measurements of influx and efflux in tissue disks. Their approach seems to be incomplete as provision was not made to include the possibility of differences in the L_p values caused by the direction of water flow. The authors also overlooked that any interfering permeation of the solute during plasmolysis experiments for water permeability determination can easily be detected by observation of changes in the degree of protoplasmic contraction after osmotic equilibrium is reached.

MEMBRANE STRUCTURE AND PERMEATION

One of the most active areas of cell membrane research involves interpretation of the constantly increasing amount of information on membrane structure. The situation is presently characterized by a multitude of new concepts, yet these often lack general applicability since they often were developed hastily and were based on results from a few isolated experiments without considering and synthesizing data from other research areas of investigation on cell membranes. Since some aspects of cell membranes seem to be basically alike for plant, animal, and human cells, references on the latter also are included. A comprehensive review on membrane structure is given in this volume by Branton.

A great need exists for an updated and rational concept on structure and dynamics of cell membranes which will explain the large amount of recently obtained data on membrane characteristics. This concept must also synthesize the observations leading to the membrane concept of Overton, Collander, and Danielli. It must allow for the diversity found in individual membranes of cells of different species, and must treat the plasmalemma and the tonoplast as surface layers of the living protoplasm.

Basic information for theories of cell membrane structure and permeability are derived primarily from three sources: (*a*) studies of passage of solutes and water through the more or less intact or altered membranes of living cells and of electrical qualities of membranes; (*b*) other biophysical and biochemical analyses of cell membranes and their components, or of artificial membranes, as well as mathematical evaluations at the molecular level; (*c*) electron microscopy.

Membrane structure. While most recent concepts of membrane structure are still based on the lipid nature of membrane they differ widely with respect to membrane architecture.

Danielli's bimolecular leaflet model was further developed and an organization of locked hexagonal subunits was suggested for the two layers of protein coats. Also proposed was a hexagonal icelike water structure between the protein and bimolecular lipid layer as well as a membrane model containing globular proteins with an aqueous phase.

Lucy mentioned a micellar model of membrane organization with globular micelles of lipids (diameter about 50Å). Between the globular micelles, globular protein molecules might be located. This micellar structure was thought to be in a dynamic equilibrium with the bimolecular leaflet structure, so that certain areas of the membrane temporarily were in one phase or the other.

Rectangular micelles ("pillars") in membranes and their reversible transition to the leaflet structure were assumed by Kavenau. However, it is questionable if Kavenau's concept can be applied to a structure as thin as the plasma membranes.

The welcomed unit membrane hypothesis is in need of major revision since an increasing number of investigations on cell membranes show a diversity of membrane thickness and composition, which hardly can be brought into agreement with the original concept.

Surface views of the plasmalemma of *Saccharomyces cerevisiae* cells, made possible by the freeze-etching technique, show more or less regular patterns of different elements. Branton interprets the oblique fractions of membranes seen in the electron microscope pictures to explain details of the interior structure of the membrane.

Permeability series. The differences in the permeability series at least partially reflect variations in membrane composition or structure. Causal analysis of these differences may become an important tool for investigating intact cell membranes.

Previous observations of differences in the ratio of permeability constants for urea and glycerol in different series led to a concept of amidophilic and amidophobic types of protoplasm. A new arrangement of 38 permeability series is based on relative permeability constants (glycerol arbitrarily set at 1) and the ratio of glycerol to malonamide.

Cells with the same width of the permeability series may have similar membrane composition but different degrees of sieve effect. Variation in width may be caused by chemical or structural differences between membranes. However, recalculation of many of these permeability constants seems to be necessary before well-founded conclusions about membrane structure can be made from comparing permeability series.

Plasmalemma. The outer plasma membrane is known to protect the cell against an unfavourable environment. Qualitative differences in the resistance of cells to harmful Na_2CO_3 solutions are interpreted as differences in the qualities of the plasmalemma. In cells of higher plants the plasmalemma usually is rather delicate and plasmolysis with Na_2CO_3 solutions damages the cell. In desmids of bogs, however, the plasmalemma often is resistant to damage and the cells show normal plasmolysis with Na_2CO_3 solutions. Hypertonic solutions of $MnSO_4$ and even aluminum salts cause normal plasmolysis which persists for many hours without causing pathological changes in protoplasm. Desmids also do not need Ca^{++} for restitution of the plasmalemma as do higher land plants.

Membrane formation. The dynamics of plasmalemma and tonoplast formation and absorption involve basically a change in area of a surface layer. There exists an immediately accessible reservoir (probably the entire ground plasma) of plasmalemma and tonoplast material in the protoplasm. This is strikingly demonstrated by the ease with which new (secondary) plasmalemma and tonoplast are formed when the protoplast increases its surface area (*e.g.*, experimentally during plasmolysis; plasmalemma formation alone occurs when Hecht's strands are formed). An increase in surface area at a rate of 23 μ^2/sec was measured for the plasmalemma and the tonoplast during deplasmolysis. Little attention has been given recently to the surface quality of the plasmalemma and tonoplast, and most new models of membrane structure do not consider the mesoplasm or its intimate connection with the membranes.

The secondary plasmalemma is very similar if not equal in characteristics to the primary plasmalemma. With certain chemical treatments (EDTA followed by plasmolysis with urea), the plasmalemma is destroyed and the uncovered cytoplasm, in direct contact with the external solutions, immediately swells. The breakdown of the plasmalemma started at a specific point on the plasmalemma surface. Breakdown then progressed gradually over the entire surface.

Site of permeation resistance. It generally is assumed that plasma membranes exhibit the major permeation resistance ; for some electrolytes the plasmalemma may be the main barrier, whereas passive uptake of non-electrolytes from external solution into the vacuole may be controlled by the tonoplast as well as the plasmalemma. Recently Hofler postulated that a continuous lipid phase existed in the ground plasm. This phase, which extends from the plasmalemma to the tonoplast, acts as a solvent for lipophilic permeators. The continuity of this lipid phase (protein-lipid complex) through the entire plasma layer is inferred from vital staining experiments as well as previous evidence. There might also be a continuous acqueous phase in the mesoplasm.

The "pores" (or the sievelike action) of the plasmalemma control passage of that portion of the total amount of a penetrating small molecular substance, which does not enter by lipid solution through the plasmalemma and the lipid phase of the plasma into the vacuole [two-pathway theory].

Levitt's criticism of the two-pathway theory is based on the hypothesis that all lipids of the cytoplasm are located in the cell membrane, and is not correct for all cell types. Formation of new plasmalemma may involve considerable surface increase. Also, naked protoplasts show high expansion capacity.

While Hofler's concept of a continuous lipid phase in the mesoplasm deserves some consideration, it is unlikely that the main portion of the permeator continues to move inside the lipid phase after it has passed through the membrane itself. The distribution coefficient definitely favours migration inside the mesoplasm in the acqueous phase.

For the "lipid phase," factors, such as membrane thickness, the number of solute molecules present in the lipid phase, and the decrease of their migratory speed while passing through the membrane may be important in the permeation process.

The water phase of the ground plasma contributes significantly to permeation resistance of the entire protoplasm layer for passage of water into or out of the vacuole from or to the external solutions. This may be inferred from the much higher speed of plasmolysis of tonoplasts when compared with plasmolysis of protoplasts in cells of the same tissue. Hofmeister compared the permeability constants for water and urea for 25 different kinds of plant cells : Cells with high urea permeability still exhibited average values for water permeability. This may indicate that at least partially different pathways are involved in urea and water permeation. Also Dainty & Hope emphasized that resistance to water movement was not confined to the tonoplast. Plasmalemma and tonoplast resistance to water permeation still reach an appreciable magnitude since microinjection of water into protoplasm of plasmolyzed cells results in a temporary swelling of the protoplasm. Microinjection of water-soluble sulfonic acid dyes was followed by rapid distribution of the dye inside the mesoplasm. Because these dyes have low lipid solubility, they cannot diffuse out through the plasmalemma or the tonoplast into either the external solution or the vacuole.

The question of whether protein layers of plasma membranes contribute to the resistance against water permeation is still open. Huang & Thompson suggested from comparison with values for the artificial lipid bilayer membrane that the protein layer can play only a minor role as permeation resistance against water. However, the water permeability values for the different cell species vary too much to make such comparisons meaningful.

Permeation process. Very little is known about the permeation process itself. Although pores or porelike structures have been assumed to exist, and thereby to account for passage of small molecules, they have never been seen in electron microscope pictures, although their supposed size is within resolution of the electron microscope. Also the equality of the half-time values for water exchange by DHO and for osmotic water withdrawal does not support the assumption of pores. Only transient rather than permanent pores could be assumed for the artificial liquid lipid bilayer membrane.

Two theories were promoted to explain the pore effect by molecular kinetics rather than by the assumption of pores. Wartiovaara compared the permeability constants of homologue series of permeaters (*e.g.*, alcohols : methanol, ethanol, propanol, butanol) and concluded that two parallel and alternative pathways for permeation of molecules through the protoplasm membranes did not exist. It seems that the sieve effect can be explained by a sievelike plate covering the membrane. The effect of such sievelike plates depends on the molecular size of the permeator molecule.

The lipid layer of the plasmalemma is assumed to consist of vertically oriented lipid molecules which form temporary holes by thermal motion. A sieve effect results when thermal motion causes the permeator molecule to come into contact with the surface of the lipid layer : They permeator molecule can penetrate only when it is or can be oriented fast enough parallel to the orientation of the lipid molecules of the membrane surface. The speed of molecular orientation depends on the moment of inertia. A close parallelism exists between the inverse moment of inertia of the permeator molecule and the sieve effect calculated from the permeability constant.

A lattice model for the cell membrane was suggested by Stein. The breaking of hydrogen bonds between permeator molecule and water during the passage of the permeator through the membrane is considered to be the important feature. The product $P.M^{1/2}$ (P=permeability constant,

M=mole weight) can be related to the number of hydrogen bonds of the permeator by using an assigned value for the number of bonds formed for individual atom groups. Thermodynamic parameters for the cell can be calculated from these data and from the temperature dependency of the permeability constant.

Wartiovaara's model attracts by its simplicity and because it can be used to explain in addition to the permeability constants of other permeators, the experimentally determined value of the water permeability constant. However, the model has not yet been elaborated so that hidden shortcomings may not as yet be apparent.

Many quantitative data are available for the lattice model, but a correction factor for CH_2 groups must be introduced and the value of $P.M^{1/2}$ for water still lies far from the other points when they are plotted against the number of hydrogen bonds. It also is difficult to explain the marked differences in the permeability constant for the same substance with different types of cells on the basis of the amount of energy needed to disrupt a hydrogen bond with the permeator molecule. This energy depends upon the structure of the permeator molecule only. Therefore it would be expected that different membranes would have the same permeability constant for the same substance, if the rupture of hydrogen bonds alone determines passage of the molecule through the membrane.

2

Plasma Membrane

The outer most limit of a cell is composed of living or non-living materials or both. Whatever their chemical constitution and structural organization, their function includes regulation of the flow of materials into and out of the cell and they provide some degree of support and maintenance of cell shape and cell size. Plasma membrane is variously called as *cell membrane, plasmalemma, plasma membrane* etc. The term plasma membrane was given by *Nageli* in 1855. On account of the technical difficulties it is not possible to investigate the structure of plasma membrane in every kind of cell. Most of the informations on the subject has accumulated through the study of mammalian R.B.C. and the myelin sheath of the nerve fibres.

This membrane is so thin that is cannot be resolved with the light microscope, but in some cells it is covered by thicker protective layers that are within the limits of microscopic resolution. For example, most plant cells have a thick cellulose wall that covers and protects the true plasma membrane. Some animal cells are surrounded by cement-like substances that constitute visible cell walls. Such layers, also called extraneous coats, generally play no role in permeability, but do have other important functions.

For the study of their structure, chemical composition and for knowing the physiological and biochemical nature of the plasma membrane or any other cytoplasmic organelle the first thing which we face is their isolation in pure state.

METHODS FOR ISOLATION

The ideal isolated membrane component is that which does not have any non-membrane contamination. The isolation of the membrane can be regarded as the removal of other cell components from the cell. This operation is somewhat less difficult in the non-nucleate mammalian red blood cells. During hemolysis the membrane structure loosen sufficiently to allow the cell contents to escape without rupturing completely. After this hemolysis the isolation of erythrocyte plasma membrane is easy. However, the product obtained depends upon the method of lysis and the pH of the liquid.

The other procedure of getting the plasma membrane is introduced by *Wolport* and *O'Neill* in 1962 by treating the *Amoeba* with buffered : 45% glycerol or 2.4 M sucrose solution, caused the cell contents to shrink away from the cell membrane, which retained its characteristic shape.

The membrane were fused from the cell contents by gently homogenization and were subsequently collected in bulk by centrifuging density.

CHEMICAL COMPOSITION

The plasma membrane is composed mainly of protein, lipid, and a small percentage (1 to 5 percent) of oligosaccharides that may be attached to either the lipids (glycolipids) or the proteins (glycoproteins). There is a wide variation in the lipid protein ratio between different cell membranes. Myelin is an exception, in the sense that the lipid predominates ; in the other cell membranes there is higher protein/lipid ratio. In myelin the area occupied by the protein is insufficient to cover that of the lipids, whereas in a red cell ghost the opposite situation is found.

Lipids

The main lipid components of the plasma membrane are the phospholipids, cholesterol and galactolipids ; their proportion varies in different cell membranes. Myelin differs, with regard to lipid composition, from other cell membranes found in the brain. The major proportion of membrane phospholipids is represented by phosphatidylcholine, phosphatidylethanolamine, and sphingomyelin, all of which have no net charge at neutral pH (*i.e.*, neutral phospholipids) and tend to pack tightly in the bilayer. This property is also shared by cholesterol. Five to 20 per cent of the phospholipids are acidic, including : phosphatidylinositol, phosphatidylserine, cardiolipin phosphatidylglycerol, and sulfolipids. Acid phospholipids are negatively charged and in the membrane are associated principally with proteins by way of lipid-protein interactions.

Recent studies demonstrate that there is an asymmetry in the erythrocyte membrane regarding not only the protein, but also the phospholipids. There are more choline phospholipids and glycolipids in the external half of the bilayer and more amino phospholipids in the inner, or cytoplasmic, half of the bilayer. It is assumed that this asymmetry is rather stable and that there is no exchange of lipids across the bilayer.

Carbohydrates

In red cell ghosts hexose, hexosamine fucose, and sialic acid are bound mainly to proteins ; the same in true of carbohydrates in liver membrane. Sialic acid is sensitive to neuraminidase and is attached to proteins by N-acetylgalactosamine on the outer surface of the membrane.

Because of the presence of sialic acid residues, as well as carboxyl and phosphate groups, the outer suface of the membrane is negatively charges ; consequently, positively charged proteins may be bound by electrostatic interactions to the plasma membrane. Only a small amount of sialic acid exists in the form of gangliosides (*i.e.*, glycolipids) in the plasma membrane of liver. However, gangliosides are important constituents of the neuronal surface and are probably involved in ion transfers. The preferential localization of gangliosides in the acetylcholinesterase rich nerve-ending membranes has been demonstrates.

Proteins

Proteins represent the main component of most biological membranes. They play an important role, not only in the mechanical structure of the membrane, but also as carriers or channels, serving for transport ; they may also be involved in regulatory or ligand-recognition properties. In addition, numerous enzymes, antigens and various kinds of receptor molecules are present in plasma membranes.

Membranes contain three different classes of proteins, structural proteins, enzymes and carrier proteins. *Structural proteins* form the 'backbone' of the cell membrane. They have little catalytic activity and are extremely lipophilic. Their amino acid composition is nearly the same. The plasma membrane consists largely of structural proteins. The average molecular weight of structural proteins is 3×104. Enzymes from the major component of many membranes and are catalytic proteins. The endoplasmic reticulum, mitochondria, and plasma membrane contain many enzymes. The structure of the enzymes varies from membrane to membrane. Some 30 enzymes have been detected in isolated plasma membranes. Those most constantly found are 5′ nucleotidase, Mg^{+}ATP ase, $Na^{+}-K^{+}$ activated $-Mg^{2+}$ ATP ase, alkaline phosphatase, adenyl cyclase, acid phosphomonoesterase, and RNAs.

Some enzymes have a preferential localization ; for example, alkaline phosphates and ATPase more abundant at the bile capillaries, while disaccharidases are present in microvillae of the intestine. A specific localization with a mosaic arrangement has been postulated for some of these enzymes. Disaccharidase forms 5 to 6 nm globular units coating the membrane of the microvillae. The plasma membrane lacks the respiratory chain and glycolytic activity.

Of all the enzymes mentioned, $Na^{+}-K^{+}$ activated $-Mg^{2+}$ ATPase is one of the most important because of its role in ion transfer across the plasma membrane. This enzyme is dependent on the presence of lipids and is inactivated when all lipids are extracted.

Carrier proteins or permeases transport substances cross the membrane against the concentration gradient. The molecular weight of structural and carrier proteins are rather similar.

The amount of protein varies in different cell membranes. Myelin contains only 20% protein, plasma membranes about 50% and the inner mitochondrial membrane 75%.

The plasma membrane proteins fall in two main categories, *intrinsic* or *integral proteins* and *extrinsic* or *peripheral proteins.* The former are firmly associated with the membrane, while the latter have a weaker association and are bound by electrostatic interaction. Human erythrocyte glycoproteins consist of three chemically distinct regions. (1) N-terminal part external to the membrane and containing all the carbohydrates, (2) a middle hydrophybic region located within the membrane and (3) a hydrophilic C-terminal portion, rich in proline but lacking carbohydrate, on the internal side of the membrane. The middle intramembranous part appears to be intimately associated with the membrane phospholipids.

The glycoprotein of human erythrocyte membrane has a polypeptide backbone to which are attached two types of oligosaccharide chains, straight chain and a longer brached one at about 10 sites.

MOLECULAR STRUCTURE OF PLASMA MEMBRANE

All biological membranes, including the plasma membrane and the internal membranes of eukaryotic cells, have a common overall structure ; they are assemblies of lipid and protein molecules held together by noncovalent interactions. The lipid molecules are arranged as a continuous double layer 4 to 5 nm thick. This lipid bilayer provides the basic structure of the membrane and serves as a relatively impermeable barrier to the flow of most water-soluble molecules. The protein molecules are "dissolved" in the lipid bilayer and mediate the various functions of the membrane ; some serve to transport specific molecules into or out of the cell ; others are enzymes that catalyze membrane-associated reactions ; and still others serve as structural links between the cell's cytoskeleton and the extracellular matrix, or as receptors for receiving and transducing chemical signals from the cell's environment.

All cell membranes are dynamic, fluid structures ; most of their lipid are protein molecules are able to move about rapidly in the plane of the membrane. Membranes are also asymmetrical structures : the lipid and protein compositions of the two faces differ from one another in ways that reflect the different functions performed at the two surfaces.

Although the specific lipid and protein components vary greatly from one type of membrane to another, most of the basic structural and functional concepts discussed in this chapter are applicable to intracellular membranes as well as to plasma membranes.

The Lipid Bilayer

The first indication that the lipid molecules in biological membranes are organized in a bilayer came from an experiment performed in 1925. Lipids from red blood cell membranes were extracted with acetone and floated on the surface of water. The area they occupied was then decreased by means of a movable barrier until a monomolecular film (a monolayer) was formed. This monolayer occupied a final area about twice the surface area of the original red blood cells. Because the only membrane in a red blood cell is the plasma membrane, the experimenters concluded that the lipid molecules in this membrane must be arranged as a continuous bilayer. The conclusion was right, but it turned out to be based on two wrong assumptions that fortuitously compensated for each other. On the one hand, the acetone did not extract all of the lipid. On the other, the surface area calculated for the red blood cells was based on dried preparations and was substantially less than the true value seen in wet preparations. Nonetheless, the conclusions drawn from this experiment had a profound influence on cell biology ; as a result, the lipid bilayer became an accepted part of most models of membrane structure, long before its existence was actually established (Fig. 2.1).

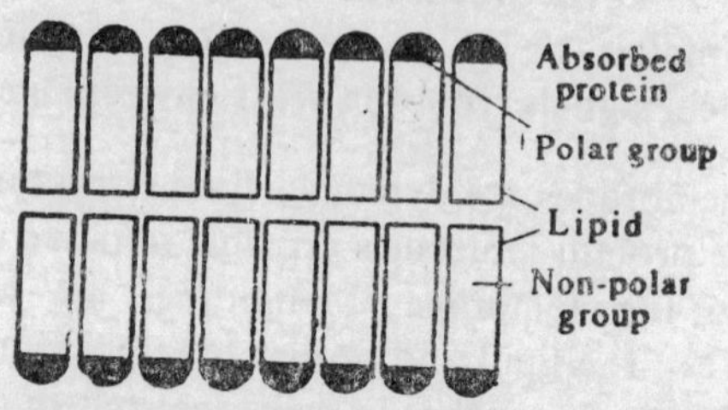

Fig. 2.1. The lipid membrane, of *Gorter* and *Grendel* (1925).

Danielli-Davson model or Protein-lipid-protein or sandwich model :

Harvey and *cole* (1931) indicated the existence of protein by studying the surface tension of cells. This led them to propose a lipo-protein model of the cell membrane. According to this model (Fig. 2.2) the plasma membrane consists of two layers of lipid molecules as shown in the lipid bilayer model. The lipid molecules have their polar regions on the outer side. Globulin proteins are thought to be associated with the polar groups of the lipids. The non-polar hydrophobic ends of the two layers of lipids face each other, whereas their polar hydrophilic ends are associated with protein molecules by electrostatic interaction. Protein-linked polar pores (Fig. 2.3) are present in the membrane. These pores are formed by periodic continuity of outer and inner layers of proteins of plasma membrane.

Modifications of Danielli-Davson Membrane Model : Several modifications of the above arrangement have been described :

(*a*) Some plasma membrane have folded β-chains of proteins on both the surfaces of lipid bilayer.

(*b*) Coiled α-chains of helical protein on the surfaces of lipid bilayer.

(*c*) With globular proteins on both the surfaces.

(*d*) With folded proteins on both the surfaces and helical proteins extending into the pores.

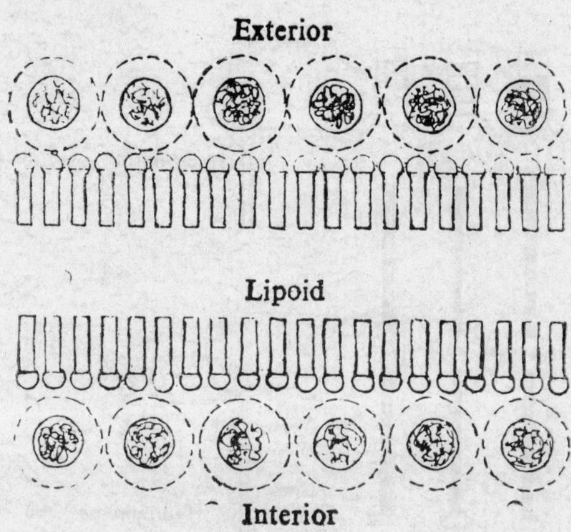

Fig. 2.2. The original Davson-Danielli model.

(*e*) With folded β-chain protein on one side and globular protein on the other side.

Robertson's unit Membrane Model

Unit membrane model was put forward in the year 1953 while studying cell under electron microscope. The basic unit membrane structure was considered to be general for a wide variety of plant and animal cells. All cell organelles such as Golgi body, mitochondria, endoplamic reticulum, nuclear membrane etc., have the unit membrane structure. The unit membrane is considered to be trilaminar with a bimolecular lipid layer between two protein layers. Two parallel outer dense osmiophilic layers of 20A° which correspond to the two protein layers. The middle light coloured osmiophobic layer is about 35A° in thickness corresponding to the hydrocarbon chains of the lipids. Thus the unit membrane is about 7A° in thickness. In this respect it resemables the Danielli-Davson model it, however differes from the Daneilli-Davson model in that the protein is asymmetrical. On the outer surface is muco-protein, while on the inner surface is non-mucoid protein. (Fig. 2.4).

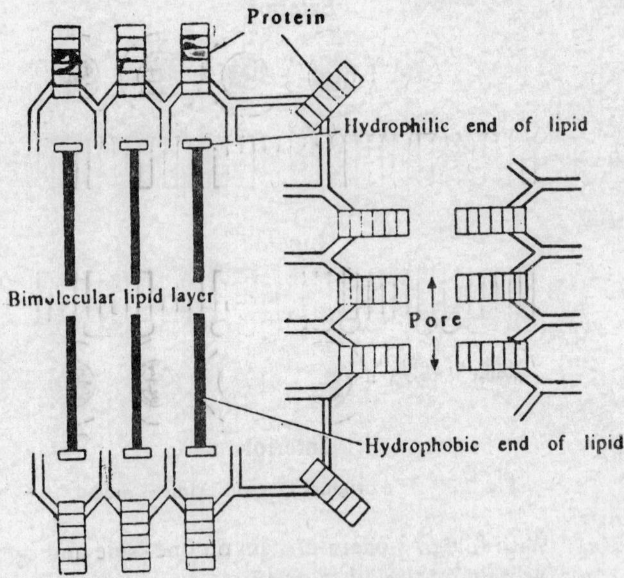

Fig. 2.3. The cell membrane with protein lined pores.

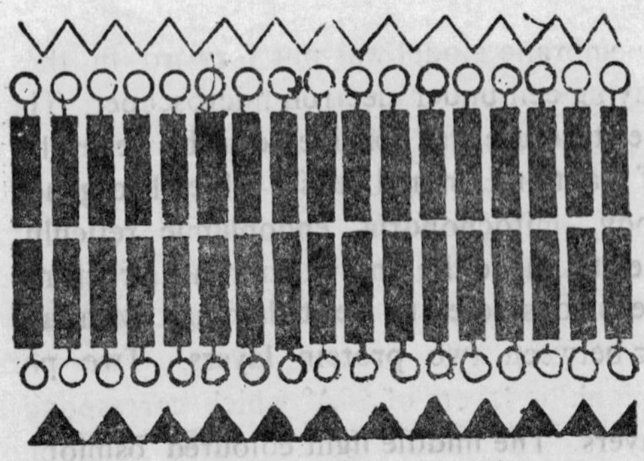

Fig. 2.4. Robertson's unit membrane model.

Objections to unit membrane theory. Objection to the unit membrane model increased during the 1960s and this led to reexaminations of lipid-protein interactions and to new models. Studies of *F.S. Sjostrand* (1963), of smooth endoplasmic reticular, mitochondrial and chloroplast membranes underscored the differences between observed features of membranes and the uniformity that was required by the unit membrane concept. Mitochondrial and chloroplast membranes contain displays of particulate units in or on the membrane. The plasma membrane did not present the same appearance as mitochondrial or chloroplast membranes. It seemed that different models might be needed to described different functional types, This unsuitable approach became unnecessary when a mosaic membrane model was proposed.

Greater membrane model

Like the trilaminar model here too the lipid layer is sandwiched between two layers of structural proteins. *Robertson* described the different nature of outer and inner surfaces of the membrane. The inner surface was thought to be covered with unconjugated protein, and the outer surface with glycoprotein, which is superimposed on the structural protein obigosaccharide chains with negatively charged sialic acid terminals are attached to the glycoprotein.

Micellar Model

An alternative interpretation of molecular structure of plasma membrane has been postulated by *Hilleir* and *Hoffman* (1953). They have suggested that biological membranes may have a non-lamellar pattern, consisting instead of a mosaic of globular subunits known as *micelles*, which have a lipid core and a hydrophilic shell of polar groups. Lipid micelles are possible building-blocks for membranes since they tend towards spontaneous association. In this model of membrane structure the protein components of the membrane may form a monolayer on either side of the plane of lipid micelles. Individual units of the micellar mosaic might be replaced by individual enzymes molecules or by arrays of enzymes with a precise-three dimensional organization, enabling specific functions to be 'built in' to the membrane structure. The spaces between the globular micelles are thought to form water-filled pores 0.4 nm (4A°) in diameter, lined partly by the polar groups of the micelles and partly by the polar groups of associate protein molecules.

Fluid Mosaic Model

This model was proposed by *Singer* and *Nicolson* (1972). According to this concept, the lipid molecules are arranged to form a rather continuous bilayer that forms the structural frame work of plasma membrane. The protein molecules are arranged in two different manners. Some proteins are located exclusively adjacent to the outer and inner surfaces of lipid bilayer and are called *extrinsic proteins*. Other proteins penetrate lipid bilayer partially or wholly and form *integral* or *intrinsic proteins*.

The lipids and integral proteins of plasma membrane are *amphipatic* in nature. The term amphipaty was coined by *Hartley*, 1936 for those molecules which have both hydrophobic and hydrophilic groups. The amphipatic molecules tend to constitute liquid crystalline aggregates in which hydrophobic or non-polar groups are situated inside the bilayer, and hydrophilic groups are directed towards the water phase. Therefore, the lipid molecules form a rather continuous bilayer. The integral proteins are interecalated in the lipid bilayer, with their polar regions protruding from the surface and non-polar regions embedded in the lipid bilayer. This arrangement explains why the active sites of enzymes and antigenic glyco-proteins are exposed to the outer surface of the membranes. The quasifluid structure of plasma membrane explains the movement of cluster of protein molecules of considerable size across the membrane.

Pores in Plasma Membrane

Plasma membrane is perforated by pores. These have a diameter of about 0.35nm (nanometer), slightly larger than the sodium ions. Less than 0.1 percent of the plasma membrane is perforated by pores while 99.9 percent of the cell surface is impenetrable for ions. Several models of structures of pores have been proposed. Some of them are :

1. *Structural pores.* These are permanent cylindrical holes that interrupt the otherwise continuous bilayer sheet.

2. *Dynamic pores.* These pores are transient cylindrical holes rather than being permanant. These appear only at the time of intake.

3. *Paving block pores.* According to this concept the pores are regarded to be the corners of the closely filled nearly hexagonal paving blocks of lipid and protein subunits.

4. *Protein channel pores.* These pores are considered to be parts of the lipid-globular protein mosaic model. These form small channels of specific proteins embedded in the membrane through which ions and small molecules can diffuse.

5. *Ionophore.* The ionophores are small polypeptides whose one end is hydrophobic and other hydrophilic. The hydrophobic (outer) end dissolves in the membrane while the hydrophilic end (inner side) picks up ions or water-soluble materials and dumps them on the other side. The ionophores help in exchange of substances from or into the cell.

SPECIALIZATION OR MODIFICATIONS

With the increased resolution of the electron microscope numerous specialization of cell surface have been recognized. Following description of *Fawceett* (1958) deals with the various specializations of plasma membrane studied topographically.

Microvilli

In the intestinal epithelium microvilli are very prominent and form a compact structure that appears under the light microscope as a striated border. These microvilli, which are 0.6 to 0.8 μm long and 0.1 μm in diameter, represent cytoplasmic process covered by the plasma membrane. Within the cytoplasm core fine microfilaments are observed which in the subjacent cytoplasm form a terminal web. The outer surface of the microvilli a covered by a coat of filamentous material (fuzzy coat) composed of glycoprotein macro-molecules. Microvilli increase ıhe effective surface of absorption. For example, a single cell may have as many as 300 microvilli, and in a square millimeter of intestine there may be 200,000,000. The narrow spaces between the microvilli form a kind of sieve through which substances must pass during absorption.

Numerous other cells, in addition to intestinal epithelium, have microvilli, although fewer in number. They have been found in mesothelial cells, in the epithelial cells of the gall-bladder, uterus, and yolk sac, in hepatic cells, and so forth.

The brush border of the kidney tubule is similar to the striated border, although it is of larger dimensions. An amorphous substance between the microvilli gives a periodic acid-Schiff reaction for polysaccharides. Between the micro-villi, at the base, the cell membrane invaginates into the apical cytoplasm. These invaginations are apparently pathways by which large quantities of flow enter by a process similar to pinocytosis.

Desmosomes or Macula Adherens

Desmosomes are cell junctions found mainly in the cells of simple columnar epithelium. These occur as specialized areas along the contact surfaces. Under light microscope the desmosomes are seen as darkly stained bodies. Under electron microscope these appear as button-like thickenings on the inner surface of plasma membranes of adjacent cells at the point of contact.

The thickenings are traversed by fine cytoplasmic fibrils called *tonofibrils*, which form a kind of loop in a wide arc. These filaments stabilise the junction and act as anchoring sites for the cytoplasmic structures.

The plasma membranes of adjoining cells in the regions of desmosomes are separated by an intercellular space about 30-35 nm. It is filled with intervening dense coating material that forms a dark line in the middle. It is formed of muco-polysaccharides and proteins.

The desmosomes are primarily concerned with cell adhe-sion, but also help in maintaining cell-shape, providing it rigidity and cellular support. The former is brought about by the intercellular coating substance and latter by the tonofibrils.

Plasmodesmata

Sometimes, the cells are joined by bridges of cytoplasm passing between pores of cell wall or plasma membrane between the adjacent cells, such connections are called plasmodesmata. They are usually simple but anastomosing plasmodesmata may also found. Their distribution and number may also very considerably. They were discoverd by *Tangl* (1879) and were named as such by *Strasburger* (1882). Endoplasmic reticulum often is closely associated with the cell surface, at the points where plasmodesmata are present. Through them cytoplasmic continuity is often maintained among the adjacent cells. They provide a mean for interaction between adjacent cells which are separated in other regions. Through them the material can pass from cell to cell.

It is not known whether all plasmodesmata are similar to each other. There exist some difference because they are not only produced at the time when cell divides but also formed spontaneously between cells that have grown into contacts with one another, e.q. tyloses in xylem vessel elements. They may occur singly or they can be aggregated into groups. In many primary walls, the plasmodesmata are usually associated with a reduced desposition of wall material, and the area then is known as primary pit or field.

Hemidesmosomes

These are found in the basal surface of some epithelial cells. Their structure is similar to desmosomes but these are represented by one half ; their counterpart is usually represented by collagen fibrils.

Terminal bars

The terminal bars are also known as intermediary junctions or *zonula adhaerens*. The terminal bars are similar to desmosomes except they lack in the tonofibrils. In terminal bar the plasma membrane is thickened and the cytoplasm of thickened area is dense. The terminal bars occur in the intermediary portion of the plasma membrane of columnar cells just below the surface. The correct identity of zonula adhaerens is still questionable.

Membrane Interactions

Another aspect of cell membranes that deserve discussion is the interaction between membranes of different cells. Intercellular communication is important in many cell functions and especially during development of the organism, when cells are constantly interacting with other cells. The nature of membrane interactions may vary from complete cytoplasmic bridges between cells to localized areas of membrane junctions that may involve an area of contact as small as a few angstroms or as large as several micrometers. The structural nature of the actual contact generally falls into one of three categories ; gap junctions, tight junctions, and septate junctions.

Gap junctions appear as multilayered structures when observed with the electron microscope. They appear to be two unit membranes closely apposed to each other with a 20 to 40A° gap between. The total thickness of the entire gap junction is 170 to 190 A°, and they are found in both vertebrates and invertebrates. They are not found in skeletal muscle fibers or red blood cells.

Tight junctions are found only in vertebrates and occur in cells such as epithelial cells. These junctions appear to be true fusions between the two membranes, and they are 100 to 140 A° thick.

Septate junctions have been found only in invertebrates. They are much larger than the other types of junctions and are characterized by electron-dense cross bridges that extend between the two cell membranes.

PERMEABILITY OF MEMBRANES

The membranes of a cell pass small ions and molecules through them. The passage of ions or molecules may occur as passive diffusion, or active transport evolving the expenditure of energy. In passive diffusion, membranes may be classed according to their degree of permeability.

(1) *Impermeable*. A membrane of this kind allows nothing to pass through it. Certain unfertilized fish eggs, such as trout, are permeable only to gases, water labelled with deuterium does not peneterate the egg.

(2) *Semipermeable*. No cell membranes are in this category. A model membrane may be constructed to allow passage of water molecules, but no solute particles.

(3) *Selectively permeable*. Most membranes of the cell belong to this category. Such membranes allow water and certain selected ions and small molecules to pass through, but prohibit other ions as well as small and large molecules.

(4) *Dialyzing membranes*. The endothelial cells and their basement membranes of the capillaries and nephron can act as a dialyzer. In this way hydrostatic pressure forces water molecules and crystalloids across the membrane down their concentration gradients while restricting the passage of colloids.

Membrane Function

In order to survive, the cell must exchange with its environment. Metabolic essentials, such as oxygen, salts, and nutrients, must get into the cell, and waste products, such as carbon dioxide and other metabolites, must pass out of the cell. All of these exchanges take place across the cell membrane by a variety of processes. Although it is really not possible to present one unifying concept of membrane transport, it is reasonable to group the membrane transport phenomena by the type of molecule being regulated.

Passive Diffusion

Passive diffusion is straight diffusion of water, O_2, or CO_2 through the cell membrane from an area of greater concentration to an area of lower concentration. The rate and speed of the diffusion is generally proportional to the difference in concentration of the substance between the two regions. When the concentration in both regions becomes equal, net diffusion ceases.

Free Diffusion

According to a large number of evidences, many substances move though membranes at rates of free diffusion that are directly proportional to their solubility in lipid. Water molecules are a notable exception to this rule, since they freely diffuse through membranes regularly and rapidly. It has been suggested that membranes contain 1 nm-wide pores that are lined by hydrophilic residues. Such openings would be large enough for water molecules but would hardly accommodate other water-soluble substances to pass through the narrow passageway. The suggestion is quite logical to explain diffusion of water through membranes, but there is little direct evidence to support the concept.

Free diffusion along a concentration gradient could be maintained for entry and exit of metabolites and waste products as long as substances were changed or washed away to keep one end of the gradient lower in concentration. If metabolites are chemically changed on entering the cell, then the concentration of the metabolite remains high outside the cell and is kept low inside the cell if it is not retained in its original form. Similarly, wastes leaving the cell could continue to move from the higher concentration inside as long as these substances were washed away or removed somehow from the immediate vicinity of the cell.

Facilitated Diffusion

Certain substances, such as glucose and amino acids, also move across the membrane as a function of a concentration differential. However, these substances do so only up to a certain concentration level. When the concentration of glucose outside the cell surpasses a certain level, the rate of diffusion no longer increases. It has been suggested that this is because certain specific reactive sites on the cell surface are saturated with the glucose molecules. This theory proposes that there are specific membrane associated carrier molecules that complex with the glucose and "facilitate" its diffusion into the cell.

Active Transport

Active transport often involves charged ions, such as Na^+ and K^+, and electrolytes. These substances do not necessarily move across the cell membrane in response to a concentration differential. In fact, active transport is a mechanism that promotes the movement against a concentration gradient and results in the buildup of a marked concentration differential. Because active transport is process that work against a concentration gradient, it is not surprising to learn that it requires the expenditure of energy. The process involves the use of carrier molecules within the cell membrane proper. These molecules apparently shuttle back and forth between the inner and outer cell membrane surfaces and either pick up or release the particular molecule being regulated. This entire process involves the expenditure of energy according to the equation $ATP \rightarrow ADP + P_1 + energy$.

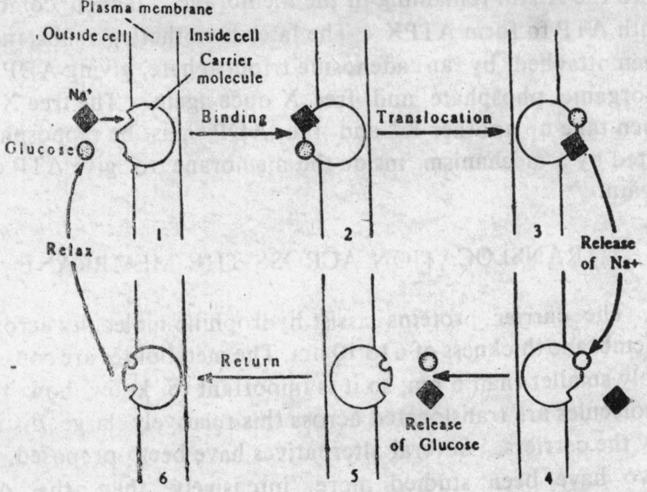

Fig. 2.5. Sodium Pump.

Ion Pump

Potassium ions, often, suggest that there must be energy available for such a "pumping" mechanism. *Hepel* observed in 1939 that in animals fed with potassium—deficient diet, the muscle cells exchange almost 50% of their K^+ by sodium (Na) even though under normal circumstances, these cell are practically impermeable to sodium ion. The passage of Na^+ could be carried out exclusively by an active process, which is sometimes called the "Sodium pump" (Fig. 2.5). Because there is a good evidence that other ions may be transported by a similar mechanism. We can speak in general of an "ion pump". Such a mechanism was postulated by *Hober* in 1911 under the name of "Physiological permeability". There is a carrier molecule 'X'.

What this carrier X might be is unknown, but in the case of Na^+ and K^+ transport, it is now thought possibly to be ATP itself or an ATPX compound. The entire mechanism could work as shown in the figure given above.

K^+ combines with X to form K^+-X, which then moves across the membrane and unloads the K^+ to the inside cell, with the X still remaining in the membrane. It then combines with ATP to form ATPX. The later hypothetical substance is then attached by an adenosine triphosphate, giving ADP and inorganic phosphate and free X once again. The free X can then take up another K^+ and the ADP must be rephosphorylated by a mechanism inside the membrane to give ATP once again.

TRANSLOCATION ACROSS THE MEMBRANE

The carrier proteins assist hydrophilic molecules across a membrane thickness of 6 to 10 nm. The metabolites are considerably smaller than 6 nm, so it is important to know how these molecules are translocated across this relatively large distance by the carriers. Several alternatives have been proposed, but two have been studied more intensively than the other possibilities.

One alternative hypothesis postulates that the carrier binds with the hydrophilic molecule and that then the entire transport protein rotates across the membrane and delivers its bound metabolite to the other side (Fig. 2.6). The second alternative proposes that the carrier is fixed in place within the membrane, and that the carrier molecule undergoes a conformational

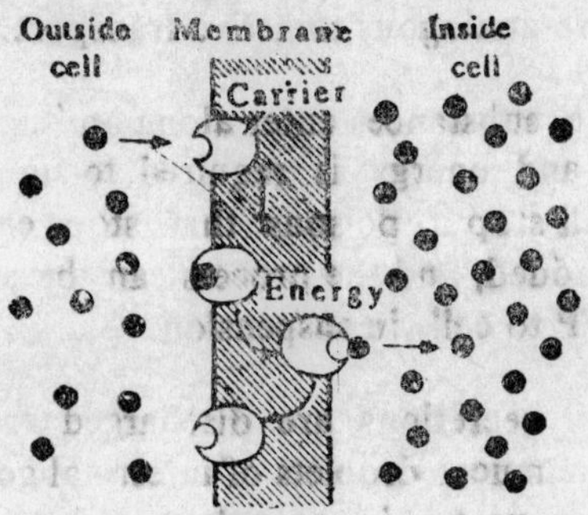

Fig. 2.6. Showing translocation across the Membrane.

change that translocates the binding site across the membrane and the bound metabolite along with it at the same time. Once the metabolite has been translocated the binding site is freed and restored to its original conformation, ready to bind another hydrophilic molecule in another transport event. This second alternative has been referred to as the *fixed-pore mechanism* (Fig. 2.7). The first alternative is known as the *carrier mechanism*.

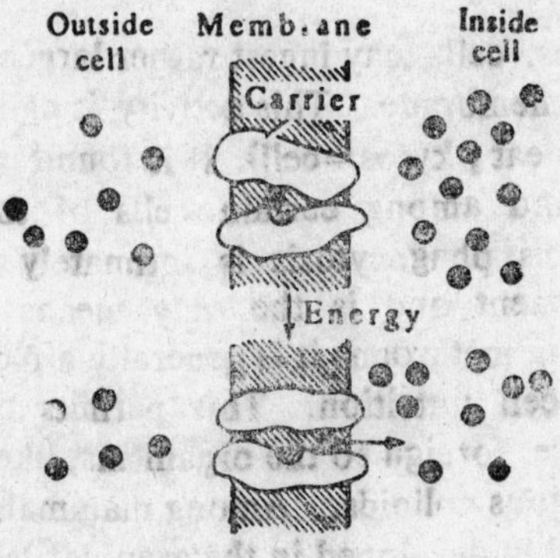

Fig. 2.7. Showing the fixed pore mechanism.

Transport by Vesicle Formation

Most cell membranes can enclose materials in vesicles and bring the substances into the cell in this way, or package materials for discharge from the cell in a reverse process. The process is called *endocytosis* when materials are brought in, and *exocytosis* when vesicle discharge takes place. Endocytosis has features that are analogous to active transport.

For example, substances enter along an "uphill" concentration gradient, and energy is required to support the process. Endocytosis will stop if poisons that stops energy production in the cell are added, and the process can be stimulated by the addition of ATP to cells in suspension.

Various cell secretions are discharged from the cell by exocytosis. The mucus droplets of intestinal goblet cells, digestive enzyme precursors in granule packages from pancreatic cells, and other cell secretions leave the cells in which they are produced by this general pathway. Although more difficult to see by light microscopy, numerous endocytotic infolded regions can be seen along the plasma membrane in suitable electron micrographs.

According to the nature of intake of food or foreign substances, the endocytosis may be classified into types :

(*i*) *Phagocytosis*

In some cases, cells may ingest rather larger, solid particles through plasma membrane. This activity is called *phagocytosis* (Gr. *Phagein*=to eat ; kytos=cell). It is found in large number of protozoans and among certain cells of the metazoans. Among protozoans, phagocytosis is intimately associated with amoeboid movement and is the only means for obtaining nutrition. Among metazoans it is generally a means of defense than serving the cell nutrition. This permits the ingestion of bodies which are foreign to the organisms, like bacteria, dust particles and various colloids. Among mammals this property is found very highly developed in the granular leucocytes. The particles become absorbed at the surface of membrane and later on they are taken into the cytoplasm by infolding of the plasma membrane which soon piched off forming vesicles consisting of the particle envelope by a membrane. The vesicle so formed may then fuse and assume various sizes, or they may fragment.

Is the contact of transportant enough to stimulate the sequence of events culminating in the formation of vesicle ? Certainly not. Whatever information available clearly indicate that the organism or substance to be phagocytized must be *cationic* in nature. There are, however, cases when the transportant is not cationic in nature (for instance eating of bacterium by W.B.C.). In such cases it has been observed that the transportant first become coated by a phagocytosis-promoting substance. The substance is called *opsonin*, which is cationic in charge.

Kinds of phagocytosis. According to the physical and chemical nature of foreign substance following types of phagocytosis have been recognised :

(*a*) *Ultraphagocytosis or colloidopexy.* The process in which plasma membrane ingests smaller colloidal particls is known as *colloidopexy* or *ultraphagocytosis*, *e.g.*, leucocytes and the macrophagic cells of *mammals.*

(*b*) *Chromopexy.* When the cell ingests colloidal chromogen particles phagocytotically the process is known as *chromopexy*, *e.g.*, some mesoblastic cells.

(*ii*) *Piocytosis*

When the ingestion of fluid material in bulk takes place by the cell through the plasma membrane, the process is known as *pinocytosis* (*Gr. pinein*=to drink). The process of pinocytosis was first of all observed by *Edward* in *Amoeba* and by *Lewis* (1931) in the cultured cells.

In the process of pinocytosis, the particle free globules of fluid are surrounded and ultimately engulfed by the clasping folds of the cytoplasm. The plasma membrane forms membranous vacuoles around the fluid globules. Such membrane bound vacuoles are called *pinosomes*. The pinosomes, later on, are transported to the interior of the cell where they are fused with the secretory granules or lysosomes. In the food vacuoles the digestion of food substances takes place and the digested food is diffused to the surrounding cytoplasm. The pinocytosis can be observed ordinarily in the cells by light microscopy.

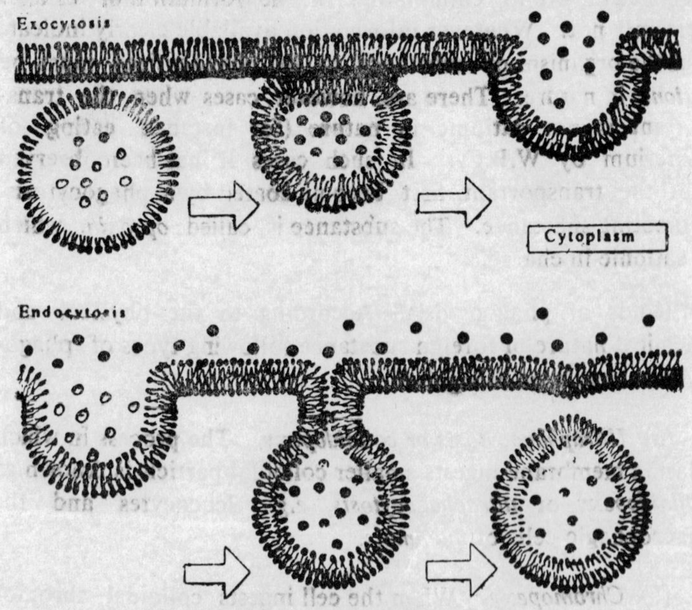

Fig. 2.8. Showing exocytosis and endocytosis

Micropinocytosis. Electron microscopic observation have been made on the pinocytotic process at sub-cellular or sub-microscopic level in the cells. The pinocytosis which occurs at sub-microsocopic level is known as micropinocytosis. In the process of micropinocytosis, the plasma membrane invaginates to form small vesicles of 650A° diameter. These vesicles have openings on both outer and inner surfaces which suggest the possible transportation of fluids through these vesicles to the cell. The micropinocytosis have been observed in endothetial cells, Schawn and satellite cells of nerve ganglion, macrophages, muscle cells and reticular cells, etc.

Emeiocytosis or Exocytosis or Cell Vomiting

The process of exudating the secretory products to outside the cell cytoplasm is known as *emeiocytosis.* In cells of pancreas, the vacuoles containing enzymes move from the interior of the cytoplasm towards the surface. Here they fuse with plasma membrane and discharge their contents to the exterior (Fig. 2.8).

Origin of Plasma Membrane

There are two interpretations to explain the nature and origin of plasma membrane :

(1) Plasma membrane may be an independent organelle which autonomously enlarges with growth.

(2) Cytoplasm may be passive and plasma membrane may be formed due to accumulation of surface active substances at the phase boundary. Such an explanation gets support from the observation that a new plasma membrane is formed wherever it is destroyed, resulting into new plasma surface. However, requisite quantities of lipoproteins must be present in the ground plasm, for the formation of plasma membrane.

A relationship between plasma membrane and other cell organelles has been postulated. In many cases, connections between plasmalemma and endoplasmic reticulum (ER) and also between plasmalemma and nuclear envelope have been shown. It has also been suggested that plasmalemma could give rise to endoplasmic reticulum. However, according *to Frey-Wyssling* and *Muhlethaler* (1955), published micrographs do not clearly settle this issue. Therefore, it is doubtful whether or not there is a direct connection between plasmalemma and ER, but such a connection with Golgi membranes has been shown.

3

The Cell Wall

In vascular plants only certain cells related to reproductive processes and early embryology are naked ; all others have cell walls. This wall, at first very thin and delicate, is modified in various ways as the cell matures. The more important changes are increase in extent and in thickness ; in chemical nature ; and in modification of grosser physical structure, such as the absorption of the end walls of vessels. The presence of a prominent wall, especially after the earliest stages of development, makes cellular structure in plants distinct—a feature in strong contrast with the condition in animals where the limits of the protoplast are less readily discernible.

In recent years the origin and structure of the cell wall have received critical attention and are now much better understood. A new interpretation is placed on the obscure earliest stages of wall formation, the structure of the wall is known in detail, and necessary changes have been made is the terminology applied to the wall.

Nature of the Wall. The cell wall is commonly looked upon as a secretion of the protoplast, laid down upon its surface. Strictly as such, the wall would be a nonliving layer. Although this is the view now generally accepted, the opinion is held by some that the wall in early stages, and the layers contiguous with the protoplast in later stages, may contain protoplasmic material and therefore form a part of the living unit. The middle lamella may also contain living substance in early stages and possibly as long as the protoplast remains active. If the wall is a living adjunct of the protoplast, certain facts of structure and behaviour of cells are more readily understood whereas others are difficult of interpretation.

Origin of the Wall. Wall formation is first evident at the close of nuclear division. As the daughter nuclei reach the telophase, the spindle fibers become stronger along the equatorial line and weaker near the nuclei (Fig. 3.1). With the

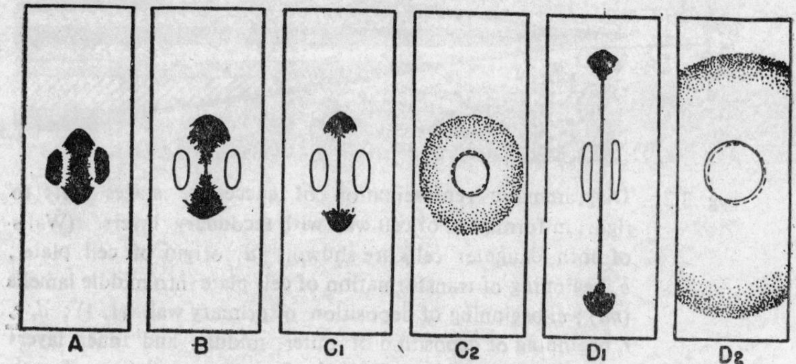

Fig. 3.1. Diagrams to show cell-wall formation during cell division of elongate cells, as seen in longitudinal sections. *A*, at late telophase stage : the phragmoplast enlarging equatorially, with droplets of cell-plate substance forming along the median line ; the spindle fibers disappearing in the polar and central regions. *B*, the beginning of the cell plate (lying at right angles to the page), formed by fusion of the droplets in the center of the phragmoplast ; new fibers forming on the equatorial periphery of the phragmoplast, the older ones progressively disappearing. *C1*, the phragmoplast, now an annular band, continuously expanding peripherally, has extended the cell plate in all directions. *C2*, section at right angles to *C1*, the phragmoplast (lying in the plane of the page), has extended nearly to the mother-cell wall in the central part of the cell ; the cell plate, a delicate membrane in the plane of page, not shown. *D1*, the cell plate and phragmoplast ring extended much farther than in *C1*, *D2*, section at right angles to *D1*, the phragmoplast ring has reached the lateral walls in the central part of the cell and disappeared there, leaving two arcs (kinoplasmasomes), which will continue to progress to the ends of the cell ; the cell plate has also reached the lateral cell walls in the center, dividing the cell in the central part, while both ends remain undivided.

surrounding, somewhat denser cytoplasm (the kinoplasm) they form a rather poorly limited, bulging-barrel-shaped structure, the *phragmoplast*. Along the median line of the figure, droplet-like particles appear, increase in size, and merge to form a fluid *plate* that divides the phragmoplast (Fig. 3.2). The plate is

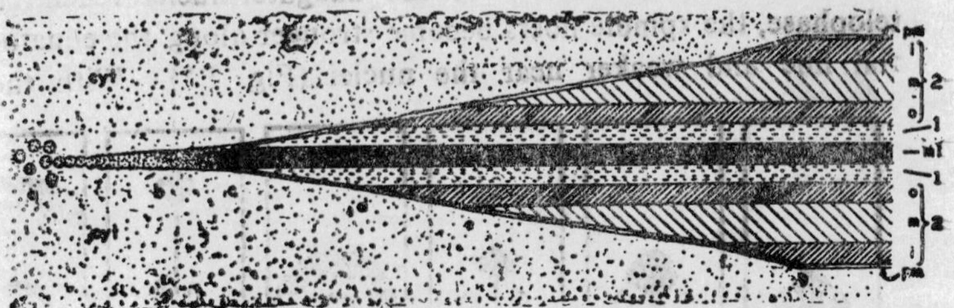

Fig. 3.2. Diagrammatic representation of successive stages (left to right) in formation of cell wall with secondary layers. (Walls of both daughter cells are shown.) *a*, origin of cell plate; *b*, beginning of transformation of cell plate into middle lamella (*ml*); *c*, beginning of deposition of primary wall (1, 1); *d*, *e*, *f*, beginning of deposition of outer, middle, and inner layers (*o*, *m*, *i*) of secondary wall (2, 2); *g*, completion of wall thickening; *pm*, plasma membrane; *cyt*, cytoplasm. (*Based on researches of W. A. Becker, I. W. Bailey, T. Kerr, and others; from Sharp.*)

first evident at the center of the phragmoplast and then extends marginally in all directions toward the mother-cell wall. As the plate forms, the central spindle fibers disappear, and with the extension of the plate, the peripheral spindle fibers also progressively disappear—the outer ones persisting longest. When the plate is complete, reaching the wall on all sides, its substance becomes less fluid, and delicate membranes appear on both its surfaces. These membranes are probably deposited by the daughter protoplasts. They constitute the first stages of the new cell walls of the daughter cells. The cell-plate material is gradually transformed into intercellular substance, which becomes an *intercellular layer*. The nature of the substance forming the cell plate is unknown; it is possibly cytoplasmic, and the spindle fibers may be "lines of flow" in the cytoplasm.

Where the mother cell is small and isodiametric and the nucleus large and central, the division figure is large, and no complex structural conditions are involved in the division of the protoplast. But cell division and the completion of the new wall are frequently much more complex. Division in cells with a large central vacuole involves predivision changes in form and position of nucleus and in structure of the vacuome (Fig. 3.3)—the cytoplasm increases in amount, extending in strands through the vacuole ; the nucleus becomes more rounded and moves away from the wall along one of the cytoplasmic strands ; the strands gradually become concentrated in a layer (the "phragmosome") in the plane in which the new wall is to be formed. Nuclear division and wall formation then take place much as in cells with abundant cytoplasm.

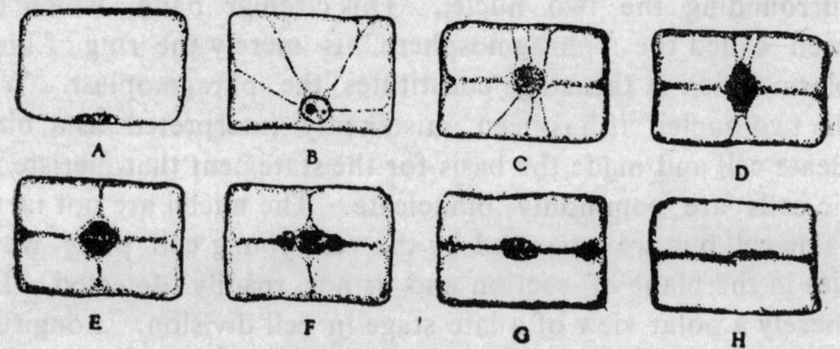

Fig. 3.3. Division in cell with large central vacuole. Cytoplasm increases in amount forming layer in position of new wall ; nucleus becomes globose and migrates to central position ; normal nuclear division with extending phragmoplast.

In cells in which one diameter is markedly greater than the others, especially in those that are very long, and when division is longitudinal, as in cambium cells, the phragmoplast persists long after nuclear division is complete and builds the cell plate far beyond the daughter nuclei and to the ends of the mother cell (Fig. 3.1 and 3.3). Its polar parts and the central fibers disappear, as in isodiametric cells, with the formation of the first part of the cell plate (that between the daughter nuclei) and the plate increases in diameter, expanding as an equatorial belt in all directions. New fibers are added progressively at the periphery as the older ones disappear. The new fibers are short, forming only close to the plate, and recurve toward the nuclei (Fig. 3.1 D). The barrel-shaped phragmoplast thus becomes a continuously expanding, annular band of short fibers and dense cytoplasm. As the band expands, it is broken up into arcs (which have been called "kinoplasmasomes") as the parts that reach the nearer walls disappear. The persisting arcs continue to progress to the farthest parts of the cell, adding to the cell plate on its margins until the cytoplasmic body is completely divided. The constant presence of the fibers as the plate is formed and their renewal on its progressing margins far from the nuclei suggest close relation of these structures to the building of the plate.

As the phragmoplast ring moves away from the daughter nuclei, it appears, when seen in face view, as a halo-like band surrounding the two nuclei. This circular band, which has been called the "phragmosphere," is merely the ring of kinoplasm which at this stage constitutes the phragmoplast. With the two nuclei, it has been mistakenly interpreted as a binucleate cell and made the basis for the statement that meristematic cells are commonly binucleate. The nuclei are not in the same cell but are separated by the very young cell plate which lies in the plane of section and is not readily detected. It is merely a polar view of a late stage in cell division. Longitudinal cross sections show the phragmoplast as two rounded or wedge-shaped masses at the ends of the weakly developed cell plate (Fig. 3.1 D).

In such cells as cambium initials, procambium cells, young sieve-tube elements, young fibers, and other elongate cells, cell-plate extension is of extraordinary degree and the wall often of unusual form. Curved and even cup-shaped walls may be built by the changing plane of the ring as the plate is formed. The breaking up of the phragmoplast ring into arcs and the slow completion of the division of the protoplast are common but neglected features of cell division. This represents not a special type of cell division but an adaptation to cell form.

GROSS STRUCTURE OF THE WALL

In mature, thick-walled cells, a concentric layering is usually evident in the wall. The layers represent successive deposits of wall substance by the protoplast upon the cell plate. The layers differ from one another in physical and chemical nature, their capacity for adaptation to changes in cell size and shape, and in extent over the earlier formed wall. They fall into two major groups, constituting the *primary wall* and the *secondary wall.* The primary wall usually consists of a single layer. The secondary wall is made up of one to many layers, most frequently of three.

The Primary Wall. The membrane formed on the surface of the cell plate is the first stage of the primary wall. As cell enlargement begins, this membrane becomes somewhat unevenly thickened but remains delicate and mesh-like in finer structure. During rapid increase in area, the wall may maintain its thickness and unevenness, or be alternately thickened and thinned. The uneven thickening soon sets off more or less distinct thinner areas, *primary pit-fields,* sometimes known as *primordial pits.* Primary pit-fields are conspicuous in cambium initials. At this stage abundant plasmodesmata, protoplasmic connections between the protoplasts separated by the new walls, are evident, clustered chiefly in the primary pit-fields. The presence of large numbers of plasmodesmata perhaps determines the location of the pit-fields. The mesh-like earlier wall is probably also freely penetrated by cytoplasmic projections, the protoplasts remaining in continuity from the division stage, but such connections have not been demonstrated. At this early stage the wall substance is possibly highly protoplasmic, as may also be the middle lamella.

The young wall remains plastic and adaptable to changing volume and shape as the young protoplast continues growth. Inequalities in thickness persist, and alternate thinning and thickening may continue. With attainment of final size and form in the cell, the primary wall takes on mature structure and may thicken further. Small, well-defined thin areas, *pits* are formed, one or more in each of the primary pit-fields. The plasmodesmata are largely restricted to the pits, but scattered, even solitary, strands may occur outside these areas. The mature primary wall varies greatly in thickness and in chemical and physical structure in cells of different types and in cells of the same type in different tissues and different plants.

The Middle Lamella. As the first stages of the primary wall are developing, the substance of the cell plate becomes, by physical and chemical changes, a firm *intercellular layer,* a *true middle lamella,* which serves to bind the cells together. The term *middle lamella* has been in common use for a middle layer in the walls separating two protoplasts, or two cell lumina, that differs from the rest of the walls in staining and light-refracting qualities. Such a layer is, however, structurally and morphologically complex ; according to this loose usage the "middle lamella" includes the true middle lamella (intercellular substance) plus the contiguous primary walls of the two cells and often also the thin, first-formed layers of the secondary wall. The intercellular layer is the middle layer of this group of three or five layers and the term middle lamella should be restricted to this. In stained preparations the true middle lamella is commonly difficult to differentiate because of its close resemblance in chemical nature to the contiguous walls. For convenience, the term "middle lamella" will doubtless continue to be used for the compound layer, but it should be remembered that this is a loose and inaccurate use.

Along the line of contact of the new wall with the mother-cell wall the new and the old middle lamellas are separated by the primary wall of the mother cell (Fig. 3.4). Union of the layers of middle lamella is secured by the development, along the line of contact of the old and new walls, of a cavity—triangular in cross section—in the mother-cell primary wall, which enlarges and extends to the middle lamella, and the extension into this cavity of the intercellular substance of the new wall. When the cavity continues to enlarge, and the intercellular substance does not fill it, an intercellular space lined by intercellular substance is formed. Commonly the formation of intercellular spaces is more complex than this. More than two walls are concerned and the development of the interwall space is preceded, accompanied, or followed by a splitting of the middle lamella of the older wall nearby. The interwall space and the space formed by the splitting merge, and the cavity so formed enlarges by further splitting and by pulling away of surrounding cells under growth stresses. Increase in size of spaces sometimes results from shrinkage of cells. When two new walls are formed opposite one another, the two triangular spaces merge—sometimes with a split which appears between them—and a diamond-shaped space is formed. In meristems where cell divisions succeed one another rapidly and all the walls are thin, intercellular spaces seem to be formed by a simple splitting apart of cells through the middle lamella.

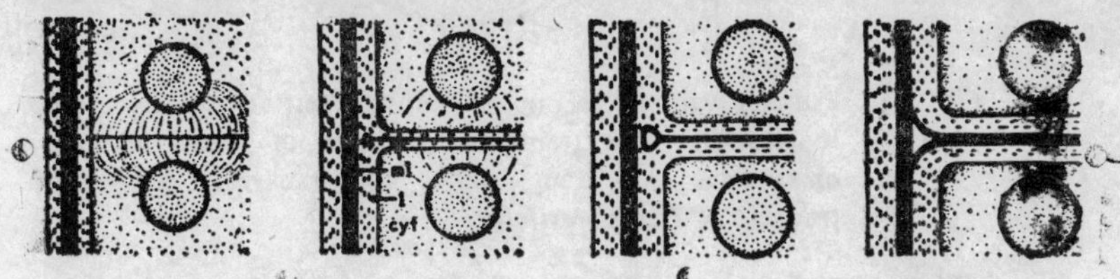

Fig. 3.4. Stages in the establishment of the connection between the middle lamella of a newly formed wall with that of the lateral wall of the divided cell, *ml*, middle lamella ; *l*, primary wall ; *cyt*, cytoplasm.

The Secondary Wall. In many cell types further wall-thickening occurs after cell size and shape are fully attained. The wall then formed is the *secondary wall.* This differs from the primary wall especially in that it is incapable of increase in area and that such modifications as occur are irreversible. The primary wall covers the protoplast except where the plasmodesmata occur ; the secondary wall is formed over the primary wall except over the pit membranes. As this wall thickens, the pit cavities become deeper, and in some kinds of cells, partly enclosed and complex in structure (bordered pits). The secondary wall tends to be more massive than the primary, and in most thick-walled cells it constitutes the major part of the wall. In the tracheids and vessels of the protoxylem the secondary wall covers much less of the primary wall ; it forms only as rings, spiral bands, and bars over the delicate primary wall.

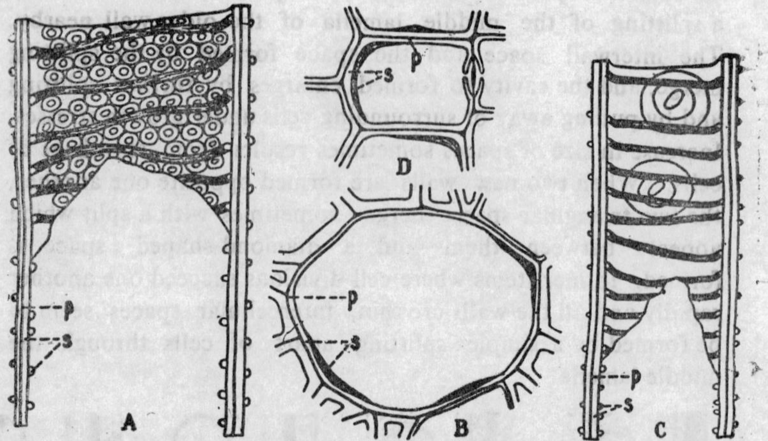

Fig. 3.5. Primary wall and secondary wall with tertiary spirals. *A,B,* longitudinal and transverse sections of vessel of *Tilia americana*; *C,D,* longitudinal and transverse sections of tracheid of *Taxus brevifolia.*

The last-deposited layer of the secondary wall of some wood elements—laid down over a wall with fully developed pits—has the form of delicate spiral bands (Fig. 3.5). Because these bands appear to represent thickening after the secondary wall is complete they have been called a *tertiary wall*, but they represent only the innermost layer of the secondary wall and are now called *tertiary spirals* or *spiral thickenings*. The term "tertiary spirals" better distinguishes them from the spiral (secondary) bands of protoxylem which they somewhat resemble. Tertiary spirals lie upon a thick secondary wall; protoxylem spirals lie upon a delicate primary wall.

MINUTE STRUCTURE OF THE WALL AND MIDDLE LAMELLA

Both primary and secondary walls are composed fundamentally of a complex mesh-like matrix of cellulose. Other substances fill the interstices of this framework and may be replaced later by different ones. In the primary wall pectic substances accompany the cellulose in early stages; the secondary wall is at first composed mostly of cellulose or of cellulose and hemicelluloses. Substances that at later stages replace, or are added to these are lignin, cutin, suberin, various inorganic materials, tannin, oils, and many others.

In minute structure the cellulose matrix of both walls is made up of aggregates of delicate fibrils that grade down in size to the limits of microscopic visibility. The fibrils are believed to be made up of *micellae*, aggregates of molecules, probably in crystalline form. The fibrils form complex, three-dimensional anastomosing systems which cohere firmly and are continuous throughout a wall. Variations in size, number, and arrangement of the fibrils give different structural patterns. The pattern may be dominantly radial or dominantly concentric. It is probably usually a combination of these—radial arrangement intensified at intervals during formation by increase in number or in size of fibrils so that a concentric layering is apparent.

The mature primary wall of many cells remains thin. When young, uneven thickening may give it reticulate structure, and when mature, abundant small pits may make it appear sieve-like. Where this wall is a place for food storage, as in endosperm and some sieve tubes, it becomes greatly thickened. Probably the thick wall of collenchyma cells, with its high percentage of water, is wholly primary. A distinction between a strongly thickened primary wall and a primary wall plus a secondary wall may be difficult to make without special study of optical characteristics.

Layering is not usually apparent in primary walls except with special treatment. In wood cells, especially in heartwood, the primary walls of adjacent cells and the middle lamella between them may be chemically closely alike and difficult to distinguish. This triple layer, together sometimes with parts of the secondary walls, has for this reason passed, in loose use, as "the middle lamella."

The secondary wall, which constitutes the larger portion of the wall in most thick-walled cells, commonly shows marked lamellate structure, the layers ranging in number up to many.

A division into three major layers is frequently seen, especially in wood cells : an outermost thin layer lying against the primary wall ; a thick median layer ; and an innermost thin layer. Constituting the "wall" between two protoplasts or lumina among such cells, there are nine major layers—six secondary layers, two primary walls, and the middle lamella (Fig. 3.6). The heavy median secondary layer itself frequently shows many minor layers varying in distinctness. The function of the thick secondary wall is primarily mechanical.

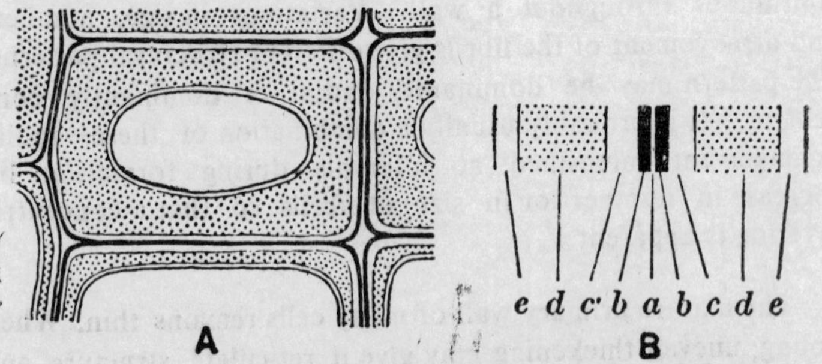

Fig. 3.6. Diagrams showing wall structure and middle lamella in thick-walled cells (wood fibers). *A*, cross section of one fiber and parts of seven adjacent fibers ; *B*, cross section of adjacent walls and middle lamella more highly magnified. *a*, middle lamella ; *b*, primary wall ; *c,d,e*, outer, median, and inner layers, respectively, of secondary wall.

The chemical and physical nature of all wall layers and of the middle lamella undergoes various changes as the walls mature, and may undergo even further changes at a later period, for example, during the transformation of sapwood into heartwood. The middle lamella, at first containing much pectic material, becomes heavily lignified in thick-walled tissues and cutinized in epidermal regions. It shows no evidence of layering. The primary wall likewise becomes strongly lignified in wood and most sclerenchyma. Much of the mineral content of the cell wall is found here. In epidermal cells of seeds and fruits this wall is often gelatinous or mucilaginous. Lignification, cutinization, and suberization are the major changes in secondary walls. Gelatinous layers are frequent in the secondary wall of wood.

Method of Wall Building. The method of addition of building material as the wall increases in area and thickness is not wholly understood. Two processes have been described : *apposition,* the placing of new particles in layers on the surface of the earlier formed wall ; and *intussusception,* the addition of new particles among those already in position. When the wall is considered a matrix to which new material is added throughout, intussusception must play a prominent part in increasing the extent and thickness of the wall. Particularly in the increase in area of primary walls, intussusception must be the chief method. The layering of the secondary wall and the varying direction of the fibrils of its successive layers suggest that the addition of particles is periodically restricted to the layers next to the protoplast. Hence, considering the major layers alone, increase in thickness is by apposition ; but each layer itself may be built by both apposition and intussusception. When the original matrix is increased by the addition internally of new micellae or fibrils, and when it is chemically transformed by replacement of particles or the deposition of new substances in the interstices of the meshwork, the method must be intussusception.

Thickenings of the cell wall. The secondary wall acquires different types of thickenings. This is specially seen in cells which are to differentiate into tracheary elements *i.e.*, vessels and tracheids. In these cells the thickening is due to the deposition of lignin. The deposition of lignin is not uniform, hence we find different types of thickenings. Chiefly the following types of thickenings are seen in xylem.

Annular. Lignin deposition is in the form of annuli or rings. These rings are unconnected. The space between the rings is unthickened (Fig. 3.7A).

Spiral. Lingnification occurs in the form of a spiral band (Fig. 3.7B).

Scalariform (*Ladder like*). Wall thickening is in the form of transverse rods, very much resembling the rungs of a ladder (Fig. 3.7C).

Reticulate. Lignin is deposited in the form of a network or reticulum. The unthickened areas of the cell wall are irregular in shape (Fig. 3.7D).

Pitted. Lignin deposition is almost uniform over the wall, except for small pores here and there. These pores are called *pits* (Fig. 3.7E,F).

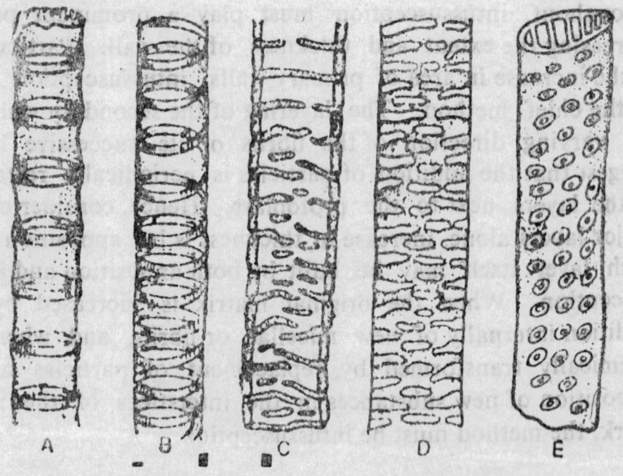

Fig. 3.7. Various thickenings of the cell wall A. Annular, B.B. Spiral, C. Scalariform, D. Reticulate, E. Simple pitted, F. Bordered pitted.

The microfibrillar organization of primary walls has been thoroughly described in recent reviews by Frey-Wyssling, Roelofsen and Wardrop and will not be considered in detail here. Since in the past few years important similarities in the organization of very diverse primary walls have become evident, in this we will emphasize general aspects of primary wall structure, noting only the more significant exceptions.

Simple multinet structure. The cellulose skeleton in the walls of some young isodiametric cells, such as the apical initials of onion roots is a rather disorganized mesh of microfibrils, although in young parenchyma cells of oat coleoptiles a net transverse orientation of microfibrils can be discerned. As elongation proceeds, a pattern becomes established in thin primary walls in which the majority of microfibrils show wide angular dispersion about a mean orientation transverse to the cell axis—the "tubular texture" of Frey-Wyssling. Closer examination of such walls reveals a difference in orientation of microfibrils on the inner and outer surfaces. Although the microfibrils are interwoven throughout the thickness of the wall and are not arranged in layers, nevertheless, on the inner surface they are transversely oriented in a fairly regular fashion whereas towards the outside they show greater dispersion and an increasing tendency to longitudinal orientation. Because of the model of wall growth proposed by Roelofsen and Houwink we shall call this type of arrangement "simple multinet structure." In this structure transverse microfibrils predominate, thereby accounting for the negative birefringence observed in the walls of many elongating cells.

Simple multinet structure was first demonstrated clearly in the walls of the sporangiophores of *Phycomyces*, staminal hairs of *Tradescantia*, cotton hairs, and hairs from the fruits of *Ceiba* and *Asclepias*. The same type of structure has since been described in the walls of parenchyma cells from oat coleoptile, onion root, celery petiole and bean stem, and in young collenchyma cells of celery. Scott *et al.* did not describe their results on onion roots in terms of multinet structure but their observations appear consistent with such an interpretation. Although the wall of the internodal cell of the alga *Nitella* is considerably thicker than the walls just mentioned, this too was found to possess simple multinet structure. All of these cells elongate over most of their length, but similar structure has also been observed in the primary walls of differentiating tracheids of conifers where growth is localized at the cell tips.

These observations indicate that simple multinet structure with predominantly transverse microfibrils is of general occurrence in thin primary walls. A possible exception is in root hairs which grow only at their tip where the wall contains a thin mesh of microfibrils with little or no preferred orientation.

This meshwork persists below the tip on the outside of the wall and a secondary wall of longitudinal microfibrils is apposited on the inside. The wall structure of a number of algae is also quite distinct from that of higher plants and has been given separate consideration recently by Preston and Roelofsen.

Longitudinal arrangements of microfibrils. In the walls of many parenchyma cells and of conifer tracheids axial ribs composed of longitudinally oriented microfibrils occur between regions of simple multinet structure. The ribs of coleoptile parenchyma are present in very young cells and persist throughout elongation. In the electron microscope face views of parenchyma walls clearly show the ribs on the outer surface, and reveal microfibrils running from longitudinal orientation in the ribs to transverse in the inter-rib regions. Transverse sections of such walls disclose two types of rib ; "integral ribs" in which microfibrils throughout the thickness of the wall are longitudinally oriented, and "outer ribs" that consist of a separate layer of longitudinally oriented microfibrils outside the microfibrils forming the simple multinet structure. Such ribs were previously regarded as secondary thickenings ; but the distinction between these structures and secondary wall is clear in oat coleoptiles and onion roots from sections of fully-elongated parenchyma cells in which layers of closely aligned, near-longitudinal microfibrils representing a secondary wall can be distinguished on the inner surface of the primary wall. The orientation of microfibrils in these secondary walls probably varies as Wardrop and Cronshaw report finding them at angles of between 30 and 60° to the cell axis in oat coleoptile parenchyma.

Another, more striking, instance of longitudinal microfibrils in primary walls is found in collenchyma cells. Preston and co-workers discovered longitudinally oriented cellulose in collenchyma walls of *Heracleum sphondyllium* and *Petasites vulgaris* with the polarizing microscope, and they were the first to describe these as "thickened primary" walls. Electron microscope observations of collenchyma in celery petioles revealed that young walls had a simple multinet structure while in progressively older cells distinct layers of longitudinally oriented microfibrils appeared at several points around the circumference of the cell. The number and extent of the layers increased with cell age and in mature cells the wall contained a large number of layers running continuously around the lumen. Thickened primary walls containing longitudinal microfibrils are also present in outer epidermal cells guard cells and sub-epidermal cells of oat coleoptiles, in epidermal and cortical cells of bean stems and in the epidermal wall of onion roots. In young oat coleoptiles the outer epidermal wall consists of 10 to 15 layers of microfibrils oriented longitudinally in the outer layers but more nearly transversely in the inner. As elongation proceeds the number of layers increases to about 25 while the microfibrillar orientation becomes predominantly longitudinal in all layers. Similar layers have been observed in epidermal cells of *Hyacinthus* and *Philodendron* but it is not certain to what extent these represent primary and secondary walls.

The above studies indicate that organized groups of longitudinally oriented microfibrils are a common feature of primary wall structure.

A generalized structural pattern for primary walls. Superficially, thickened primary walls containing numerous layers of longitudinally oriented microfibrils seem quite distinct from thin parenchyma walls with their predominant simple multinet structure. From the preceding discussion, however, it is clear that such a distinction is arbitrary and the arrangements of microfibrils in most primary walls can be regarded as variations of the same basic organization. This organization consists of two distinct mocrofibrillar components : transverse microfibrils which as a result of cell elongation form a simple multinet structure and organized layers of longitudinal microfibrils. Different types of primary wall reflect variations in the relative amounts of these two components. Thus in the extreme case of hair cells and internodal cells of *Nitella* the longitudinal components is absent and the entire microfibril arrangement is simple multinet. Ordinary parenchyma cells are intermediate in having both components present while collenchyma cells have a predominant longitudinal component.

Epidermal cells are a special case in which the longitudinal component predominates in the outer wall but is absent from the radial walls. The cortex of the elongation zone of bean stems shows that this concept is not completely artificial. In this tissue, there is a gradual transition from extensively multi-layered walls in the epidermis and outer collenchyma through walls with progressively fewer layers to typical parenchyma walls deep in the cortex. This transition is achieved by a change in the proportion of the two microfibril components which are deposited.

One clear exception to the above discussion should be mentioned. Sterling and Spit studied walls of developing fibres in asparagus and their observations showed that the inner surface of the wall consisted of a system of thin layers of microfibrils in which the orientation relative to the cell axis alternated at an angle of 55 to 60° for most cells. In young fibres the outside of the wall showed predominantly transverse microfibrils but in older fibres the outermost microfibrils were oriented longitudinally. Unfortunately this study is open to the criticism that, because of the preparative procedure employed, the inner and outer surfaces of the walls could not be conclusively identified in the electron microscope. Such cells might profitably be reinvestigated with direct sectioning techniques.

Plasmodesmata. That the relation between cell wall and protoplast is very close is evident by the presence in the wall of many delicate threads of cytoplasm which connect the protoplasts of adjacent cells (Fig. 3.8). These threads, known as *plasmodesmata*; fill minute passages which constitute the only breaks in the primary walls of the two cells. They are characteristic of living cells and are undoubtedly present in all living parts of the higher plants, maintaining the continuity of the protoplasm of the plant although its units, the cells, are sharply set apart by firm walls. Their extreme delicacy and protoplasmic nature make special technique necessary for their demonstration in most tissues. Where the fibrils of contiguous cells meet at the middle lamella there are often seen slight

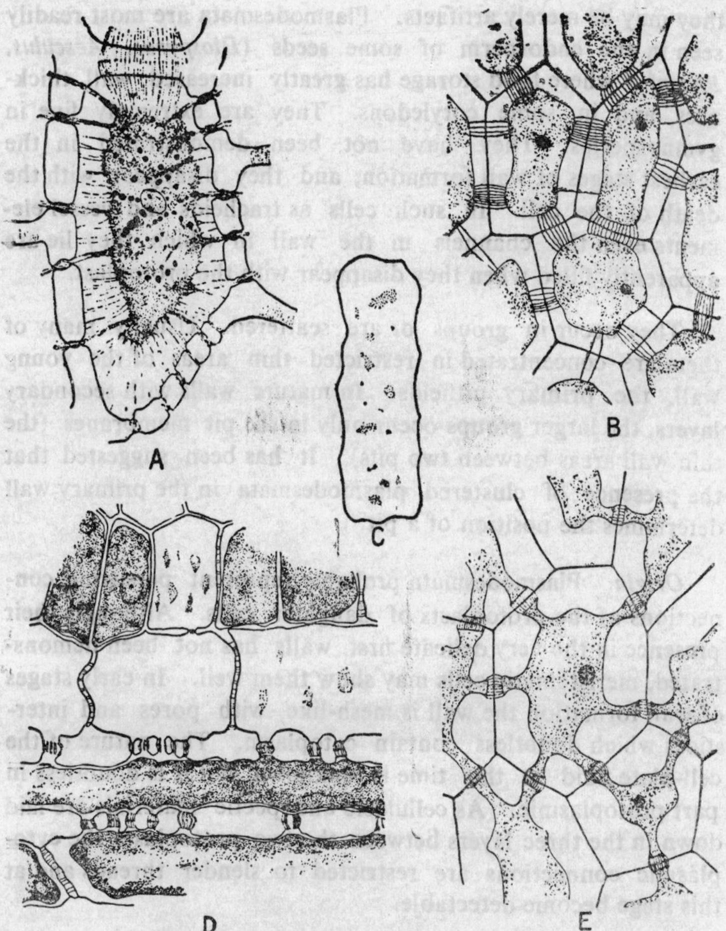

Fig. 3.8. Plasmodesmata. *A,* in a cen from the petiole of *Marattia* (the cytoplasm plasmolized); *C,* in a cortical cell of *Ophioglossum,* the strands seen "end on"; *D,* in cells of root cap and adjacent cortex of *Vicia Faba*; *B,* in endosperm of *Diospyros* (persimmon); *E,* in endosperm of *Phoenix* (date). (*A and C after Poirault*; *D after Gardiner and Hill.*)

knob-like thickenings. These may be enlargements of the threads owing their form to the nature of the intercellular substance, which in early stages may be partly protoplasmic or they may be merely artifacts. Plasmodesmata are most readily seen in the endosperm of some seeds (*Diospyros, Aesculus, Phoenix*) where food storage has greatly increased wall thickness, and in some cotyledons. They are extremely fine in gymnosperms. They have not been demonstrated in the earliest stages of wall formation, and they disappear with the death of the cell. In such cells as tracheids and vessel elements even the channels in the wall in which they lie are apparently filled when they disappear with the protoplast.

They occur in groups or are scattered. Usually many of them are concentrated in restricted thin areas of the young wall, the primary pitfields. In mature walls with secondary layers, the larger groups occur only in the pit membranes (the thin wall areas between two pits). It has been suggested that the presence of clustered plasmodesmata in the primary wall determines the position of a pit.

Origin. Plasmodesmata probably represent persisting connections of the protoplasts of daughter cells. Although their presence in the very delicate first walls has not been demonstrated, meristematic cells may show them well. In early stages of wall formation the wall is mesh-like, with pores and interstices which doubtless contain cytoplasm. The nature of the cell-plate fluid at this time is unknown but it is doubtless in part protoplasmic. As cellulosic and pectic materials are laid down in the three layers between the two protoplasts the cytoplasmic connections are restricted to slender threads and at this stage become detectable.

As the wall increases in area with growth of the cell, the number of plasmodesmata seems to increase. The increase possibly results from the splitting of the original strands. New connections between protoplasts are apparently not freely made but when cells during growth establish new contacts, secondary plasmodesmata may form in these areas. During symplastic growth the strands are distributed over larger areas, accompanied perhaps by multiplication. In intrusive growth they are ruptured and when the new-contact areas of both walls are young enough are replaced by new strands. It has been claimed that no secondary strands are formed ; that evidence for this is seen, for example, in their absence on the new-contact walls of wood fibers and developing vessel elements. But where such new contacts of enlarging vessels are with parenchyma cells new plasmodesmata are formed. The presence of plasmodesmata in graft hybrids and periclinal chimaeras between cells of different genetic origin seems to be sufficient evidence that plasmodesmata may develop secondarily.

Function. It has been suggested that plasmodesmata are "channels of translocation of solids," that they "conduct stimuli," but no demonstration of such functions has been made. Their existence is doubtless important in many ways because they unite the otherwise isolated protoplasts of the plant body into a single protoplasmic structure.

SCULPTURE AND MODIFICATION OF THE WALL

Pits. During early stages in the thickening of the primary cell wall, small, sharply defined areas in the primary pit-fields are left thin and uncovered. As thickening continues, recesses or cavities are left in the wall at these places. These cavities, together with the walls that surround them—the exposed area of primary wall that forms the "bottom" or external wall, and the side walls and roof—are known as *pits.* (The term pit is often applied to the cavity alone but best covers not only this but the modified wall area in which the pit lies.) Pits are conspicuous features of mature cell-wall structure. They vary in type, in form, and in abundance. They are characteristic of all kinds of cells and are probably lacking only in cells with very thin walls. Functionally, pits seem to be diffusion areas, regions where better interchange between cells may occur. That pits may be areas of interchange of another type also is apparent from the fact that in living cells groups of plasmodesmata are frequently or always present in the pits in larger numbers than elsewhere (Fig. 3.8).

Pits are structural features either of the primary wall alone or of the primary plus the secondary wall. Those that involve both wall layers are more readily seen ; those of the primary wall alone are conspicuous only where the wall is fairly thick.

Pit-pairs. A pit has a complementary or opposing pit in the wall of the contiguous cell. The two pits form a structural and functional unit, a *pit-pair.* (The use of the term "pit" for both the pit and the pit-pair has long caused difficulty and confusion in descriptions. The adoption of "pit-pair" for the double unit corrects the morphological error.)

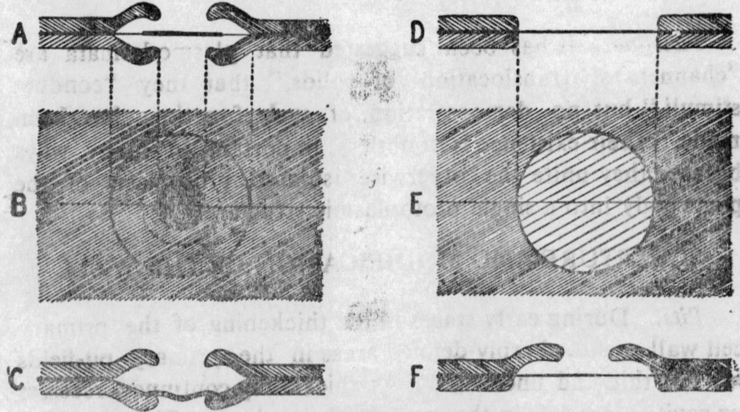

Fig. 3.9. Diagrams of three kinds of pit-pairs. *A,B,* section and face view of bordered pit-pair, showing the overarching borders of secondary wall which enclose the pit cavities, between which passes the closing membrane whose central portion is thickened to form the torus ; *C,* section of same showing the closing membrane in a lateral position, the torus closely appressed to the pit aperture ; *D,E,* section and face view of a simple pit-pair in which the secondary wall does not overhang the pit cavities and the closing membrane has to torus ; *E,* section of half-bordered pit-pair, the secondary wall overarching only one pit cavity, no torus. (*A—C, from Sharp, after Bailey.*)

The recess of the pit is the *pit cavity* ; the membrane that separates the two cavities of a pit-pair is the *pit membrane*, or *closing membrane* of the pit-pair ; the opening or mouth of the pit at the inner wall surface is the *pit aperture.* A pit-pair has two cavities, two apertures, and one closing membrane. The closing membrane consists of three layers—the primary wall of each of the two cells (different in thickness and chemical nature from adjacent parts of the primary walls) and, between these, the middle lamella which also is different from its adjacent parts in this area.

Two types of pits are recognized : *simple* and *bordered* (Fig. 3.9). These form *simple pit-pairs* when both members of the pair are simple ; *bordered pit-pairs* when both are bordered ; and *half-bordered pit-pairs* when one pit is simple and the other bordered.

Simple pits are those in which, as the wall is built up in thickness, the cavity remains of the same diameter, or is enlarged, or is only slightly constricted, and the closing membrane is simple in form and structure. In section, such pits show the side walls of the pit at right angles to the primary wall, sloping away from the margin of the exposed closing membrane, or sloping only slightly and gradually toward the center (Fig. 3.10). Simple pits with chambers slightly narrowing to the aperture somewhat resemble some forms of bordered pits—showing evidence of a narrow border in face view—but the closing membrane is without a torus.

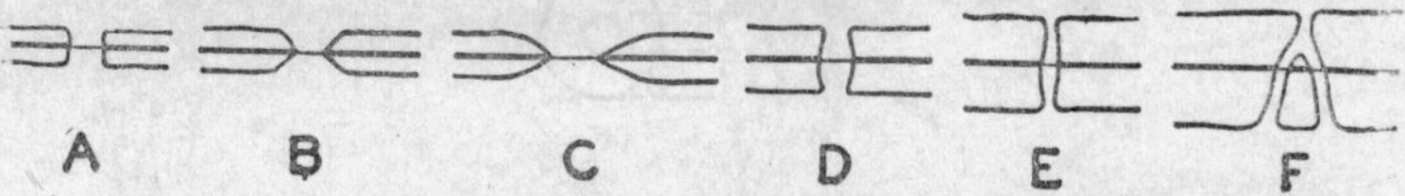

Fig. 3.10. Diagrams showing varieties of simple pit-pair structure as seen in section. *A*, the walls of the cavities at right angles to the closing membrane ; *B, C*, the walls sloping away from the closing membrane ; *D, E*, the cavities narrowing slightly to the apertures ; *F*, two pit-pairs, with two pits fused to form one cavity at one aperture. In no form is the pit cavity differentiated into cavity and canal as in bordered pits.

In *bordered pits*, the cavity, as the wall thickens, is so constricted that a rim or roof is built, overarching the recess. The closing membrane is elaborate in form and structure (Fig. 3.11). The pit is called "bordered" because in face view the rim forms a border around the aperture. In section such pits show the cavity more or less nearly enclosed by the secondary wall (Fig. 3.11). Where the secondary wall is thick, the pit cavity is divided into parts, an external *pit chamber*, and a passageway leading from the lumen to the chamber, the *pit canal*; where the secondary wall is thin (Fig. 3.9), no distinction of chamber and canal can be made. The pit canal in bordered pits has an *outer aperture*, the opening into the pit chamber, and an *inner aperture*, the opening into the cell lumen. The cavity of a simple pit is not divided into chamber and canal, but deep simple pits in thick walls have "*canal-like cavities.*" Such a pit has only one aperture.

In *half-bordered pit-pairs* both pits vary in type as do the pits of simple and bordered pairs. The closing membrane may have a weak torus or no torus.

The bordered pit-pair is complex not only in the form of its cavity but in the structure and behaviour of its membrane. The membrane of a bordered pit-pair (consisting of parts of two primary walls and the middle lamella) has a thickened central part, the *torus*, surrounded by a delicate marginal area. The triple nature of this membrane is not readily evident

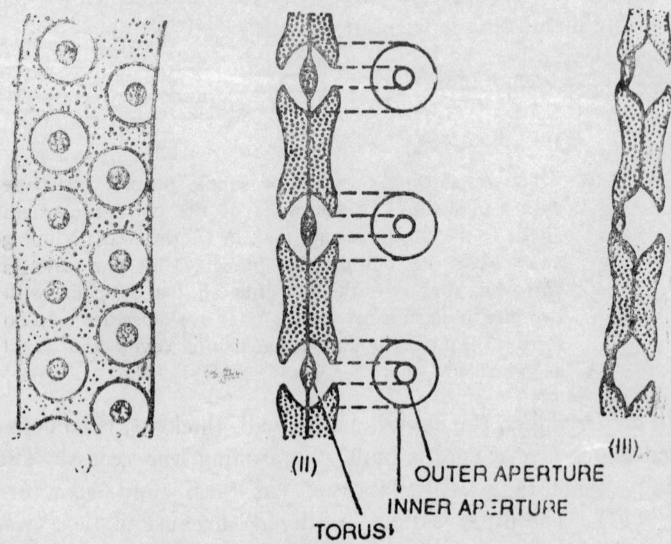

Fig. 3.11. Surface and Sectional views of Bordered pits.

because all the layers are much alike chemically (mainly cellulosic) and structurally. In some of the gymnosperms the thin area of the closing membrane is perforated by minute openings, or even by larger pores, so that the torus is suspended by a meshwork or, rarely, by a few strap-like pieces. The existence of these perforations has been clearly demonstrated by the passage of minute solid particles (finely ground India ink) from tracheid to tracheid.

The closing membrane of many bordered pit-pairs may change position within the cavities, apparently because of changes of pressure within the cells. The torus may occupy the median position (Fig. 3.9) ; or lie close against the aperture on either side (Fig. 3.9). When in the lateral position it may be distorted by the forcing of its central part into, or even partly through, the aperture. The diameter of the torus is always greater than that of the aperture so that, when the torus is in the lateral position, passage through the pit-pair can be only by diffusion through the torus. With the torus in the median position, the thinner margin provides passage of a perhaps different type. In the conifers with actual perforations, free passage is possible when the torus is in median position, but this has not been demonstrated in other groups. Membranes capable of these changes of position are characteristic of living spring wood. Those of summer wood are inflexible (in some woods 20 to 60 per cent are said to retain median position) and all membranes of heartwood are "frozen" in position by the changes incident to heart-wood formation. These pit-membrane characteristics are important from the standpoint of impregnation of wood under pressure with preservatives and other substances. Under pressure the membranes of heartwood and those of the summer layers of sapwood, held rigidly in position (perhaps in heartwood often cracked by shrinkage), can be ruptured to permit the passage of fluids ; those of living spring wood are forced into the lateral position with the torus jammed into the aperture and cannot be ruptured.

The simple pit-pair has no torus ; the half-bordered pit-pair has either no torus or a weakly developed one. The closing membrane of simple pit-pairs retains its median position; that of half-bordered pit-pairs may bulge well into the lumen on either side when the apertures are large.

Departures from normal pit-pair structure are few. A large pit in one cell may have two or more smaller pits opposite it in the adjacent cell. This is *unilateral compound pitting*. A pit without a complementary pit is a *blind pit*. Such pits are possibly halves of pit-pairs that have been split apart in the readjustment of cell contacts during the cell growth. (The pits without complementary pits commonly seen in sections of wood cells and other thick-walled cells are halves of pit-pairs that have been cut obliquely in sectioning.)

Simple pit-pairs, with the canals slightly and gradually narrowing toward the aperture, are frequent in thick-walled parenchyma, in some sclereids, and elsewhere. Some of these have been incorrectly called "bordered." These do not have the pit membranes of bordered pits or the distinction of cavity into chambers and canals. The bordered pit, as a morphological structure, is characteristic of and restricted to water-conducting cells and their phylogenetic derivatives. If pits with slightly overhanging borders are called "bordered," important morphological structure is ignored.

Simple pits occur in walls that are wholly primary and those that are both primary and secondary; bordered pits occur only in primary plus secondary walls. Simple pits occur in many kinds of cells; bordered pits are characteristic of water-conducting cells and of kinds of cells morphologically derived from these cells. In a given cell or kind of cells, the type is constant.

Variations in Size and Structure of the Bordered Pit. The bordered pit-pair varies greatly in size and structure (Figs. 3.12, 3.13, 3.14). The prominent modifications of its typical form and size are those associated with the specialization and reduction of the pit as it occurs in fiber tracheids and fibers. In the typical bordered pit-pair, the cavities, which are circular in outline, are large, and the apertures are also circular (Figs. 3.11 and 3.15 A). Where such pit-pairs occur in thin walls,

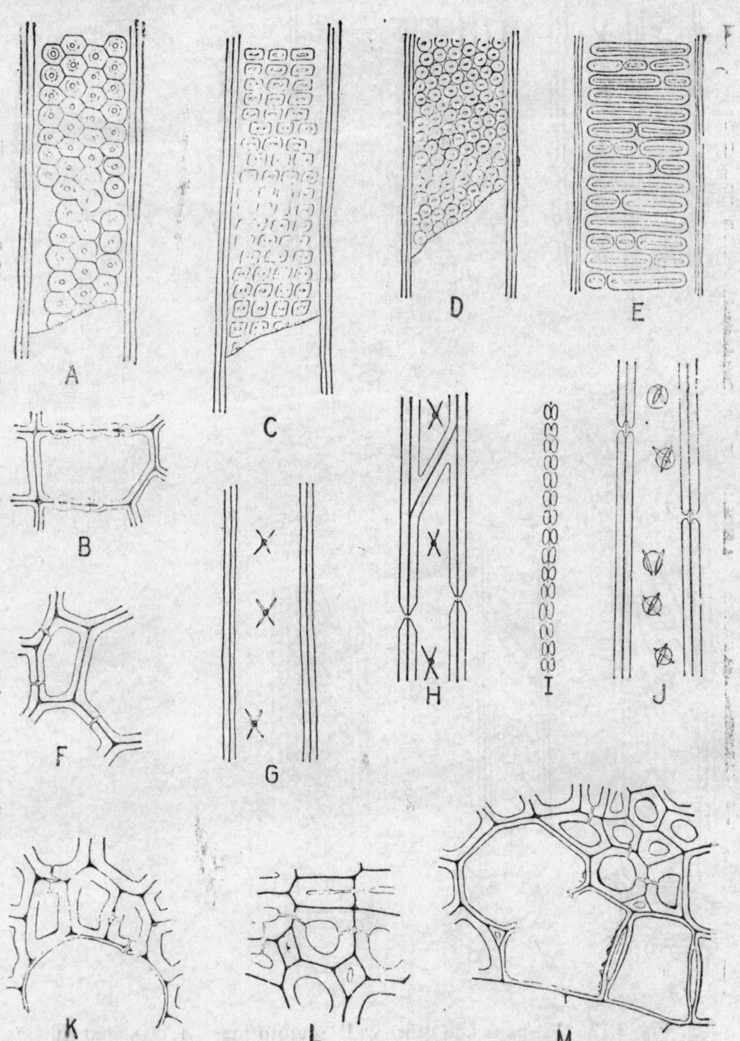

Fig. 3.12. Bordered pit-pairs. *A, B,* crowded angular pit-pairs in face view and in section in tracheid of *Agathis australis* ; *C,* rectangular pit-pairs, face view, in horizontal rows in vessel of *Liriodendron* ; *D,* crowded pit-pairs, face view, in spiral rows in vessel of *Acer rubrum* ; *E,* elongate pit-pairs in face view in vessel of *Magnolia acuminate*—the same pit-pairs in vertical section in *I,* and in transverse section in *M* ; *F, G,* pit-pairs with narrow. inner apertures, in section and face view in thick walls of wood fiber of *Populus tremuloides* ; *H,* similar pit-pairs in wood fibers of *Magnolia,* face view and longitudinal section—the same pit-pairs in transverse section in *L* ; *J, K,* pit-pairs with narrow, unsymmetrically placed inner apertures in wood fiber of *Malus pumila* in face view and in longitudinal and transverse section ; *M,* large, elongate pit-pair (*E*) in vessels and small pit-pairs in fibers, both in transverse section in *Magnolia.*

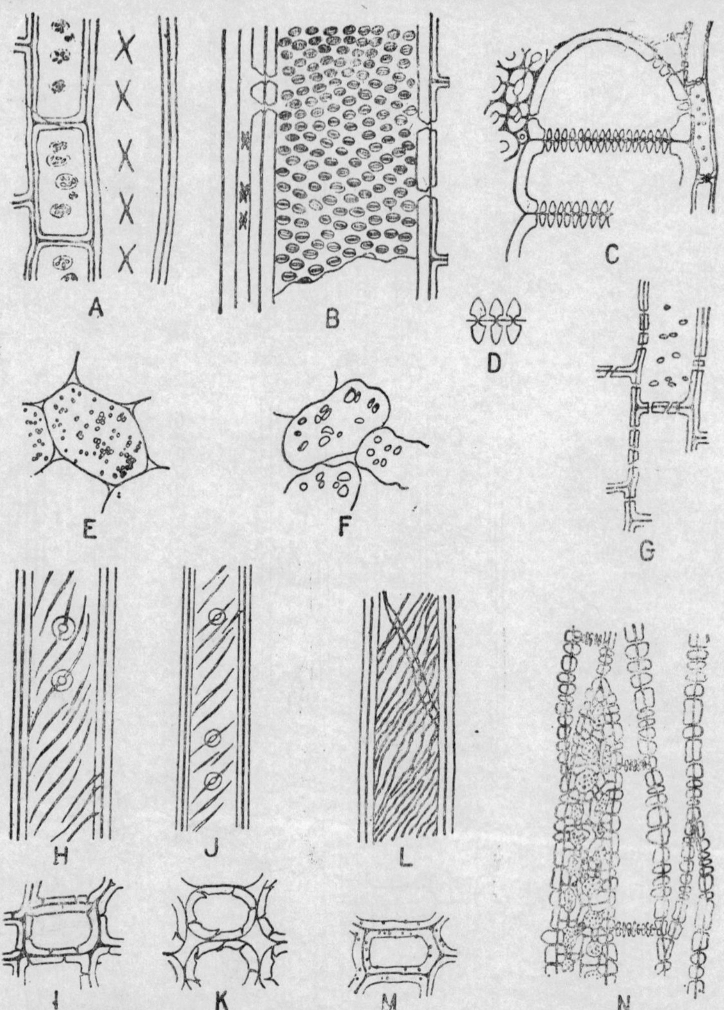

Fig. 3.13 Pit-pairs and other wall sculpturing. *A*, clustered simple pit-pairs in wood parenchyma, and very small bordered pit-pairs in fibers, face view, in *Fraxinus americana*; *B*, bordered pit-pairs of two forms, face view, in vessel and in fiber, and half-bordered pit-pairs, in section, between vessel and parenchyma cell, in *Diospyros* (ebony); *C*, the same pit-pairs as in *B*, in transverse section, also simple pit-pairs, face view and section in ray and parenchyma cells; *D*, detail section of bordered pit pairs between vessels, as shown in *C*; *E* and *F*, simple pit-pairs in thin-walled parenchyma cell—*E*, in pith of *Chenopodium*, *F*, in pericarp of *Citrullus*; *G*, simple pit-pairs in thick-walled parenchyma of pith of *Clematis*; *H* and *I*, *J* and *K* microscopic checking, face view and section, of secondary wall in tracheids of *Pinus Strobus* and *Sequoia sem pervirens*—the cracking extending through pit apertures simulates elongate apertures; *L*, *M*, erosion of secondary wall by fungus hyphae, simulating checking, face view and section, in tracheid of *Podocarpus*; *N*, simple pit-pairs in thick-walled ray cells and wood parenchyma, face view and section, in late summer wood of *Magnolia*.

the borders may arch and bulge into the lumen, as in the spring-wood tracheids of *Larix* and species of *Pinus* (Figs. 3.14E, 3.15E). These pit-pairs are large in comparison with the cells to which they belong (Fig. 3.11) and, in section, may resemble small cells (Fig. 3.14E). Where the wall is thicker the presence of a pit causes no mounding or bulging of the wall and the aperture opens into the pit canal, not directly into the chamber.

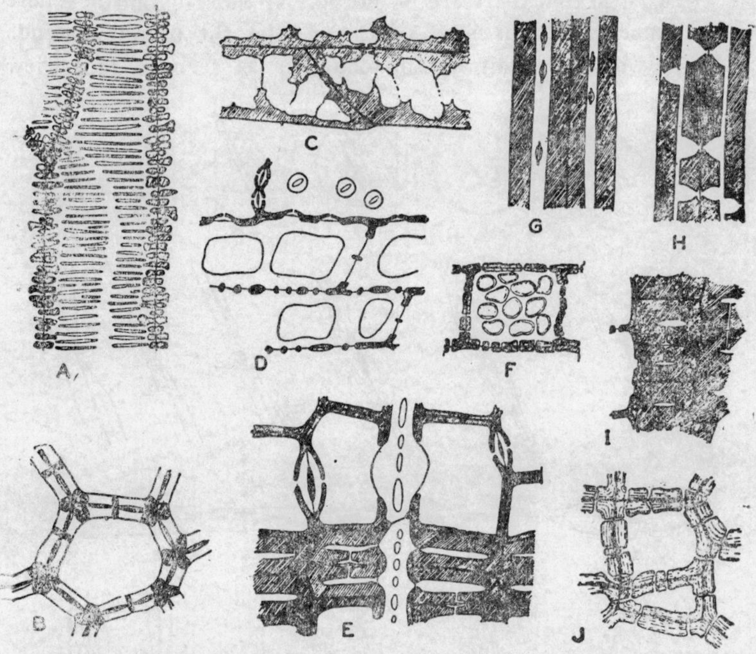

Fig. 3.14. Pit-pairs. *A,B,* elongate bordered pit-pairs of tracheid of *Osmunda Claytoniana,* face view and section ; *C,* bordered pit-pairs and "dentations" of wall in section in marginal ray cell of *Pinus Banksiana* ; *D,* full-bordered, half-bordered, and simple pit-pairs, face view and section, in wood ray cells, radial section of *Pinus Strobus* ; *E,* similar, in part the same, pit-pairs, in face view and section, in tracheids and ray cells in transverse section of wood of *Pinus Strobus* ; *F,* simple pit-pairs in section and half-bordered pit-pairs in face view in ray cell of wood of *Salix alba* ; *G,H,* and *I,* small bordered pit-pairs with narrow, vertical inner apertures coinciding in position, and long, narrow pit-canals between pit chamber and lumen, face view, longitudinal section, and transverse section, respectively, in wood fibers of *Eucalyptus globulus* ; *J,* simple pit-pairs in stone cells of phloem of *Platanus.*

In general, the thicker the wall in bordered pit-pairs, the smaller the chamber and the smaller the aperture. In thick walls, although the chambers and outer apertures remain always circular in outline, the inner apertures become narrowed and elongated, being elliptical to slit-like in outline (Fig. 3.15). When only slightly narrowed, the inner aperture, seen in face view, may extend only part way to the margin of the cavity (Fig. 3.15*A*), or may reach the very limit of the cavity (Fig. 3.15*A* 3), so that there are no borders at those points. Where the inner aperture is much narrowed, the slit usually extends far beyond the limit of the chamber as seen in face view

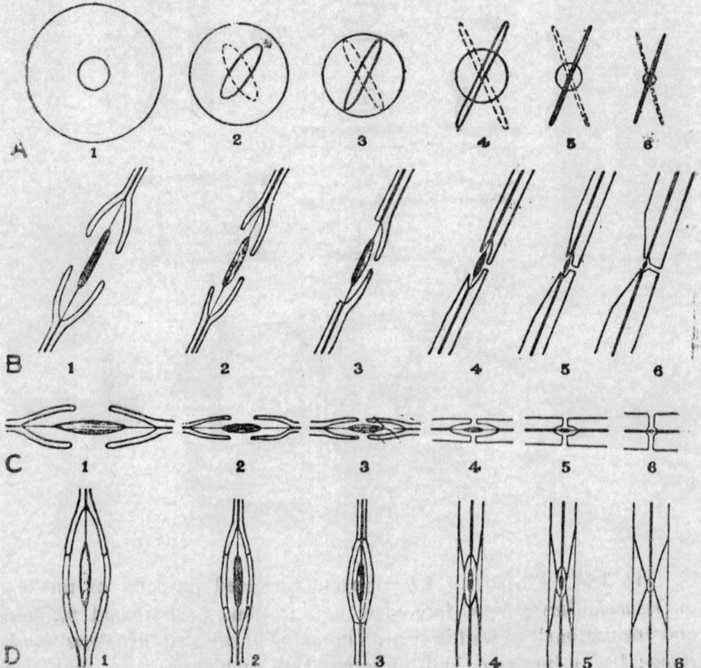

Fig. 3.15 Diagrams of various types of bordered pit-pairs, showing changes of from accompanying reduction in size and function, and occurrence in walls of various thickness. *A*, face views ; *B*, sections along plane of an elongated aperture ; *C*, transverse sections ; *D*, view, "edge on," of entire pit-pair lying in wall—the view of small pit-pairs commonly seen, because of their small size—in longitudinal sections of wood.

(Fig. 3.15 *A* 4—6). In reduction of pit size in thick walls, the greater the reduction in diameter of the chamber, the greater the increase in length of the inner aperture. These much elongated apertures occur only in thick walls, and hence the pit canal through the thick wall to the small inner chamber is in shape like a much flattened, or laterally compressed, funnel (Fig. 3.16). The outer aperture of this funnel-like canal remains circular. Where the inner pit apertures are circular, the two apertures of a pit-pair lie directly opposite one another (Fig. 3.15 *A* 1), coinciding in position ; where the inner apertures are elongate, the apertures are commonly crossed, usually symmetrically (Fig. 3.15 *A* 2—6). (The angles at which the inner apertures lie are apparently controlled by the orientation of the strands of fibrils that make up the major layer of the secondary wall.) These crossed pit apertures are conspicuous features of pit marking in the walls of wood cells (Figs. 3.12,*G*,*H*,*J*, 3.13 *A*). The pit cavity in such pits is complicated in form (Figs. 3.15, 3.16). With extreme reduction the borders of the pits nearly disappear, and slit-like pit-pairs, essentially simple in structure, are formed (Fig. 3.15 *A* 6). Morphologically, such pit-pairs are, however, clearly bordered pit-pairs. The pits of libriform wood fibers are often wholly vestigial, being simple and hardly more than markings on the wall. In much reduced bordered pit-pairs the closing mem-

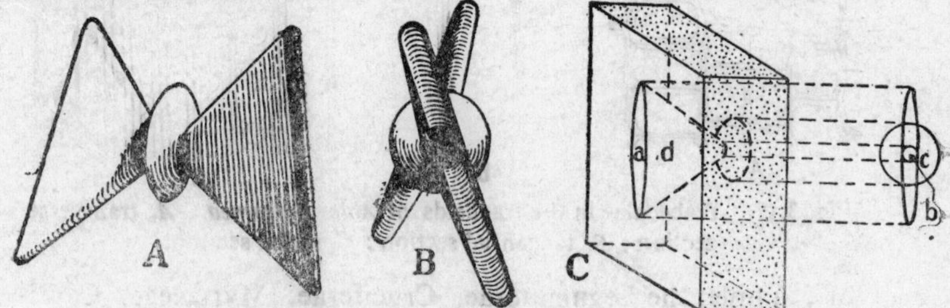

Fig. 3.16. Diagrams showing bordered-pit structure of the type shown in Fig. 3.15 *A*, 4. *A*, *B*, showing the form of the cavities of a pit-pair—the inner apertures are slit-like; the small chambers, flat-dome shape ; the canals like flattened funnels. *C*, a pit shown in position in a thick cell wall. *a*, the inner aperture ; *b*, the outer aperture ; *c*, the chamber ; *d*, the canal.

brane is of simpler structure, with no torus and no capacity to change position. Such pits are probably physiologically functionless or nearly so.

Vestured Pits. Pits that have minute outgrowths from the surface of the side wall of the pit chamber or along the rims of the apertures are known as *vestured pits.* The projections that form the vesture are of many types—filamentous, papillose, coralloid, branched and anastomosing, or simple. They may be scattered or closely clothe the surface of the wall. In some pits they cover also the surface of the wall of the vessel in which the pits occur. When long, they may fill the pit cavity and project into the cell lumen. Only bordered pits are vestured ; in half-bordered pit-pairs the simple pit is without vesture.

The refractive and deep-staining qualities of these minute structures give a punctate or reticulate appearance to pits seen in face view. For this reason such pits were formerly called "cribriform" and "sieve-like" in the mistaken interpretation that the closing membrane was perforated like a sieve plate.

Vestured pits are probably restricted to the tracheary elements of the secondary wood of certain angiosperm families,

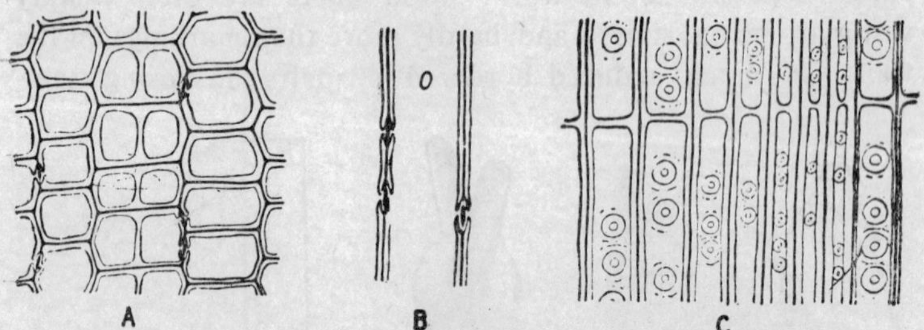

Fig. 3.17. Trabeculae in the tracheids of *Abies balsamea*. *A*, transverse section ; *B*, tangential section ; *C*, radial section.

for example, the Leguminosae, Cruciferae, Myrtaceae, Caprifoliaceae. When present, they are constant throughout the tissue in the species or genus and probably also in subfamilies and families, and so may be important in systematic and phylogenetic studies. No function has been suggested for them. Their presence is associated with highly specialized wood types and they appear to indicate a highly specialized pit type.

Other Wall Sculpturing. The modeling of the inner surface of the cell wall is in large measure due to the presence of pits. Pits and performations of vessel walls form the conspicuous structural modifications. Also prominent are the secondary-wall thickenings of protoxylem—rings, spirals, and reticulations. Other internal thickenings or projections occur—the "dentations," really ridges, of the ray tracheids of hard pines (Fig. 1.14C) ; the Casparian strips of some endodermal cells ; the trabeculae of conifer tracheids (Fig. 3.17) ; the supporting bars or meshes of cells of the velamen and another wall.

Trabeculae. Rod-like or spool-shaped wall projections crossing the lumen radially are known as *trabeculae*. They are frequent in the secondary wood of conifers, rare elsewhere. They usually occur in rows radially, and the rows may extend across the cambium and into the phloem.

Crassulae. Structural variations within the wall itself may appear like surface features. Bar-like or crescent-shaped thickenings of the primary wall and middle lamella that separate or partly encircle individual bordered pits (Fig. 3.11) or small groups of pits may be visible through the secondary wall when properly stained. These *crassulae* are the persisting marginal rims of the primary pit-fields of the young cell wall. The secondary wall covers but does not form a projection over these primary ridges so that they are not features of the sculpture of the wall surface. The names "bars of Sanio" and "rims of Sanio" have long been applied to crassulae but these terms have been confused in usage and in part applied also to trabeculae. The new term "crassulae" should replace both older ones.

Centrifugal Thickening. In cells that have free surfaces the wall is apparently sometimes thickened centrifugally. Some hairs and other epidermal cells, some sclereids, and certain cells abutting on large intercellular spaces have minute nodules, granules, or ridges on the outside of the wall. Some of these are parts of the wall ; others are deposits on the surface, as is the cuticle. The external wall layers and surface projections of spores and pollen grains are formed in part by tapetal fluid or mothercell cytoplasm.

Chemical Nature of the Cell Wall. The chemical nature of the early stages of the cell wall have already been briefly discussed. Later modifications and additions are of many kinds. These are brought about by infiltration of the interstices of the original wall matrix by new substances, or perhaps in part by replacement or transformation of part of that first-deposited wall. *Lignification*, infiltration by lignin, is prominent in these changes. It takes place in both primary and secondary walls of many types of cells, especially those of wood. Primary walls of wood cells, together with the intercellular layer between, are characteristically heavily lignified ; the secondary walls of the same cells are more lightly lignified. In angiosperm wood the so-called "middle lamella"—intercellular layer, primary walls, and thin outermost layers of the secondary walls—is usually the most heavily lignified part of the cell wall. The secondary wall, or its inner layers, is *gelatinous* in the fibers of many angiosperm genera. The function of such gelatinous wall layers is not understood. Gelatinous or mucilaginous cell walls are found also in other tissues, such as phloem and especially the outer layers of seeds and fruits. A general breakdown of wall structure, known as *gummosis*, may occur even in woody tissues under certain pathological conditions.

Cutinization and suberization are processes in which the wall and often the middle lamella are infiltrated by chemically allied substances, *cutin* and *suberin*, respectively. Cutin and suberin are mixtures of substances, chiefly fatty acids, that are impermeable to water. These waterproof the cells in which they are deposited. Cutinization is found chiefly in the epidermis, and suberization in the periderm, but both occur less prominently elsewhere.

"*Mineralization*," the deposition of large amounts of inorganic salts in the wall is only rarely a prominent feature of wall modification. All mature walls contain small amounts of mineral matter, sometimes in the form of visible minute crystals. Silica and the oxalate and carbonate of calcium are commonly present ; the first heavily infiltrates the walls of the outer tissues of grasses, sedges, horsetails, and some other families. Large crystals may become attached to, or even embedded in, the secondary wall by the formation about them of the later layers of wall substance.

Other changes in the wall involve resins, gums, oils, tannins and tannin-decomposition products, various aromatic and coloring substances, and many others of minor importance. Tannin, together with other substances, in responsible to a large extent for the durable and useful qualities of certain woods, and of heartwood as contrasted with sapwood. These substances are present not only in the walls, but often fill the lumina and pit cavities as well. They are formed in part from the contents of the living cells of the immediate region, but doubtless also in part from additional material brought to the region.

The modifications discussed above may involve the middle lamella and both chief wall layers or any of these alone. The primary wall is perhaps less changed than the secondary.

Cystoliths. In a few dicotyledonous families peculiar structures known as *cystoliths* project into the protoplasts in large specialized cells. These structures are lime-encrusted or -infiltrated, stalked projections of the wall. The foundation of the cystolith is a stalked, stratified, cellulosic body which arises early in cell development as a local wall thickening. With the addition of large amounts of calcium carbonate this becomes an irregular body which may nearly fill the cell. In shape, cystoliths vary greatly in different genera and families. They are characteristic of the leaves of the Urticaceae, Moraceae, and a few other families and may appear even in the wool rays and phloem of these groups. Cystoliths are not common or important features of cell structure.

All these changes are in nature either those of direct chemical transformation or molecular replacement ; or those of apparent change brought about by infiltration, or by deposition of large amounts of new, different substances, within the original wall, so adulterating the latter that the first substance is obscured. Pseudochanges are secured by the addition of new material in lamellae over the old wall, completely burying the latter. Such modifications may accompany the maturation of the wall in cell ontogeny, or may take place long after cell maturity, for example, when sapwood becomes heartwood, and when cortical photosynthetic parenchyma cells or phloem parenchyma cells become sclereids after a long period of activity in their first condition.

Polysaccharide components. The microfibrillar component of walls usually consists of cellulose or chitin while the amorphous matrix is composed predominantly of polysaccharides which can be classified simply as hemicelluloses and pectic substances. Native cellulose appears to be similar in both primary and secondary walls of higher plants but almost all knowledge of non-cellulosic matrix substances is derived from studies of mature secondary walls and is useful here only to indicate possible components of primary walls.

Cellulose occurs naturally as microfibrils 100 to 200 A wide, within which long parallel polymers of predominantly β-1,4 linked D-glucose residues form a well-ordered core surrounded and interrupted by regions of less order. A model for the ordered arrangement of molecules in native cellulose was proposed by Meyer & Misch in which neighbouring molecules run in opposite directions, *i e.*, are all parallel, although the evidence for this is not conclusive. This model, which was later modified by Frey-Wyssling to provide better agreement with the observed behaviour of microfibrils, has not been supplanted although recent studies have found it inadequate.

Celluloses from different sources differ in microfibril width degree of crystallinity, and in the traces of non-glucose residues found associated with them (1). There is no evident correlation between these features, but Preston and Roelofsen suggest that non-glucose residues are concentrated in the less-ordered regions of the microfibril. It has not been proved, however, that such residues are covalently linked to cellulose and it is dossible that they are merely adsorbed since customary extraction procedures are known to be incapable of separating artificial mixtures of cellulose and hemicellulose.

Chitin, a polymer of N-acetylglucosamine, replaces cellulose as the microfibrillar component in many fungal walls. In the well-founded model for chitin proposed by Carlstrom, the molecules are antiparallel. Instances of other microfibrillar polysaccharides have been reported, notably among the algae, but they do not appear to occur generally and their significance is not clear at present.

The hemicelluloses, which have been discussed recently by Aspinall and Whistler and Sannella, are a heterogeneous group of compounds containing D-xylose, L-arabinose, D-galactose, D-glucose, D-mannose, and D-glucuronic acid residues. Generally one or two residues predominate in a backbone chain, allowing four relatively distinct groups of compounds to be recognized : xylans, mannans and glucomannans, galactans and arabogalactans and glucans. The xylans and arabogalactans show considerable branching, branches of the former usually containing mixtures of several residues, while glucomannans are relatively unbranched. Pure galactans, mannans, and glucans are rare in higher plants but the latter two are common in fungi. Preliminary analyses of primary walls of oat coleoptiles indicate the existence of several hemicellulose fractions and the presence of all of the above component residues, suggesting that these walls contain several general types of hemicelluloses.

The pectic substances are basically straight clain polymers of α-1,4 linked D-galacturonic acid residues which are usually partially esterified with methyl groups. Molecular weights ranging from 25,000 to 360,000 have been reported. Much pectin of walls occurs as relatively insoluble protopectin, which may represent very long polygalacturonic acid chains or molecules extensively cross linked. Protopectin of oat coleoptiles contains a relatively low percentage of methyl esters, thereby increasing the possibility of carboxyl interaction. Henglein presents an extensive discussion of all aspects of pectic substances.

Table 3.1. Summary of Quantitative Chemical Analyses of Primary Walls.
[Values represent percentages (rounded off) of total dry wall material.]

Object	*Reference*	*Cellulose*	*Hemicel-luloses*	*Pectic Substances*	*Protein*	*Lipid*
Corn coleoptile— 9 mm.	(174)	38	46	10	—	7
Corn coleoptile—32 mm.	(174)	33	34	10	—	23
Corn coleoptile—50 mm.	(174)	35	30	13	—	21
Corn coleoptile	(107)	28	22	—	30	7
Corn coleoptile	(95)	27	—	8	2.5-5	—
Corn mesocotyl	(133)	30	50 (Hemicelluloses + Pectic Substances)		3	2
Oat coleoptile	(158)	42	38	8	12	—
Oat coleoptile	(18)	25	>51	<1	10	4
Oat coleoptile	(88)	—	—	5	—	—
Pea stem	(38)	26	40	20	13	1
Young wheat leaves	(30)	30	11	22	—	—
Young wheat roots	(30)	38	15	16	—	—
Sunflower epicotyl	(78)	37	4	53	—	—
Sunflower hypocotyl	(78)	38	9	46	—	—
Patato path	(78)	—	—	5-6	—	—
Patato tuber	(33)	29	—	16	—	—
Young cotton hair	(159)	50	—	10	13	8
Young cotton hair	(169)	—	—	5-6	—	—
Young cotton hair	(98)	33	20	23	—	24
Pine cambium	(6)	20-25	—	10-20	—	—
Black locust cambium	(8)	—	—	<13	—	—

Quantitative studies. Analyses of a number of primary walls are summarized in Table 3.1. Most of the data were obtained by selective extraction coupled with gravimetric or relatively non-specific chemical analyses. Because of the reasonably distinct solubility properties of cellulose, the values given for this substance are probably fairly accurate and indicate that it usually forms only 25 to 35 per cent of dry walls. The figures for the preponderant matrix substances are less reliable since chromatographic analyses reveal considerable cross contamination between fractions usually designated as hemicellulose and pectin. In oat coleoptiles at least, hemicelluloses are the most abundant matrix materials and only small amounts of pectic substances are present. Most earlier analyses probably underestimate hemicelluloses and overestimate pectic substances.

The actual amount of pectin in oat coleoptiles is in doubt. Using two different extraction procedures, Bishop *et al.*, found less than 1 per cent galacturonic acid, but characterized it by several direct methods. On the other hand, with the pectinase-carbazole method Albersheim and Bonner obtained 3.5 to 4 per cent and Jansen *et al.* 5 per cent anhydrouronic acid. Galston and Purves argued that the latter method, being gentler, is more likely to give accurate quantitative values. Kivilaan *et al.* also obtained less galacturonic acid from corn coleoptile walls by chemical hydrolysis and chromatography than Albersheim found, presumably by pectinase-carbazole, and suggested hydrolytic destruction to account for the discrepancy. The destruction argument is reasonable but before the pectinase-carbazole values are taken as final this method should receive more critical evaluation. We have found that, even after extensive purification, commercial pectinase preparations readily hydrolyse hemicelluloses and may liberate sugars which could upset quantitative uronic acid determination.

Table 3.1 shows that an appreciable amount of nitrogen, presumably from protein, accompanies primary walls. Probably much of this is due to plasmodesmata and cytoplasmic contamination, but the 2.5 to 5 per cent protein remaining in well-washed preparations of isolated walls may represent enzymes bound in the wall. Pectin methylesterase has been shown to be bound to cell walls outside the cell membrane; and in the light of recent evidence that cell wall deposition can occur remote from the cytoplasm, it is probable that other enzymes involved in cell wall metabolism may be found in walls. Ginzburg also suggested that cells may be cemented together by a complex of protein and non-cellulosic substances. His evidence for this is indirect and derived from cells of the extreme tips of roots where the walls are tenuous and contain abundant plasmodesmata. The discussion of the middle lamella, below, suggests that a proteinaceous cement in elongating cells is unlikely.

Lipid constituents are commonly found associated with primary walls (Table 3.1) and are probably derived mainly from cuticle. It appears, however, that some waxes also occur within walls although their nature and function is obscure. Electron microscopy of cells fixed with osmic acid and permanganate has not consistently revealed any structure that could represent organized lipids within primary walls.

The above discussion concerns dry wall material, but it is well known that the native primary wall is highly hydrated. A figure of over 90 per cent water in young walls is often quoted but Roelofsen disagreed with this and obtained a more reasonable value of 60 per cent water for isolated walls of corn coleoptiles.

Composition of the middle lamella. The paucity of pectic substances in walls as typical as those of coleoptile parenchyma raises the question of the nature of the middle lamella. Kerr & Bailey defined this structure as the amorphous intercellular material between adjacent primary walls and electron microscope observations on sections of extracted walls confirm the exsitence of such a cellulose-free region. It has also been widely accepted that the middle lamella is principally composed of pectates, but the evidence needs re-examination as it rests mainly on non-specific extraction and staining procedures.

Even if all of the 1 to 5 per cent polygalacturonic acid of coleoptile walls were localized in one layer between cells it probably could not be resolved in the light microscope. A systematic electron microscope study of the middle lamella has not been performed, but no discrete intercellular layer can be discerned consistently in either unshadowed or shadowed sections. In the latter type of section, the texture of the matrix appears uniform throughout the wall. Concerning extraction of the middle lamella, Ginzburg has pointed out that conditions needed to separate root-tip cells with EDTA (ethylenediamine tetra-acetic acid) are far more drastic than expected if simple chelation of calcium and solubilization of pectic acid were involved. Similarly, we have consistently observed both with coleoptiles and with the elongation zone of roots, that treatments required to separate cells readily also thoroughly extract matrix substances from the entire wall revealing clean microfibrils in the electron microscope. These considerations suggest that the middle lamella may merely be a region of the polysaccharide matrix where cellulose is absent, rather than a region of pectate localization. On the basis of electron microscope examination of walls of pea root tips extracted with EDTA and stained with uranium, Klein and Ginzburg also concluded that there was no specialized cementing layer between walls.

Chemical interactions and wall structure. The physical properties of walls must depend largely on interactions between the various wall substances. If wall structure is likened to reinforced concrete with microfibrils corresponding to reinforcing steel, it is clear that, for strength, interactions must occur between microfibrils, between matrix substances, and between microfibrils and matrix. Furthermore, in the primary wall such interactions must be labile enough to permit controlled wall expansion and deposition.

The nature of such interactions is largely unknown but several possibilities exist. Carboxyl groups of pecitic substances and hemicelluloses containing uronic acid residues possess an obvious capacity for reaction through ionic bridges and ester bonds. Such bonds are commonly invoked in theories of wall structure but since some growing walls contain only small amounts of uronic acid residues this type of interaction may be of limited importance. It is possible but unproven that pectic substances are strategically dispersed through the wall and act as key bridging agents with the hemicelluloses. Certainly the common occurrence of insoluble protopectin and the presence of arabans and galactans in pectic preparations suggests that pectic substances are closely linked to other components of the matrix.

Of the matrix polysaccharides, the preponderant hemicelluloses probably make the main contribution to primary wall structure. These substances possess less obvious capacity for covalent bonding than the pectic substances and they probably interact largely through hydrogen bonds and van der Waal's forces. In discussing cellulose structure it was pointed out that hemicelluloses bind readily to cellulose. Difficulties in purifying individual hemicelluloses suggest that they also interact with each other. The branched structure of some hemicelluloses would increase the chance of molecular entanglement and interaction. Recently Jensen suggested that hemicelluloses in walls hold cellulose under tension since their removal with alkali caused 30 per cent shrinkage of 10μ longitudinal sections of onion roots. It seems unlikely that the relatively short segments of microfibrils present in such sections would be under tension and the shrinkage was probably due to mercerization of the cellulose. We have obtained shrinkage of sections with alkali extraction but it is not apparent when hemicelluloses are removed with enzymes.

BIOSYNTHESIS OF WALL POLYSACCHARIDES

It will become evident in later sections that knowledge of the biosynthesis of wall components is basic to an understanding of both the structural development and metabolic responses of walls. Unfortunately, although intermediary carbohydrate metabolism is extensively elaborated the nature of the reactions leading to polysaccharides of walls is largely unknown. This section summarizes available evidence on the subject and discusses some of the outstanding problems.

Cellulose. Knowledge of cellulose synthesis in plants is derived solely from studies using isotopically labelled substrates and essentially consists of one well-established, somewhat unsurprising, fact, *viz.* the intact glucose carbon skeleton is used directly in some form or other for cellulose synthesis.

Several workers administered specifically ^{14}C-labelled glucose to cotton bolls and wheat plants and found that it entered cellulose with little skeletal rearrangement. The small amount of rearrangement found by all workers except Greathouse indicated some degradation and resynthesis of hexose by Embden Meyerhof glycolytic reactions. Besides glucose, Neish and co-workers supplied wheat plants a large number of other specifically ^{14}C-labelled monosaccharides and related compounds and their results consistently indicated that glucose was the most efficient precursor of cellulose and that other compounds such as pentoses, sedoheptulose, and succinic acid were converted to hexose before entering cellulose.

The mechanism of glucose polymerization is unknown but sugar phosphates and nucleotides have been implicated. In particular Neish focused attention on uridine diphosphate glucose (UDPG) since known enzymatic reactions involving this compound could explain distribution of administered isotope into other cell wall polysaccharides (see below.) Furthermore, UDPG has been shown to be a substrate for enzymes producing a number of polymers containing glucose : glycogen β-1.3 linked glucan, type III pneumococcal capsular polysaccharide, and hyaluronic acid. However, attempts to demonstrate an enzyme from plants which will synthesize cellulose from UDPG have been unsuccessful although Glaser and Brown found an enzyme, chitin synthetase, in a particulate fraction of *Neurospora crassa*, which converted UDP-N-acetylglucosamine to chitin. The relative ease with which this enzyme was obtained, compared with the failure to demonstrate an analogous cellulose synthetase in higher plants, suggests that the systems for chitin and cellulose synthesis may be distinctly different despite similarity of the products.

Certain bacteria, notably *Acetobacter xylinum*, produce extracellular cellulose as a "by-product," allowing study of cellulose synthesis without problems of cell wall deposition. Schramm *et al.* investigated the carbon metabolism of this organism and showed that glucose was rapidly converted to gluconate and that a well-developed pentose-phosphate pathway was present by which 6-phosphogluconate was decarboxylated at carbon-1 and resynthesized to hexose phosphate. Results from supplying glucose, specifically labelled with ^{14}C in different positions, indicated that cellulose was synthesized from such hexose phosphate and that as in higher plants the intact hexose skeleton was used in the polymerization reaction. Great-house also administered specifically labelled glucose but although the results indicated direct utilization of the glucose skeleton in cellulose synthesis they were inconsistent with operation of the pentose-phosphate pathway. The reason for this discrepancy is not clear.

Several attempts have been made to obtain cellulose synthesis in cell-free preparations of *A. xylinum*. Although Gromet *et al.* detected many other enzymatic activities, they could not demonstrate cell-free production of cellulose. Colvin used an homogenate supplemented with glucose and ATP and by electron microscopy detected very low levels of microfibril production. Glaser demonstrated a small amount of net cellulose synthesis using a particulate cell fraction, UDPG-^{14}C and cellodextrin primer. Recently, Klungsoyr claimed to have enhanced cellulose synthesis through this reaction by adding excess UDP (uridine diphosphate) and cellodextrins to the homogenate. Unfortunately this author used alkali insolubility as the only criterion for cellulose and Colvin has shown that amorphous non-cellulosic material insoluble in alkali may be obtained from A. *xylinum.* Greathouse described a cell-free preparation from *A. xylinum* which, when incubated with glucose, ATP, and yeast extract, gave 15 per cent yield of cellulose from glucose. This yield is remarkable as the methods used for preparing the extract were not essentially different from those of other workers. Since the methods used to characterize the product as cellulose were not given these results must await independent confirmation. Finally, Colvin reported that ethanol extracts of *A. xylinum* contained substances that formed microfibrils when water and a heat-labile catalyst from the culture medium were added. He could not detect cellobiose, related oligosaccharides, UDP, or UDPG in the extract, suggesting that an entirely new mechanism is involved in cellulose formation. It is too early to explain the differences between Glaser's and Colvin's results as neither obtained more than trace amounts of cellulose, but as the former author notes, more than one pathway to cellulose may operate in this bacterium.

Aside from the biochemistry, formation of cellulose by *A. xylinum* has some peculiarities which deserve comment. Electron microscope studies indicate that the microfibrils grow at their tips without contact with the surface of the cells, a finding of considerable importance in relation to microfibril growth in walls. There are two difficulties with such a system : first, well-washed cells immediately form cellulose without the lag period expected if polymerizing enzyme must first be formed and liberated into the medium ; second, if the immediate precursor of cellulose is a glucoside, it is improbable that the cells could lose large amounts of the aglycone after the glucose entered cellulose. The first objection would be satisfied if microfibrils were initiated on the bacterial surface and then separated from the cell, carrying polymerizing enzymes bound to their tips. If such is the case, gently prepared, short microfibrils should act as primers for cellulose synthesis in cell-free systems. The second problem is more difficult. Possibly much of the aglycone could be reabsorbed by the cells, and some form of surface diffusion along the microfibrils might reduce necessary external concentration of the precursor.

Roelofsen discusses the additional problem in cellulose synthesis of the relation of simple polymer formation to microfibril organization. He concludes that glucosan chains and microfibrils are formed simultaneously by addition of individual glucose residues at the microfibril tip rather than by organization of pre-formed glucosan molecules into microfibrils. On the basis of the following evidence we would tentatively agree : (*a*) dispersed molecules of cellulose will not reform into native microfibrils ; (*b*) under the electron microscope the tips of growing microfibrils in *A. xylinum* cultures appear smooth and tapered, not frayed, and no evidence of an amorphous precursor is found ; (*c*) abundant oligoglucosides are not found in *A. xylinum* ; and (*d*) microfibrils are formed readily by *A. xylinum* in stiff gels of carboxymethyl cellulose which would presumably interfere with migration of large cellodextrin molecules. However, simultaneous formation of the β-glucosan chain and microfibril presents an enzymatic problem. If glucosan chains in the microfibril run in both directions, *i.e.*, antiparallel, both free carbon-1 and carbon-4 will be present at both ends of the microfibril and two enzymes would be required for growth, Alternatively, with a parallel arrangement of glucosan chains one enzyme could bring about growth at one end of the microfibril only. It is impossible to choose between these systems at present, other than to note that the antiparallel arrangement of molecules in chitin favours the two enzyme hypothesis.

Non-cellulosic carbohydrates. The only available information on synthesis of the non-cellulosic carbohydrates of walls concerns metabolism of the constituent monosaccharides. The utilization of many specifically ^{14}C-labelled precursors in a variety of plants has been studied : Neish and co-workers followed incorporation from several hexoses and pentoses into xylan of wheat ; Ginsburg and Hassid analysed both xylan and araban of wheat following administration of labelled sugars ; Seegmiller and co-workers administered specifically ^{14}C-labelled glucose and galactose to boysenberries and strawbarries, and analysed the uronic acid and arabinose of a pectin fraction; Loewus *et al.* (101) supplied ^{14}C-labelled glucuronate, glucuronolactone, and galactose, and analysed polygalacturonic acid of strawberries. The results from all of these studies are in general agreement and indicate that the hexose skeleton remains predominantly intact during formation of all of the 6-carbon components of wall polysaccharides. Further, little rearrangement of the first 5 carbon atoms occurs during the formation of xylose and arabinose from either hexoses or hexuronic acids. This result suggests that the pentoses are formed by oxidation and decarboxylation of carbon-6 of the hexoses. In support of this Slater and Beevers showed that in corn coleoptiles carbon-6 of exogenously supplied glucuronate and glucuronolactone was rapidly lost as CO_2. Feeding of pentoses indicates that arabinose can be used directly for pentosan formation, but in wheat and strawberry xylose must be converted to hexose before entering wall pentosan.

The interconversions of labelled compounds in the above studies are consistent with the scheme in Figure 3.18 which is based on those of Neish and Hassid *et al.* The individual reactions are catalysed by the following types of enzymes pyrophosphorylases ; C-4 epimerases ; dehydrogenase ; decarboxylase.

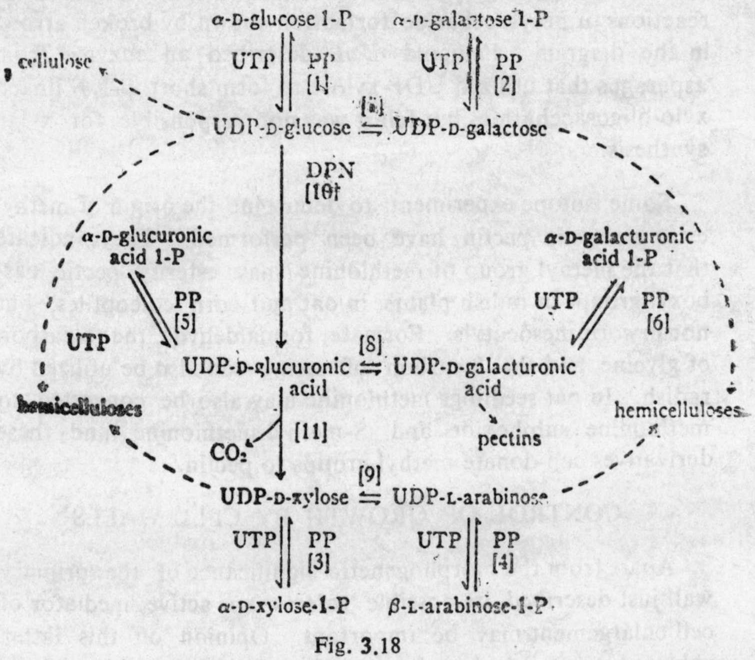

Fig. 3.18

Extensive evidence for such a scheme in higher plants comes from studies by Hassid and co-workers on mung bean seedlings. They isolated all of the sugar nucleotides shown in the diagram and demonstrated enzymes that catalysed all numbered reactions. In addition, the enzyme for reaction has been isolated from pea seedlings and for reaction from asparagus, radish, and spinach. The apparent failure of exogenous xylose to enter directly the xylans of wheat and strawberry suggests that reaction is not effective in these plants and that xylose is rapidly drawn into the pentose-phosphate pathway and partially converted to hexose phosphate. Apparently arabinose is readily utilized through reaction but there is difficulty in explaining the conversion from the pyranose form used in reaction to the furanose form found in hemicelluloses, UDP-mannose is not included in the scheme since it has not been reported, but guanosine diphosphate mannose has been found in yeast where cell wall mannan is present. There is as yet no evidence for the direct utilization of the UDP derivatives in the reactions of polysaccharide formation shown by broken arrows in the diagram. Feingold *et al.* described an enzyme from asparagus that utilized UDP-xylose to form short β-1,4 linked xylo-oligosaccharides but felt it was not responsible for xylan synthesis.

Some isotope experiments to determine the origin of methyl ester groups in pectin have been performed. They indicate that the methyl group of methionine may esterify pectic carboxyl groups in radish plants, in oat and corn coleoptiles, but not in corn mesocotyls. Formate, formaldehyde, the α-carbon of glycine, and the β-carbon of serine can also be utilized by radish. In oat seedlings methionine may also be converted to methionine sulphoxide and S-methyl-methionine, and these derivatives can donate methyl groups to pectin.

CONTROL OF GROWTH BY CELL WALLS

Apart from the morphogenetic significance of the primary wall just described, its possible role as an active mediator of cell enlargement may be important. Opinion on this latter subject has varied, but interest has been reawakened by a number of recent studies. Although most of the evidence is indirect, by implication it mainly concerns the continuous phase of the wall, the polysaccharide matrix.

Mechanisms of cell expansion. Three obvious physiological processes occur during normal growth of the cell ; water enters, new wall material is deposited, and the wall increases in area. Growth might conceivably be controlled by direct regulation of any of these reactions. Studies on water metabolism have been reviewed by Galston and Purves and by Thimann and the evidence indicates that osmosis rather than "active" water uptake plays the major role in water movement. Experiments with growth regulators further indicate that control of growth by direct regulation of osmotic properties of cells is unlikely and although such properties generate turgor pressure, which serves as a driving force in growth, the mechanisms of control reside elsewhere in the cell, perhaps in the cell wall. Attention is thus directed to the latter two growth processes, deposition and expansion of walls.

Heyn outlined three possible mechanisms of wall expansion : (*a*) increase in surface area by active synthesis of new wall ; (*b*) elastic extension of the wall by turgor pressure, made permanent by deposition of new wall substance ; (*c*) direct plastic stretching of the wall by turgor pressure. In the first case control of growth is exercised by direct regulation of mechanisms of wall synthesis while in the other two schemes, controlled variation of physical properties of the wall could regulate growth.

The first mechanism, active wall growth, has been discarded on the basis that it requires strict stoichiometry between deposition and elongation and that uncoupling of the processes can be demonstrated. For example, Preston and Clark showed that in intact oat coleoptiles longer than 1 cm. weight of wall per unit length decreased steadily during elongation, resulting in about 30 per cent "thinning" of the walls at cessation of growth. Similar situations where rate of expansion exceeds rate of deposition have been noted in isolated segments of oat coleoptile and pea stem. Using optical techniques Green found an initial rise followed by a sharp drop in mass per unit area of *Nitella* internodal cell walls in early stages of growth, and then a slow increase during the main portion of elongation. However, although such partial non-correspondence of deposition and expansion suggests that deposition is not the immediate cause of growth it is not conclusive proof of this. Active growth on the inside of the wall, for example, could promote elongation while the outer regions are passively thinned.

Several other lines of evidence, however, indicate that active wall deposition is not the basic driving mechanism in cell elongation. Bonner found small amounts of elongation without deposition of wall in coleoptile segments at 2°C. Several workers showed that the regulatory action of auxin can be separated in time from the act of growth. The presence of simple multinet structure, discussed earlier, indicates that in thin parenchyma walls intussusception of new microfibrils does not occur throughout the wall but rather deposition of cellulose is primarily by apposition. Finally, there is abundant evidence to indicate that walls become more plastic in response to growth regulators. As a result of considerations such as these, the theory of growth by wall deposition has been largely dismissed in favor of theories involving stretching of the wall. However, as will be discussed below, growth and deposition are not fortuitously related processes. Furthermore, direct evidence on mechanisms of wall growth is scanty and it is possible that enlargement by active wall growth may occur in some cells.

Growth through wall plasticization. During the 1930's Heyn performed extensive experiments to demonstrate the importance of wall plasticity in growth. In particular he treated coleoptile segments with auxin under conditions of much reduced growth and noted that plasticity of the walls, as revealed by bending under weight, varied in relation to the capacity for elongation. He also studied the elastic properties of the wall and found that they were not correlated with capacity for expansion but depended on the growth history of the segment. From these and other experiments Heyn concluded that control of growth is achieved by varying the plasticity of the wall and that plastic extension, produced by turgor pressure, is the primary mechanism of cell growth.

Recently several workers have repeated and extended Heyn's observations. Tagawa and Bonner confirmed that auxin produced a relatively rapid increase in plasticity of coleoptiles as measured by plastic bending and observed that calcium and magnesium ions, which retard elongation, caused a stiffening of the wall while potassium ions, which stimulate elongation, increased plasticity. Adamson and Adamson using bending techniques, also found increased plasticity of coleoptile walls with auxin and a decrease with calcium chloride. They also noted that although anaerobiosis and low temperature increased rigidity of the tissue auxin still increased wall flexibility under these conditions. Cleland refined Heyn's experiments by applying auxin in the presence of 0.28 *M* mannitol, which osmotically prevented elongation, and concluded with Heyn that growth stimulation by auxin was directly related to increased plasticity of the wall. Objections to determinations of wall plasticity by bending measurements have been met by Preston and Hepton who measured physical extensibility by applying a force in line with the axis of the coleoptile. By this method it was demonstrated that auxin caused a noticeable increase in extensibility of walls even when growth was limited by mannitol or absence of water. In contrast to other workers, Preston and Hepton concluded that elastic properties of the wall may be important in growth. However, the coleoptiles used in many of their experiments elongated without added sugar and may have shown changes in elasticity due to thinning of walls.

Additional support for Heyn's plasticity theory comes from studies in which action of growth substance is separated in time from the period of growth. Thimann retarded the growth of potato discs with mannitol and found that on subsequent removal of the mannitol the rate of cell expansion exceeded that of controls. A somewhat more definitive experiment was performed by Kobayashi *et al.* who incubated oat coleoptile segments with and without auxin in 0.25*M* lactose, which inhibited growth, and then transferred all segments to water without auxin. Segments which had received auxin with the lactose grew significantly more in water than ones treated with lactose alone. This experiment was further refined by Cleland and Bonner who gave an aerobic pretreatment with auxin in a concentration of mannitol which inhibited growth and then allowed expansion to occur in water under argon, a condition which normally inhibits auxin stimulation of growth. They observed a significant increase of elongation under argon caused by auxin in the pretreatment. All of these results are readily interpreted on the basis that auxin produces a plasticization of the wall which persists, and then results in increased elongation when turgor pressure is restored.

The evidence therefore strongly supports Heyn's notion that plasticity of the cell wall is an important factor in cell growth and that it can serve as the basis of at least one growth regulating mechanism. However, there are some ambiguities concerning the role of metabolism in the process. At low temperature, increases in extension and flexibility produced by auxin have both been reported. Preston and Hepton however could find no increase in physical extensibility at low temperature and Cleland and Bonner noted that elongation under argon following auxin treatment in mannitol was also prevented by low temperature. Furthermore, auxin-induced growth is prevented by anaerobiosis and all auxin effects on plasticity and extension are abolished by cyanide and other metabolic inhibitors. Yet auxin stimulation of plasticity under anaerobic conditions is reported. The inhibitor evidence would definitely suggest that plasticization of the wall must be a metabolic process.

Chemical nature of wall plasticization. Bennet-Clark in reporting growth inhibition by divalent ions, particularly calcium, suggested that they might reduce wall plasticity by increasing ionic binding between pectic carboxyl groups [a theory also advanced by Carlier and Buffel] and that growth stimulation by ethylenediamine tetraacetic acid was due to its decreasing such cross-linking by chelating calcium. Interaction of divalent ions with pectin was also advanced by Tagawa and Bonner to explain decreased plasticity of coleoptile walls treated with calcium and magnesium. Additional evidence for involving pectin in wall plasticity came from studies of pectin methyl esterase (PME). Bryan and Newcomb found a large apparent increase in the amount of this enzyme in tobacco pith cells stimulated to enlarge by auxin. Similarly, Yoda noted that inhibition and stimulation of growth of pea stem segments by various substances was closely paralleled by changes in PME activity in tissue homogenates. In a number of *in vitro* and *in vivo* studies Glasziou showed that auxin can increase binding of PME to walls and proposed that such binding prevented deesterification of pectin, thereby decreasing the amount of calcium bridging. This mechanism has been invoked to explain auxin effects on wall plasticity at low temperatures and in anaerobic conditions (2). Other authors have taken the effects of commercial preparations of pectinase on growth of roots to indicate that pectin is involved in wall rigidity.

Although such indirect evidence is suggestive, the hypothesis that calcium-pectin interaction controls wall plasticity is not supported by recent studies. The low amount of pectin found by Bishop *et al.* in oat coleoptile, if accurate, seriously restricts such a mechanism. Furthermore, the calcium content of primary walls appears to be very low. In wheat coleoptile walls Carr and Ng detected only 0.1 per cent calcium on a dry weight basis, a figure corresponding to about 1 per cent calcium pectate. Similarly, Jansen *et al.* found only enough calcium in oat coleoptile walls to form the salt of one-fifth of the non-esterified anhydrouronic acid residues which in turn represented only 3 per cent of the dry wall. More directly, Cleland used a sensitive assay with calcium and showed that in both corn mesocotyls and oat coleoptiles auxin had no effect on the amount or distribution of calcium in growing walls. Earlier, Carr and Ng were unable to find any removal of calcium by ethylenediamine tetraacetic acid under conditions where growth stimulation would be obtained. The general importance of pectin methyl esterase in growth is doubtful since Jansen *et al.* found no evidence of auxin-mediated binding or stimulation of this enzyme in oat coleoptiles. It is also difficult to see how the binding mechanism could allow auxin to produce a rapid increase in growth once considerable calcium cross linking had occurred. Experiments using commercial pectinase to influence growth have been justifiably criticized by Jackson on the basis that such enzyme preparations are very impure.

A more direct test of the pectin hypothesis in growth might be expected from comparative chemical analyses of walls grown with and without auxin. Several such analyses exist but the picture is confused. Using fairly specific methods for polygalacturonic acid. Wilson and Skoog found a 50 to 100 per cent increase in the concentration of pectic substances in tobacco pith segment streated with auxin, whether or not growth resulted. Carlier and Buffel noted a somewhat smaller increase in concentration of pectic materials in potato tuber discs supplied with auxin. On the other hand, Jansen *et al.* claimed that auxin caused no change in the pectic composition of oat ooleoptile walls. Using less specific methods, Burstrom also concluded that relative concentrations of wall materials do not change significantly in normal growth of roots.

Several studies on patterns of ^{14}C incorporation have been performed, but again the results are equivocal. Boroughs and Bonner supplied ^{14}C-labelled acetate and sucrose to coleoptiles and found that auxin increased incorporation of isotope into the fraction soluble in 4 per cent NaOH (hemicelluloses and decreased incorporation into other wall fractions. Surprisingly, although auxin stimulated growth it caused a slightly decreased total incorporation of ^{14}C into the wall from both substrates. Wightman and Neish noted increased incorporation into both hemicelluloses and cellulose of oat coleoptiles and pea stem segments due to auxin treatment. There was, however, no difference in the relative incorporation into the two fractions. In excised wheat roots Perlis and Nance used auxin as a growth inhibitor and obtained an increased relative incorporation of ^{14}C from pyruvate and acetate into cellulose. As they pointed out, this was probably related to synthesis of secondary wall and in fact multilayered secondary walls are common in the parenchyma of roots. Nance also supplied acetate-^{14}C to pea stem segments and noted decreased isotope incorporation into non-cellulosic wall materials due to auxin. This result was apparently due mainly to alterations in intermediary metabolism rather than immediate effects on synthesis of cell wall. Ordin *et al.* administered glucose-^{14}C to coleoptiles prevented from elongating by mannitol and found little difference in incorporation due to auxin. Recently, however, Albersheim and Bonner using labelled glucose have shown that auxin stimulates the rate of synthesis of pectic material soluble in cold water. They suggest that this stimulation could be a basis for wall plasticization, but since the fraction forms only about

0.2 per cent of the total wall material and is presumably not very closely bound in the wall its significance in growth seems questionable.

One reasonably clear cut, although again quantitatively small, metabolic response of pectin during growth involves methyl esterification. Ordin *et al.* supplied methyl-^{14}C-methionine to coleoptiles and found an auxin-increased rate of incorporation of ^{14}C into methyl ester of pectic substances soluble in hot water. Subsequently, Jonsen *et al.* showed that this increased incorporation was due to increased turnover, not net synthesis of methyl ester. However, Cleland was unable to demonstrate that auxin stimulated incorporation of ^{14}C from methyl-labelled methionine into pectic substances of corn mesocotyls and suggested that either auxin-stimulated methylation of pectin is not a general mechanism or a different methyl donor is used in mesocotyls. Cleland also found in oat coleoptiles that, although ethionine rapidly prevented incorporation of methyl groups from methionine and at the same time reduced elasticity of the wall, inhibition of growth was not directly related to these processes.

Taking the evidence as a whole, no consistent major change in the metabolism of pectic substances in relation to growth has been demonstrated. Thus, although the hypothesis that pectic substances determine plasticity of the wall is attractive theoretically, it remains unsubstantiated. Unfortunately, the studies that failed to demonstrate a role for pectin also failed to show an obvious growth response of other wall substances, suggesting that the basis of plasticity of the wall may reside in qualitative, rather than quantitative, changes in the preponderant and complex hemicelluloses. That slight structural changes are capable of changing the properties of these compounds is indicated by studies such as those of Perlin, who showed that the solubility of xylans is directly related to the frequency of arabinose side chains.

Relations between cell elongation and wall deposition. Although plastic stretching appears to be important in growth, wall deposition normally accompanies cell elongation and even if, as noted above, the two processes do not always remain perfectly in step it is impossible to separate them for long periods. Thus, coleoptile segments may elongate 15 or 20 per cent without wall deposition at low temperature or without oxygen, but sustained growth under these conditions is impossible. Limiting substrates for wall synthesis will also limit growth and galactose may inhibit growth of coleoptiles by direct inhibition of cellulose synthesis. General metabolic inhibitors will of course inhibit both processes but simple osmotic inhibition of elongation also prevents wall deposition. Growth inhibition by lack of auxin is accompanied by decreased wall synthesis. These facts lead to the conclusion that although wall deposition may not be the direct cause of elongation, the two processes are closely interrelated and essential for each other during continued growth.

Dependence of wall deposition on extension might be visualized as some process by which stretching of the wall frees active surface areas required for addition of new materials. Such a mechanism could explain reduction of wall deposition in the absence of auxin but the situation is more complex in the case of osmotic inhibition of growth. By manipulating external concentrations of mannitol and diffusible salts. Ordin obtained evidence that synthesis of wall in oat coleoptile segments depends directly on turgor pressure rather than on elongation. A similar proposal was made by Adamson and Adamson to explain results from Jerusalem artichoke. Ordin also found that synthesis of different wall fractions shows a variable sensitivity to turgor although the main trend is for reduced synthesis with reduced turgor. In this connection it might be noted that, without entering the cell, mannitol could exert direct effects on polymerization reactions occurring outside the plasma membrane.

The alternative relation, dependence of wall elongation on deposition, might be explained on the basis that a given amount of wall material is only capable of a certain amount of stretching by turgor, and for continued expansion to occur, new material capable of being stretched must be added. Auxin would mediate the stretching process but could only function in the presence of wall material which was not fully stretched. The plastic material would presumably be a component of the matrix and would be added by intussusception throughout the wall while the transverse microfibrils were apposited on the inner surface. Such a concept would not be in entire agreement with the simple multiner theory in its present from, but the site of deposition of matrix substances in parenchyma walls is as yet unknown. If these substances are added only by apposition they must be capable of indefinite stretching and a more complex feedback mechanism would be required for expansion to be limited by wall synthesis.

To this discussion should be added the ideas of Busse that both plastic substance and material capable of stiffening walls are added during deposition and that the rate of elongation depends on a balance between the two. He has presented evidence that stimulation of coleoptile growth by cobalt results from an inhibition of synthesis of the stiffening material. Auxin-independent stiffening of coleoptile walls has also been noted by other workers and Cleland has invoked two types of bonds, auxin sensitive and auxin insensitive, to explain the fact that elongation of coleoptiles supplied with auxin shows a sharp change in rate at an external mannitol concentration of about 0.12 M. The stiffening material would presumably be a component of the matrix since structurals studies indicate that the microfibrils, although reasonably rigid themselves, are able to slip past each other in the matrix of growing walls. Once stiffening of the matrix stopped growth, further reinforcement could be achieved by deposition of secondary wall.

As noted by Busse, considerations of this type are quite speculative and can serve only as working hypotheses. However, they have several consequences. First, growth may be experimentally regulated by interference with at least four processes : water uptake, plasticization by auxin, synthesis of plastic materials, and synthesis of stiffening materials. Second, a change in the composition of the matrix substances should be detectable as the proportions of stiffening and plastic materials change. Jermyn and Isherwood have shown that quite complex changes can occur in walls of parenchyma cells during ripening of fruit and it seems reasonable to expect that changes of a similar nature take place during the course of growth. Recent cytochemical analyses of walls by Jensen and Ashton and Fuller suggest that fairly complex changes occur in root meristems but more detailed chemical study of matrix substances will be necessary to test these ideas fully. Here, studies of lower plant forms which appear to have walls of relatively simple composition may be useful.

THE CUTICLE

The surface of the aerial parts of vascular plants is covered by a layer of cutin, the *cuticle*. The cuticle covers the epidermis closely, following the contour of all cells (Fig. 3.19). Cutin is impermeable to water and the cuticle contributes to the control of water loss from the underlying cells. On young parts the cuticle is thin and pliable ; as growth continues, it is stretched and added to in thickness. When growth of the tissues beneath is complete, the cuticle becomes firm and highly resistant to stretching and to breakdown by any agent. (Cuticles persist, little changed, even in coal.) In thickness and in character, the mature layer varies greatly ; it is thin in plants of shade and moist habitats, generally thick in those of dry and sunny situations. Its thickness on the leaves and fruit of the same plant varies somewhat with position on the plant and from season to season with weather conditions. On ephemeral organs and aquatic plants it remains very thin and may not be readily detectable. Typical roots have no cuticle.

The only breaks in the continuity of the cuticle are the openings of stomata and lenticels. The cuticle extends for varying distances into the stomatal apertures (Fig. 3.20A,B), and a thin layer of cutin may coat the stomatal chamber. The walls of large air chambers may have a thin cuticle, and small intercellular spaces between the epidermis and mesophyll cells may also be coated with cutin.

In origin, the cuticle is a secretion of the epidermal cells and perhaps also of underlying cells. The cutin, in apparently a liquid or semiliquid form, is passed through the outer wall in an unknown manner. On the surface it forms a continuous layer and becomes hard and tough. According to one view, droplets of cutin pass through exceedingly minute openings of the outer wall ; according to another, they are formed in the

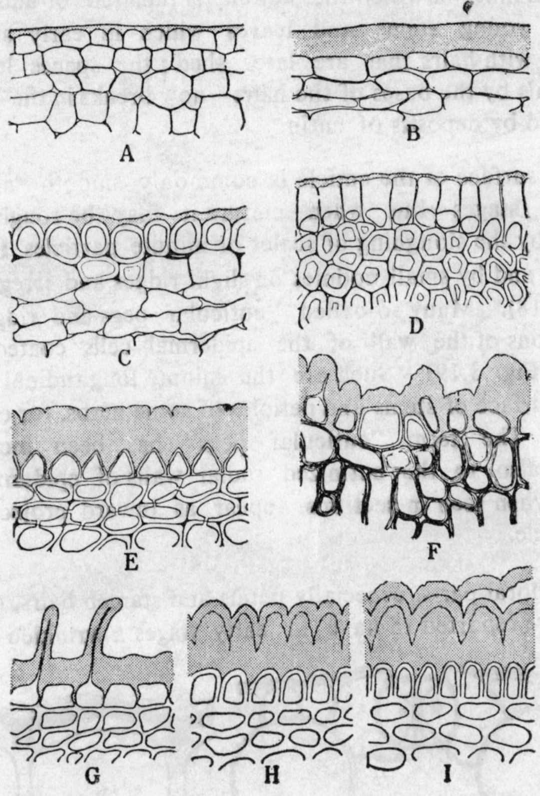

Fig. 3.19. The cuticle. Sections of the epidermis and adjacent tissues, showing various degrees of thickness and of extent of the cuticle between epidermal cells. *A*, *Citrus sinensis* (*orange*), fruit ;*B*, *Malus pumila* (var. Ben Davis), fruit ; *C*, *Dracaena Goldieana*, stem ; *D*, *Dasylirion serratifolium*, leaf ; *E*, *Acer pennsylvanicum*, stem, the older, outer layer cracking and disintegrating ; *F*, *Smilax rotundifolia*, stem ; *G*, *Vaccinium corymbosum*, stem ; *H* and *I*, *Cornus circinata*, stem, showing two stages, the outer, older layers being replaced by new ones formed below.

deeper lying cells and migrate along the walls, reaching the surface through the anticlinal walls of the epidermis. It is clear that the cuticle is thickened from below, although in early stages—and perhaps at some other times—new material is apparently dissolved in the preexisting layer. An example of the manner in which the cuticle is repaired or added to is seen in young stems and leaves which in early stages are clothed with hairs that are later shed ; the spaces left in the epidermis by the bases of the hairs—now breaks in the cuticle—are filled by deposits of cutin.

The surface of the cuticle is commonly smooth—always so in early stages—but when mature it may be roughened by cracks, by the sloughing of scales or minute particles from the surface, and by small nodules or slight ridges and irregularities (Fig. 3.18). Many so-called "cuticular pegs and ridges" are projections of the wall of the epidermal cells coated by the cuticle (Fig. 3.19). Such are the minute longitudinal flutings of the surface of stems and petioles of some herbs, for example, *Rumex*. The term "cuticular pegs" has been incorrectly applied also to the cutinized radial walls of epidermal cells which, when seen in sections, appear as inward projections of the cuticle.

The floral parts, especially petals and stamen hairs, of some plants are reported to have in early stages a wrinkled cuticle,

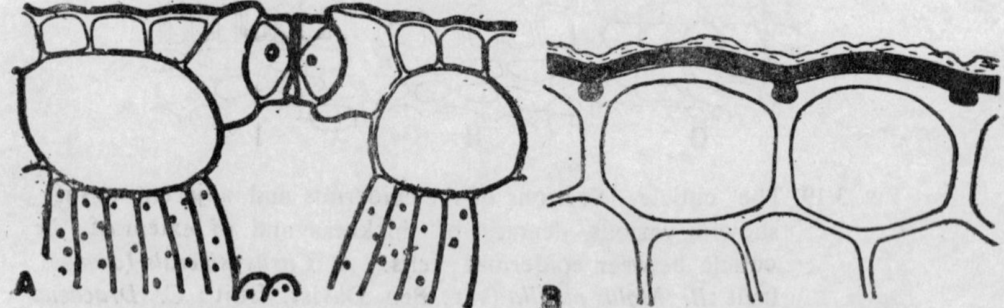

Fig. 3.20. Cuticle of the leaf of banana (*Musa*). *A*, section of upper epidermis through stomatal region, showing extent of cuticle over guard cells ; *B*, section through pulvinar band showing "pegs" of cuticle over anticlinal walls of epidermal cells, freedom of cuticle between pegs, and waxy layer on surface of cuticle.

the wrinkles appearing as minute folds when the deposition of cutin is rapid and growth of the epidermis in area is slow. The folds are said to form a "reserve surface," and later, when growth of the organ is very rapid, are stretched out or reoriented. Over the mature epidermis the only folds remaining are those whose axes lie in the direction of greatest growth of the cells beneath. The radiating striations of the cuticle of papillate epidermal cells of petals are said to represent this condition.

The cuticle commonly adheres closely to the surface of the epidermis but has been reported to lie free and loosely over the walls in some floral hairs and to be loose, but pegged down to the epidermis over pulvinar swellings, where the surface undergoes great changes in curvature, as in the leaf of banana (*Musa*) (Fig. 3.20). The separation of the cuticle from the epidermis by an air space is said to be responsible for the silvery appearance of some leaves.

The development of the cuticle, *cuticularization*, is accompanied, or followed in many plants by *cutinization* of the cells below. The outer walls of the epidermal cells are first modified. These may be increased greatly in thickness becoming many times as thick as the other walls, and as the cuticle itself (Fig. 3.19*F,G*). They may be so heavily impregnated with cutin that they are separable only with great difficulty from the layer of pure cutin, the cuticle. The radial and inner walls may similarly be cutinized and even the walls of the subjacent cells. Where the entire walls of the epidermal cells become cutinized, the cells may die, but frequently they remain alive, with normal pits and plasmodesmata. The term cuticle is unfortunately sometimes incorrectly applied in a functional sense to the entire outer "waterproofing" layer, the cuticle plus all cutinized cells. The layer which is stripped from leaves as "the cuticle" is usually the cuticle plus cutinized parts of underlying layers.

In some woody stems where the epidermis persists for more than one year—for example, species of *Acer*, *Cornus*, *Kerria*—

the original cuticle is cracked or ruptured and is repaired by secretion of new material from beneath, or a complete new layer is formed (Fig. 3.18*H*,*I*).

Many fruits, such as plum, tomato, persimmon, and some varieties of apple (Fig. 3.19*B*), have, when ripe, a thick cuticle. This reaches full thickness only as the fruit attains mature size. The heavy layer aids in the conservation of water in the fruit after the supply of water ceases. Some of the keeping qualities of fruits are due to the presence of a heavy cuticle. Varieties of apples with thick cuticles keep better than those with thinner cuticles. Fruits grown in sunshine have heavier cuticles than those grown in the shade ; and apples grown in a "dry year" have heavier cuticles than those borne on the same tree in a "wet year." In the drying of prunes the cuticle is chemically or mechanically perforated to hasten the drying.

Minute particles or rods of wax on the surface of the cuticle are responsible for the glaucous character of stems and leaves and constitute the bloom of fruits. Wax may occur in larger particles, as in the bayberry, *Myrica*, or may heavily coat the leaves, as in the wax palms.

4

Structure and Function of Mitochondria

Kolliker (1880) was the first who observed the granules mitochondria) in muscle cell of insects. *Flemming* (1882) named the mitochondria as *fila*. *Altmann* in 1894 observed tham and they were called *Altmann's granules bioplasts*. The term *mitochondria* was applied by *Benda* (1897-98) to these granules which were described as *cytomicrosomes* by *Velette St. George*. *Benda* stained mitochondria with alizarin and crystal violet. *Kingsbury* (1912) related them to cellular respiration and *Warburg* (1913) observed the presence of respiratory enzymes. In 1934 *Bensley* and *Horr* isolated mitochondria from liver cells and *Porter* and *Palade* described their electron microscopic structure.

Distribution. Ordinarily mitochondria are evenly distributed in the cytoplasm. They may, however, be localized in certain regions. In proximal convoluted tubules of the kidney they are found in the basal region of the cell, opposite the renal capillaries. In skeletal muscles they lie between the myofibrils. In insect flight muscle several large mitochondria are in contact with each fibril. In cardaic muscle the mitochondria are situated in clefts between the myofibrils, numerous lipid droplets are associated with the mitochondria. In many sperms the mitochondria fuse into one or two structures which lie in the middle piece of the tail, surrounding the axil filament. In columnar or prismatic cells they are oriented parallel to the long axis of the cell. In leucocytes they are radially arranged.

Orientation. Mitochondria may have a more or less definite orientation. For example in cylindrical cells they are generally oriented in the basal-apical direction, parallel to the main axis. In leukocytes, mitochondria are arranged radially with respect to the centrioles. It has been suggested that these orientations depend upon the direction of the diffusion currents within cells and are related to the submicroscopic organization of the cytoplasmic matrix and vaculolar system.

PLASTICITY OF MITOCHONDRIA IN CELL

Lewis and *Lewis* (1914-15) concluded that the mitochondria are extremely variable bodies, which are continually moving the changing shape in cytoplasm. There are definite types of mitochondria, as any one type may change into another. They appear to arise in cytoplasm and to be used up by cellular activity. The shape may change fifteen to twenty times in ten minutes ; it can be changed by heat, hypertonic and hypotonic media or by acids, fat solvents, potassium permanganate and osmotic changes. *Frederic* (1958), *Littre* (1954), *Tobioka* and *Biesels* (1956) studied the effect of large number of chemical and physical agents on mitochondrial behaviour. Some material, such as detergents, show some effect *in vivo* as on mitochondria isolated from homogenates.

MORPHOLOGY

Shape. The shape is variable but is characteristic for a cell or tissue type, this too is dependable upon environment or physiological conditions. In general, they are *filamentous* or *granular.* They may swell out at one end to become *club-shaped* or hollow out at one end to assume a shape of *tenis-racket.* They may become *vesicular* by the appearance of central clear zone. *Rod-shaped* mitochondria are also observable. Fig. 4.1

Size. The size of mitochondria also varies. In majority of the cells, which is relatively constant, about 0.5μ, but the length varies and, sometimes, reaches a maximum of 7μ. The size of cell also depends on the functional stage of the cell. Very thin mitochondria, about 0.2μ, or thick rods 2μ are also seen.

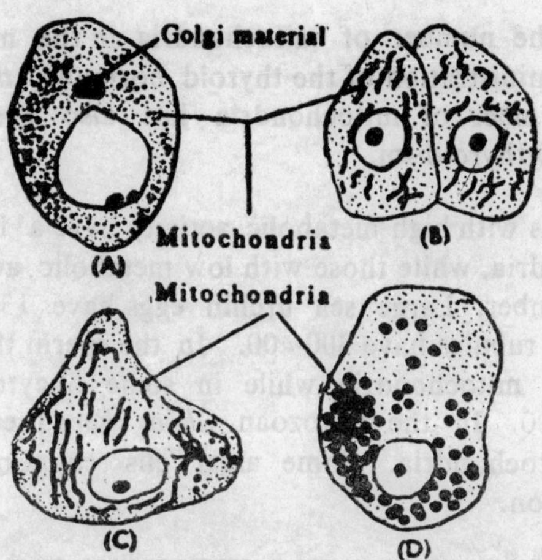

Fig. 4.1. Different shapes and sizes of mitochondria.

The size and shape of the fixed mitochondria are determined by the osmotic pressure and pH of the fixative. In acid mitochondria are fragmented and become vesicular. Mitochondria, in the rat-liver, are usually 3.3μ in length ; in mammalian exocrine pancrease, they are about 10μ in length and in oocytes of Amphibia, they are approximately of the length of 20 to 40μ.

Number. Mitochondria are found in the cytoplasm of all aerobically respiring cells with the exception of bacteria in which the respiratory enzymes are located in the plasma membrane. The mitochondria content of a cell is difficult to determine, but in general, it varies with the cell type and functional stage. It is estimated that in liver mitochondria constitute 30 to 35 per cent of the total protein content of the cell and in kidney, 20 per cent. In lymphoid tissue, the value is much lower. In mouse liver, homogenates there are about 8.7×10^{10} mitochondria per gram of fresh tissue. A normal liver cell contains about 1000 to 1600 mitochondria, but this number diminishes during regeneration and also in cancerous tissue. This last observation may be related to decrease oxidation that accompanies the increase to anacrobic glycolysis in cancer. Another interesting finding is that there is an increase in the number of mitochondria in the muscle after repeated administration of the thyroid hormone, *thyroxin*. An increased number of mitochondria has also been found in human hyperthyroidism.

Thus cells with high metabolic activity have a high number of mitochondria, while those with low metabolic activity have a lower number. Large sea urchin eggs have 13,000-14,000, while renal tubules have 300-400. In the sperm there are as few as 20-24 mitochondria while in some oocytes there are about 300,000. In the protozoan *Chaos chaos* there are about 500,000 mitochondria. Some algal cells contain only one mitochondrion.

STRUCTURE OF MITOCHONDRIA

A typical mitochondrion is sausage shaped with an average diameter of about 0.5μ. When it is properly fixed in osmimum containing fluid and studied under electron microscope which reveals that there is hardly any difference between plant and animal mitochondria. In both the cases the mitochondrion is bounded by two membranes, the *outer membrane* and the *inner membrane*. The space between the two membranes is called the *outer chamber* or *inner-membrane space*. It is filled with a water fluid, and is 40-70A° in width. The space bounded by the inner chamber is called the *inner chamber* or *inner membrane space*. The inner membrane space is filled with at *matrix* which contains dense granules (300-500A°), ribosomes and mitochondrial DNA. The granules consist of insoluble inorganic salts and are believed to be the binding sites of divalent ions like $Mg^{++}$and Ca^{++}.

In some cases they apparently contain polymers of sugars. The side of the inner membrane facing the matrix side is called the *M-side*, while the side facing the outer chamber is called the *C-side*. Two to six circular DNA molecules have been identified within mitochondria. These rings may either be in the open or in the twisted configuration. They may be present free in the matrix or may be attached to the membrane. The enzymes of the Krebs' cycle are located in the matrix.

The inner membrane is thrown up into a series of folds, called *cristae mitochondriales*, which project into the inner chamber. The cavity of the cristae is called the *intercristae space*, and is continuous with the intermembrane space.

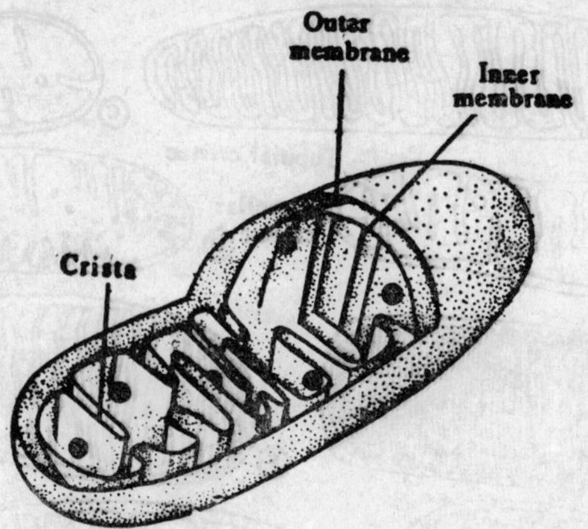

Fig. 4.2. Showing the structure of a mitochondria.

MITOCHONDRIAL CRISTAE

The space and arrangement of crests is variable and may be of the following types (Fig. 4.3).

(*i*) Parallel to the long axis of mitochondria as in the neurons and the striated muscle cells.

(*ii*) Concentrically arranged as in the matrix of certain spermatids.

(*iii*) Interlaced to form villi as in *Amoeba*.

(*iv*) Cristae in the form of vesicles which form a network of interconnected chambers as in the cells of parathyroid gland and W.B.C. of man.

(*v*) Arranged in a tubular fashion but perpendicular to mitochondrial axis as in the cells of adrenal gland.

(*vi*) Haphazardly distributed as in the cells of kidney of insects and hepatic cells.

(*vii*) Cristae extremely small and irregular as in the interstitial cells of *Opossum*.

(*viii*) Rarely the mitochondrial wall is smooth with no cristae. The number and size of cristae in a mitochondrion directly affects its efficiency. The greater and larger are the cristae, the faster is the speed of oxidation reaction.

(*ix*) Perpendicular to the long axis of mitochondria.

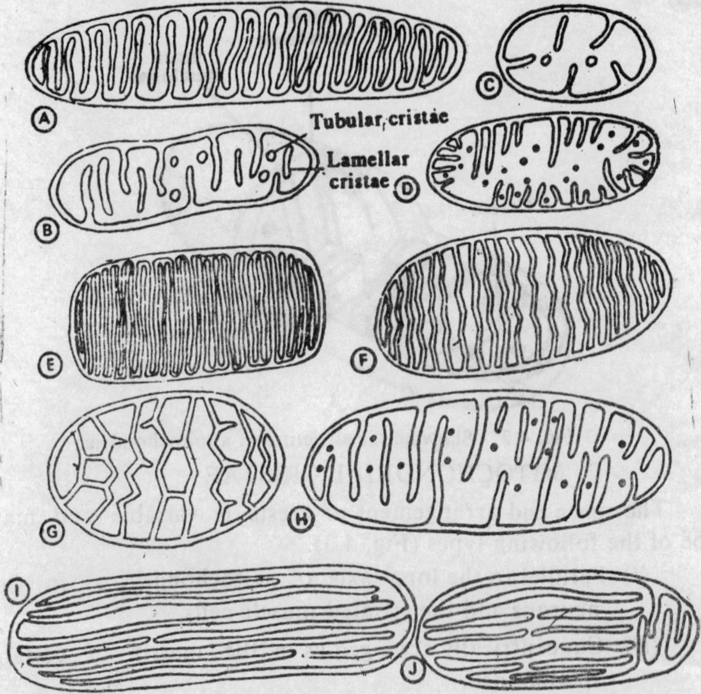

Fig. 4.3. Structural variations in mitochondrial cristae, A. Tubular cristae, B. lamellar cristae, C. Cristae in salamander, D. Plate like cristae, E. Parallel cristae, F. Cristal with sharp angulation, G. Honey corubs cristae, H. Usual cristae, I. Longitudinal cristae, J. Transverse and longitudinal cristae.

MITOCHONDRIAL PARTICLES

According to earlier descriptions the outer surface to the outer membrane and the inner surface of the inner membrane were supposed to be covered with thousands of small particles, (Fig. 4.4). Those on the outer membrance were described as being stalkless and were called the *subunits of Parson.* There may be as many as 10,000 to 100,000 particles per mitochondrion. The stalked inner membrane particles were called the *subunits of Fernandez-Moran, Elementary particles, F_1 particles* or the *oxiosomes* or *ETP* or *electron transport particles.* These particles are about 85A° in diameter and are regularly spaced at intervals of 10 nm on the inner membrane. There may be as many as 10^4 to 10^5 elementary particles per mitochondrion.

ORIGIN OF MITOCHONDRIA

Our understanding of the process where by mitochondria are produced, is still very incomplete. *Lehninger* (1964) classified the various theories of possible routes of mitochondrial genesis into three main groups :

(1) Formation from other membranous structures in the cell.
(2) Growth and division of pre-existing mitochondria.
(3) *De novo* synthesis from submicroscopic precurors.

(1) *Formation from other membranous structures in the cell.* The formation of mitochondria by "pinching off" or budding from pre-existing cell structures has been suggested

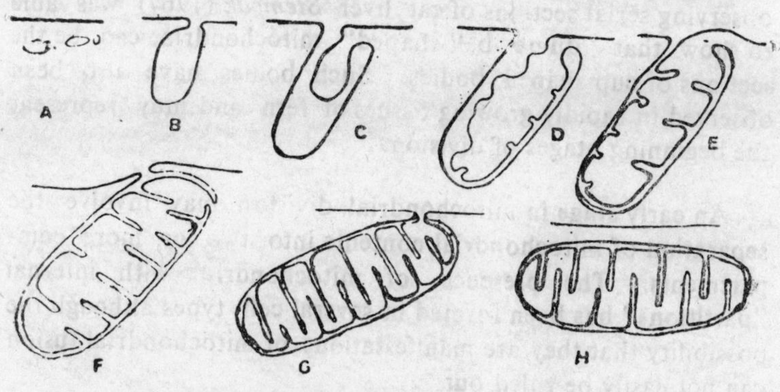

Fig. 4.4. A possible mechanism for the formation of mitochond ria from the plasma membrane.

for a range of cell membranes including those of the plasmalemma, endoplasmic reticulum, nuclear envelope, and Golgi-complex. But the support for such evidence, in the absence of supporting biochemical data, can not be wholly conclusive. Part of the problem undoubtedly lies in our fragmentary knowledge of the structure and composition of, and differences between cell membranes in general. Indeed the similarities discussed in the litterature between the mitochondrial membrane and the endoplasmic reticulum, could lend weight to the idea that probably mitochondria are formed when cytoplasm pushes into a cavity surrounded by an internal membrane, which then 'pinches off' and separates from the continuous system.

(2) *Growth and division of pre-existing mitochondria.* Electron microscopic evidence for mitochondrial division by fission, although plentiful is difficult to assess the danger of producing artifacts is very real because of the harsh chemical and physical agents brought to bear on the test material during processing. Interpretation is not made easier by the ability of mitochondria to undergo extreme changes in shape *in vivo* which may or may not be associated with mitochondrial fission. There are numerous reports of mitochondria connected to each other by narrow bridges of membrane, especially in rapidly melopolizing tissue, and it is thought that such figures may represent mitochondria in an early stage of fission. By observing serial sections of rat liver *Stempak* (1967) was able to show that "dumb bell-shaped" mitochondria can be the sections of cup-shaped bodies. Such bodies have also been observed in rapidly growing tissues of fern and may represent the beginning stages of division.

An early stage in mitochondrial division may involve the separation of mitochondrial contents into two or more compartments. The presence of mitochondria with internal "partitions" has been formed in several cell types although the possibility that they are manifestations of mitochondrial fusion can not easily be ruled out.

(3) *De novo synthesis.* The possibility of *de-novo* synthesis of mitochondria arose with experiments in the early part of the century, when mitochondria containing larvae were seen to develop from sea urchin egg cytoplasm which had apparently been freed of mitochondria by centrifugation. Using the greater resolving power of the electron microscope, it was later shown, that mitochondria could not be dislodged by

centrifugation of the egg. In the earlier experiments, mitochondria had probably been present in the "centripetal end" of the egg cell after all, and these mitochondrial could have served as precursors in subsequent mitochondrial production.

In the above description, a number of views were described, regarding the generation of mitochondria in the cytoplasm of different types of cells. But it is perhaps unwise to group the evidence to suit one or other of a limited number of clear cut methods by which mitochondria could replicate. The actual situation is probably complex, and it may well be that different methods of replication take place in different tissues, and at different stages in development. One could imagine the early mitochondria being formed from membrane structures in the developing embryo concentration of mitochondria around the nuclear membrane have been noted in embryonic tissues from several phyla and formation of mitochondria from this membrane could involve the transfer of nuclear genetic information essential for subsequent mitochondrial growth and multiplication by division. The maltiplication of mitochondria could then proceed by the incorporation of large-pre-fabricated molecules and association of molecules, with division by fission, when the mitochondria reached a critical stage.

Prokaryotic Origin of Mitochondria. This fact that mitochondria can grow, divide and are capable of mutations support a long-held view that mitochondria originated with their host. Bacteria would have originated the mitochondria and blue green algae, the chloroplasts.

There are numerous homologies between mitochondria and bacteria. In bacteria the electron transport system is localized in the plasma membrance which can be compared with the inner membrane of mitochondria. Some bacteria even have membranous projections extending from the plasma membrane which are comparable to mitochondrial crests since both contain the respiratory chain. The inner membrance and matrix, it has been postulated may represent the original symbiont which may become enclosed within a membrane of cellular origin (ER). Further mitochondrial DNA is circular;

it replicates and divides like that of bacteria. The ribosomes are also found which are, however, smaller than those of the bacteria. In mitochondria and bacteria, the protein synthesis is inhibited by chloramphenicol. From these similarities, one can easily conceive of mitochondria as being evolved from an ancient prokaryote, possessing all the attributes of an independent, probably aerobic organism. However, with the adaptation over a long period it became an essential and dependent symbiont, lost some of its identity to the cell, and conversely, the host cell lost some of its functions, deriving it now from the endosymbiont or mitochondrion. As a results, both became obligatory symbionts to each other.

This symbiont hypothesis for the origin of mitochondria and plastids has achieved wide popularity, but all biologists do not necessarily accept it. *Raff* and *Mahler* (1972) concluded that "while the symbiotic theory may be esthetically pleasing, it is not compelling." They presented a lot of evidences and proposed that mitochondrial arose by inward blebbing from plasma membrance, by the acquisition somehow of an outer membrane, and by the additional acquisition of a DNA genophore from the DNA of the protoeukaryote in which the evolution of mitochondrion occured. *Borst* (1972) proposed an *episome theory* and supposed that the DNA of mitochondrion left the 'nuclear' DNA by a sort of amplification to become mapped within a membrane containing the respiratory chain.

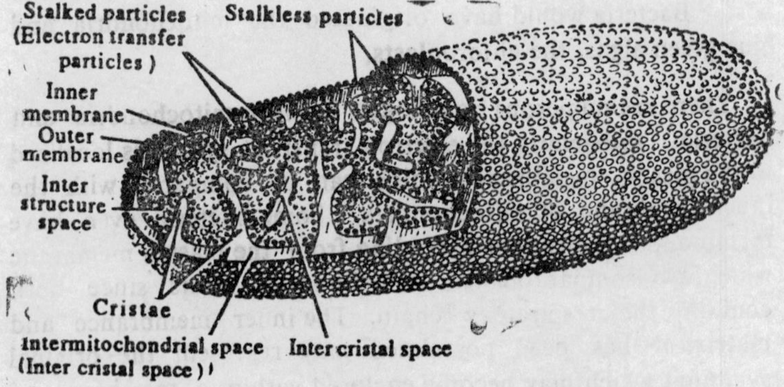

Fig. 4.5. Showing mitochondrial particles.

BIOCHEMISTRY OF MITOCHONDRIA

Lindberg and *Ernster* (1954) have given the data of chemical composition of mitochondria as follow : proteins 70 to 75%, Lipids 25-30%, and RNA 5% of the dry weight. But the recent biochemical analysis shows the following components :

(*i*) *Proteins*. The proteins are the main constituent which is insoluable in water. The outer limiting membrane of mitochondria contains less than 10 per cent of the total protein. There are about 14 different proteins having molecular weight from 12,000 to 22,000. The inner membrane contains about 60% protein having molecular weight varying from 10,000 to 90,000. The protein composition of mitochondrial membranes is not fully known.

(*ii*) *Lipid*. The lipid forms about 1/5th of the weight of the membranes. It is present almost entirely in the form of the molecules known as *phospholipid*. It has been reported by *Meluick* and *Packer* in 1971 that the outer membrane fraction has a 40% lipid contents as compared to 20% in the inner membrane.

(*iii*) *Enzymes*. About 70 enzymes and 12 co-enzymes have been recognized in the mitochondria. Enzymes lie in a non-aquous membranes region as solid arrays, with perhaps as many as 5000 to 20,000 such assemblages in a single liver or heart mitochondria.

Mitochondrial DNA. Recently the DNA is also reported from mitochondria. The mitochondrial DNA is double-stranded like the nuclear DNA. Each mitochondrion may contain one or more DNA molecules depending on its size. If the mitochondrion is larger then that may have more DNA molecules, having a circular shape.

Mitochondrial DNA differs from nuclear DNA in several respects. The GC content is higher in mitochondrial DNA, and consequently the buoyant density is also higher. Another difference is the higher denaturation temperature of mitochondrial DNA and facility with which it denatures. The amount of genetic information carried by mitochondrial DNA is not sufficient to provide specifications for all the proteins and enzymes present in this organoid. The most likely possibility is that mitochondrial DNA codes for some structural proteins.

Yeast mitochondria have been shown to contain DNA polymerase and more recently *Kalf* 1968 has succeeded in isolating the enzymes from rat liver mitochondria. Mitochodrial DNA polymerase appears to be involved in DNA replication rather than repair and possesses properties which are different from those of the nuclear enzymes. These include a differing requirement for metal ion. *Yeast* mitochondrial DNA polymerase appears to be smaller than its nuclear counter part, and is active at different stages of the cell cycle.

Mitochondrial DNA does not appear to have histones associated with it as nuclear DNA of higher organisms. In this respect mitochondrial DNA resembles with the bacterial DNA.

Mitochondrial RNA. South and *Mehlar* (1968) suggested that the amount of mt-RNA is about 10 to 20 times that of mt-DNA. All sorts of RNA have been identified in mitochondria. The present evidences point out conclusively that mitochondria contain complete set of t-RNA aminoacyl RNA synthetases, as well as ribosomal RNA. All these components differ from their respective counterparts in the groundplasm. The presence of m-RNA, transcribed from mitochondrial DNA is still uncertain. However, there are authorities who suggest its presence. These ribosomal RNAs' are coded for by mitochondrial DNA and are thus apparently synthesized within the mitochondria by a mitochondrial DNA-dependent RNA polymerase system.

Mitochondrial Ribosomes. Mitochondria appear to contain ribosomes which are smaller in diameter than cytoplasmic ribosomes. *Yeast* mitochondria contain RNA species of 23S and 16S which would correspond to a 70S ribosome of bacterial type rather than the 80S ribosome of the cytoplasm. Ribosome like particles with sedimentation values of 81S and 55S have also been reported, and the extent of degradation suffered by the particles during isolation is not yet clear. Polysome like aggregation of ribosomes have been observed in sections of yeast mitochondria by *Vignais, Huet* and *Andre* in 1969. High molecular weight and RNA species associated with mitochondria which differ in sedimentation value from cytoplasmic ribosomal RNA have been reported in *Yeast, Neurospora* and *He-La cells.* Mitochondrial ribosomes require a higher concentration of Mg^{++} ions to maintain their integrity than do cytoplasmic ribosomes.

Protein Synthesis. In general, mitochondria can code and synthesize protein, but the DNA present therein is not sufficient to code for all the proteins. It is suggested that mitochondria can synthesize the proteins of structural nature (cytochrome oxidase), but much of the proteins if not all, of the soluble proteins of the matrix as well as the proteins of the outer membrane and a number of proteins located in the cristae are under the control of nuclear DNA. Of the proteins coded by nuclear DNA, it is generally agreed that the m-RNA derived from nucleus are translated in the cytoplasm, and the resulting proteins are then transported into the mitochondria. How then protein enter the mitochondria ? Two methods have been proposed :

(1) The precursors enter the mitochondrion and inside are changed into the end products, thereby affecting a undirectional flow of material into the mitochondrion.
(2) There is the synthesis of lipoprotein vesicles which then merge and combine with the growing mitochondrion.

THE SWELLING PHENOMENA

It has been recognized for some time that the structure of mitochondria is altered when they are exposed to hypotonic solutions and the swelling so induced has been shown by Tedeschi and Harris, also, to be related to membrane permeability. There is an increase in total volume as well as an increase in the volume of intercristeal space. This swelling is related functionally to certain of the "opening phenomena" described. These phenomena involve the appearance of certain enzyme activities hidden in the intact mitochondrion. The primary enzyme function that appears coincident with swelling of beef heart mitochondria is the ability of the mitochondrion to oxidize external DPNH and citrate (2). De Duve and co-workers have described other "openings" in liver mitochondria. An opening effect of this type might best be related to a permeability change that permits the substrate to get to an active site. However, the alternative hypothesis that these enzyme functions appear because of an uncoiling or unfolding of the surface lipoprotein structure has not been eliminated. Pauly and Packer have examined the electrical capacity (by measuring the dielectric constant of mitochondrial suspensions) of mitochondrial membranes and find that it does not change over a fourfold range of mitochondrial size induced by osmotic swelling. Conductivity studies indicate that swollen mitochondria tend to equilibrate with the external medium, whereas shrunken mitochondria tend to maintain a constant ionic composition regardless of changes in external ionic strength. This indicates a relatively large, osmotically active space in the swollen condition, but a predominant binding of charge by macromolecules in the shrunken condition. These authors suggest constant replacement of the inner membrane by material from the matrix during swelling.

More extensive "opening phenomena" are encountered as mitochondria are treated with detergents. The classic opening effect of deoxycholate on the cytochrome-*o* reductase of the electron transport system from beef heart mitochondria is an example of an effect that is accompanied by extraction of material from the mitochondria (lipoprotein in this case) and can best be interpreted in terms of exposure of an active site, something like scraping butter off toast.

With our meager knowledge of lipoprotein structure and the effects of external materials on that structure, all opening effects certainly cannot be assumed to be related to permeability changes in a bag holding a soup of enzymes. Watson and Siekevitz, in their electron microscope study of progressive degradation of mitochondria under the influence of detergents, indicate that inner contents are lost after extensive expansion of the membrane. It has been our experience that there is a successive extraction of enzyme units with an increasing concentration of detergent.

SWELLING AND CONTRACTION ASSOCIATED WITH PHOSPHORYLATION

Oxidative phosphorylation has been regarded until recently as a rather labile process closely allied to mitochondrial "integrity." Harmon and Fiegelson were able to show that loss of integrity paralleled swelling of mitochondria. Subsequently it was found that ATP was lost during swelling and the addition of ATP and magnesium ions would reverse degradative swelling of mitochondria. Since these initial studies, extensive work in a number of laboratories has introduced numerous swelling agents into the field and in some instances the results seem to imply that several mechanisms for swelling exist. A detailed discussion of all of these effects and the conditions required to induce and inhibit swelling cannot be attempted here. It will be more useful for the purposes of the review to discuss certain recent papers that in themselves present a fairly good outline of the present status of the problem.

Lehninger has outlined the various factors that induce swelling in rate liver mitochondria. These include thyroxine, phosphate, arsenate, calcium ion, exposure to hypotonic media, incubation at 37°C. (spontaneous swelling), *p*-chloromercuribenzoate, phlorizin, *U* factor [which apparently is an unsaturated fatty acid released by lipase action from mitochondria since oleate shows similar effects], glutathione, carbon tetrachloride, digitonin, and zinc ion. He has been able to show that the swelling induced by most of these agents can be reversed by ATP, magnesium ion, serum albumin, and ethylenediaminetetraacetic acid in various combinations. Only swelling induced by glutathione, carbon tetrachloride, and digitonin could not be reversed by some combination of the above reagents. It is proposed that the reversal of swelling by ATP, together with the additional factors mentioned, involves a single fundamental mechanism.

Hunter *et al.* have examined the swelling produced by an oxidizable substrate and oxygen, especially with reference to the agents that will inhibit this swelling. Their results all point to the general conclusion that an oxidation-reduction process at one or more points in the electron transport chain is associated with the swelling process whether or not phosphate is present. These authors, as well as Lehninger, point out that swelling induced by glutathione, cysteine, and ascorbate is manifestly different than that associated with other substrates, although oxygen is necessary in all instances. Their discussion of metal chelation and the oxidation of sulfhydryl groups in the mitochondrial membrane seems highly pertinent to possible mechanisms for the swelling reaction. Some reports, however, indicate that reducing agents prevent swelling.

Chappell and Greville showed that swelling induced by substrate oxidation in the presence of phosphate does not directly involve coupled phosphorylation. This conclusion is primarily based on the relative inhibitory effects of oligomycin, amytal, and antimycin-*A* on swelling induced by electron transport from substrates to ferricyanide or to oxygen.

Packer studied the swelling induced by substrate oxidation in the presence of phosphate and showed that ADP causes shrinkage of the swollen mitochondria, and that the shrinkage is followed by expansion as the ADP is converted to ATP. Lehninger reported ADP effective in reversing only the swelling induced by hypotonic media, or by *U* factor or oleate and suggested that these conditions are auspicious for adenylate kinase activity to produce ATP. Packer's studies develop a further point of interest in that reduction of cytochrome-*b* is shown to precede swelling, whereas oxidation accompanies shrinkage. Overall electron transport increases when ADP is added as shrinkage begins. Thus the primary act leading to swelling may involve a reduced rather than an oxidized state of the key compound. More recently, Packer and Tappel have extended to mitochondrial fragments the swelling studies associated with phosphorylation. Lehninger and Gotterer have reported a protein fraction required for mitochondrial contraction.

These extensive studies in mitochondrial swelling and contraction have been concentrated on animal tissue. Studies with plant mitochondria would provide useful comparative data and could help to emphasize aspects of general significance. It would also be intriguing to know if plants possess a contractile protein in the mitochondrial membrane or if permeability changes alone account for swelling phenomena.

Knowledge of swelling changes in plant mitochondria might also apply to physiological studies. The type of external conditions that effect ageing changes in potato slices as reported by Hackett *et al.* are similar to the conditions that affect mitochondrial swelling. Thus they report that the ageing changes are inhibited by cyanide, dinitrophenol, and anaerobiosis. Mitochondria from the aged tissue show remarkable increases in the activity of certain enzymes that are suggestive of permeability changes. Key *et al.* show swelling of mitochondria and increased enzyme activity after 2,4-dichlorophenoxyacetic acid treatment of soybean seedlings.

ATP ase activity is also often associated with swelling of mitochondria. Maley and Johnson however, have shown that agents that cause swelling do not always increase ATPase activity. There remains a group of compounds, however, that cause both swelling and an increase in ATPase activity. The possibility that the effects of these compounds relate to a site common to both processes cannot be ignored. Calcium, thyroxin, fatty acids, and cysteine seem to be the compounds of primary interest in this regard. Since DNP stimulates an ATPase reaction but inhibits swelling it cannot fit into the above group. This is consistent with the indication by Lipsett and Corwin that the effects are independent. Forti has shown that calcium ion stimulates ATPase activity as well as swelling in pea mitochondria.

The relation of ATPase and phosphate-ATP exchange reactions in mitochondria to the process of oxidative phosphorylation has been reviewed and further discussion will be left to the review of phosphorylation reactions.

BINDING OF PYRIDINE NUCLEOTIDES

Intact mitochondria of animal tissue contain considerable amounts of bound pyridine nucleotide. The bound pyridine nucleotide can be released from or destroyed in mitochondria by exposure to aerobic conditions or to phosphate buffer.

A supply of ATP or an anaerobic condition is necessary to maintain the nucleotide in a bound form. Also, the presence of ethylenediaminetetraacetic acid in the incubation medium prevents loss of bound phosphopyridine nucleotide.

In a study of conditions necessary for rebinding DPN, Hunter *et al.* find that the oxidized form rebinds in the mitochondrial structure better than does DPNH, but it is not clear whether the same binding sites are involved as in the original attachment of DPNH. They point out that a more precise localization of binding sites will be necessary to understand the mechanism of the process.

FUNCTION

Citric Acid Cycle Oxidations

Evidence for operation of the tricarboxylic acid cycle in plant mitochondria has been reviewed several times. This evidence continues to be extended to tissues in which difficulties had been encountered and to new material. Results so far are strongly in favor of the usual tricarboxylic acid oxidation pattern in plant mitochondria.

Oxidative phosphorylation steadily improves in plant preparations until now in many cases it has reached the levels found in animal systems that closely approximate theoretical values. In most cases the system acts like the animal system, but extensive study of oxidative phosphorylation in plant particles (as conducted with liver and heart mitochondria) has not been done. In some instances the inhibitors of the system encountered in animal studies are not effective.

Inhibitory effects of CO_2 on castor bean mitochondria have been reported. Hackett *et al.* studied the comparative effects of inhibitors on phosphorylation and electron transport and have shown that cyanide uncouples phosphorylation in potato mitochondria.

Low sensitivity to DNP uncoupling has been reported and needs further examination. Usually DNP has acted as an efficient uncoupler with plant mitochondria and Gaur and Beevers present evidence that many substituted phenols function as uncouplers in carrots. It should also be of interest to know if the animal hormone thyroxin uncouples phosphorylation in all plant mitochondria as it does in yeast.

The action of ribonuclease in uncoupling phosphorylation in both plant and animal mitochondria raises a question of polynucleotide function in phosphorylation, as suggested by Pinchot or of a possible uncoupling action of the products after digestion.

ELECTRON TRANSPORT

The electron transport system of plant mitochondria appears to be similar to that of various animal systems. Spectrophotometric evidence indicates that flavoproteins and cytochromes of the *b*, *c*, and *a* type participate in the oxidation of DPNH and succinate. In most cases the system is susceptible to the same inhibitors as animal systems. Antimycin and heptyl quinoline N-oxide, azide, cyanide, and carbon monoxide usually cause inhibition at expected levels. Certain plant systems, however possess a cyanide-insensitive respiration that is indicative of other than the usual cytochrome-a-a_3 terminal oxidase system. Perhaps the best interpretation so far advanced to explain this situation is based on the existence of a special auto-oxidizable *b* cytochrome that can link the non-phosphorylating electron transport from substrate to oxygen. This may not be the final answer, however, since Chance and Hackett presented evidence that no unusual pathway branches off in skunk cabbage. It is this author's conviction that the components of plant mitochondria are not sufficiently inventoried to preclude surprises in the nature of unknown electron carrier molecules. Even in animal mitochondria the studies of Thorn and Jackson indicate separate unknown sites for the action of antimycin-*a* and BAL (2,3-dimercaptopropanol).

Much remains to be accomplished in isolating the members of the electron transport chain from plant mitochondria. Although there is general agreement that plant systems possess cytochromes of the *a*, *b*, and *c* series, it is not clear how closely related they are to the corresponding cytochromes of mammalian tissues.

Lundegardh presented spectra from wheat roots indicating, by different reduction rates for cytochromes with absorption bands at 550, 553, and 563, the presence of cytochromes c, c_1, and b respectively.

On the other hand, Chance and Hackett in a study of *Symplocarpus* mitochondria, found a cytochrome-*c* band at 551mμ, a *b* cytochrome with an α band at 558mμ. After dithionite treatment, they find indications of an α band at 567mμ that cannot be equated with any known cytochrome-*b*.

Bendall and Hill indicated in their study of *Arum* that absorption bands of cytochrome-b_7 (557.5) and b (563) were present in that tissue. Evidence for *c*, by low temperature treatment in *Arum* mitochondria, has been reported. Sisler and Evans reported cytochromes of the *b* and *c* type in tobacco root preparations and Baker and Lieberman found evidence for four cytochromes in sweet potato mitochondria that would seem very close in absorption maxima to mammalian cytochromes except for a slight shift of the cytochrome-*c* and the 558 band, indicative of the presence of b_7 rather than c_1. In potato mitochondria Hackett *et al.* found difference spectra similar to those mentioned above with a b_7 type band rather than c_1.

Of course the unusual respiratory characteristics of *Symplocarpus* and *Arum* spadices are appropriate to unusual cytochromes especially if the presence of b_7 type compounds can explain cyanide-insensitive respiration.

In comparing mitochondria from fresh cut potato slices with those from 24-hour-old slices, Hackett *et al.* indicated that development of cyaniderestistant respiration in this tissue during ageing depends on a metabolically induced modification of the respiratory chain. They show development of cyanide insensitive succinoxidase in mitochondria during ageing, as well as a large increase in cytochromes-c and -b_7 relative to the cytochromes-a and -a_3.

A system analogous to the cyanide-insensitive respiration in plants is found in the flavin-linked oxidative enzymes of *Lactobacillus* which have been studied by Strittmatter. A new terminal oxidase, cytochrome-o, has been reported by Castor and Chance in certain bacteria.

Table 4.1. Extraction of Cytochromes.

Cytochrome	*Source*	*Reagents Used**
c	pig heart	acetone and salt
c_1	beef heart	detergents
c_4	*Azotobacter*	butanol
c_5	*Azotobacter*	butanol
b	beef heart	detergents
b	beef heart	detergents
b_2	yeast	autolysis
b_5	rabbit liver	lipase digestion
b_3	leaves	soluble
a	beef heart	detergents and thioglycollate
f	parsley	alkaline ethanol

*Reagents used to extract cytochromes in procedures that lead to a highly purified enzyme. Other methods used for isolation of cytochrome-c in plants have been reviewed by Hill and Hartree.

It would seem that the apparent lack of the usual cytochromes in certain plant tissues would call for a further development of this subject. If new cytochrome combinations exist in plant tissues, careful and precise spectrophotometric examination of new plant tissues, as well as isolation of the cytochromes, will be necessary to establish their identity. The isolation of various cytochromes from animal tissue and microbes (cf. Table 4.1) indicates that the isolation of corresponding compounds from plant tissue can be accomplished, especially if the operation is done on a sufficiently large scale.

There seems to be general agreement that the cytochrome a-a_3 group with bands at 445 and 603 mμ exists in plants although, as Chance and Hackett point out, cytochrome oxidase activity seems out of proportion to the amount of cytochrome-a_3 present.

In studies of cytochrome oxidase in plants it would be worthwhile to consider the evidence that copper is associated with the purified oxidase in animal mitochondria it has been shown by Wainio *et al.* to be present in proportion to the oxidase activity during purification. Evidence for oxidation-reduction of the copper has been presented. It may well be that the copper unit in the oxidase system is present in plants in larger amounts, relative to cytochrome-a or -a_3, than in animal mitochondria. In the isolation attempts so far reported it looks as if the cytochromes-a and -a_3 in plants [perhaps better called oxidase unit until a_3 is shown as a discrete compound cf. Ambe and Venkataraman are not easily extracted at least by the procedures that are so successful with animal mitochondria.

Giuditta and Singer have studied the relative efficiency of various electron acceptors with animal preparations containing succinic dehydrogenase activity and find considerable variation in the rate of reduction with different acceptors. Phenazine methosulfate was the only one to react at a rate comparable with that of the overall oxidase system. These results are pertinent in any attempt to assess overall *in vivo* dehydrogenase activity by use of *in vitro* acceptor systems.

Ziegler, Green and Doeg have described a lipoflavoprotein diaphorase from heart mitochondria that reacts more efficiently with ferricyanide than other diaphorases or cytochrome-c reductase. There is also an accumulation of evidence indicating that certain diaphorase activity may more accurately represent lipoyldehydrogenase. The evidence that fatty acyldehydrogenases can be released from heart mitochondria by sonic treatment would indicate that the electron transferring flavoprotein and its accompanying diaphorase may also be expected in mitochondria. Thus activity designated as diaphorase activity may represent the presence of at least three different enzymes in mitochondria (Fig. 4.6). In addition, Chance and Hackett report an unusual amytal-insensitive system.

Cooperstein has described direct reduction of cytochrome oxidase and cytochrome-*b* by a DPNH—cytochrome-*c*— reductase preparation. His preparation of DPNH—cytochrome-*c* —reductase also contains the mysterious 408 compound (oxidized max. 408 mμ, reduced 430 mμ) which has turned up in many studies of DPNH—cytochrome-*c*—reductase systems (reported as cytochrome-*c*) but has usually been dismissed as a degradation product. In his studies Cooperstein also does not take into account the evidence that a lipid-cytochrome-*c* complex is the active form of cytochrome-*c*.

Sisler and Evans have reported the purification of a diaphorase from the soluble fraction of tobacco root homogenates

Succinate – – – → FAD_{III}
cyt-*b*
α-Ketoglutarate→lipoate→FAD_I– – – →DPN→FAD_{II}→CoQ cyt-*c*→cyt-*a*→Cu→O_2
Pyruvate
cyt-c_1
Other pyridinoprotein-linked enzymes
TPNH ?
Fatty acyl CoA derivatives– – – – – – – – →FAD_{IV}– – –FAD_V
DPNH TPNH

Fig. 4.6. The flavoprotein situation in mammalian mitochondria, showing the pathways of probable electron or hydrogen flow from substrates to oxygen. This is a summation of pathways reported from diverse mitochondria (see text). The following symbols are used : FAD_I (diaphorase, also lipoyldehydrogenase) ; FAD_{II} (DPNH-cytochrome-*c* reductase) ; FAD_{III} (succinic dehydrogenase) ; FAD_{IV} (acyldehydrogenases) ; FAD_V (electron transferring flavoprotein) ; FAD (flavoprotein) ; CoQ (coenzyme Q) ; (cytochrome with individual cytochromes designated by letter) ; Cu (copper) ; and O_2 (oxygen). Other flavoproteins such as α-glycerophosphate dehydrogenase have been omitted.

that is equally active with either DPNH or TPNH as the electron donor. In this respect it resembles the electron transport flavoprotein diaphorase from pig liver mitochondria. It is noteworthy that they find other diaphorase activity remains in the insoluble (mitochondrial ?) fraction, but they do not indicate whether the particle-bound diaphorase is active with both nucleotides.

LIPID CYTOCHROME-*c* AND THE INTEGRITY OF ELECTRON TRANSPORT SYSTEMS

When the electron transport capacity for the oxidation of succinate or DPNH is examined in mitochondria and mitochondrial fragments from various sources, it is immediately apparent that some preparations require addition of cytochrome-*c* for maximum activity whereas other preparations show no cytochrome-*c* requirement. The electron transport particle of beef heart mitochondria and that from *Azotobacter vinelandii* contain tightly bound cytochrome-*c*, whereas in other systems such as liver cauliflower, tobacco, etc., cytochrome-*c* is easily removed. Evidence is accumulating that the active form of cytochrome-*c* is a lipoprotein, both from the studies on cytochrome-*c* extraction from electron transport particles and from the requirement for phospholipid supplement to activate purified cytochrome-*a* or cytochrome-*c* oxidase preparations. The effects of lecithinases on liver, kidney, and brain mitochondria are consistent with this interpretation and studies on insect mitochondria by Wojtczak and Wojtczak, as well as corn mitochondria by Kahn and Hanson, indicate that endogenous lecithinases are present in mitochondria.

In some instances, therefore, the binding of proteins in mitochondria can be accomplished by formation of proteolipid or lipoprotein structures. The release of materials such as cytochrome-*c* or ATPase from mitochondria would then depend on the action of endogenous or exogenous phospholipase or on the presence of activators (*e.g.*, calcium) for the hydrolytic enzymes, Lehninger and Remmert showed enzymic release of an uncoupling factor that is most likely a fatty acid.

FUNCTIONAL LIPID MOLECULES

Coenzyme *Q* is apparently present in plant mitochondria and may be presumed to function in electron transport as shown for animal mitochondria. However, evidence for this function is meager so far.

Solvent extraction procedures have been immensely useful in studying coenzyme *Q* function. The same type of process was successfully used by Lynch and French to extract lipids from chloroplasts and show restoration of activities when the lipid was re-added. A specific requirement for plastoquinone has been demonstrated by Bishop in the chloroplast system. Tocopherol and othes lipid materials have also been functionally implicated in rat liver mitochondria by Nason and coworkers. Certain of the restorative effects of tocopherol and other lipids are undoubtedly related to a release of solvent inhibition in one way or another, as shown by Pollard and Bieri. Others have also shown reversal of solvent inhibition by added lipids. On the other hand, Corwin and Schwarz in a study of respiratory decline in vitamin E-deficient rat liver found no effect of vitamin E deficiency on DPNH—cytochrome-*c*—reductase, but they did find a decline of succinoxidase activity when DPN was present and attributed this to an accumulation of oxalacetate. An alternative interpretation of vitamin E function in the respiratory systems involves its antioxidant function. The fact remains that tocopherol is present in animal mitochondria although the chance for a direct role in electron transport is lessened, as Deul *et al.* point out by stating that there is no evidence for oxidation-reduction change. This fact does not exclude tocopherol function in phosphorylation. It might be well to ask if the tocopherol in plants is localized in mitochondria.

Not all lipids that restore activity in solvent-extracted mitochondria act nonspecifically to reverse solvent inhibition. Ambe and Crane have shown that the restorative effects of coenzyme *Q* for succinic—cytochrome-*c*—reductase activity are specific to the quinone moiety. Certain other lipid factors have also been found to be rather specific in their function, and induction of a requirement correlates with an appearance in the extracting solvent. It would seem that destruction of a lipoprotein unit within the mitochondrial membrane can lead to successive extractions of lipid materials originally responsible (by concerted action) for the function of the unit.

FACTORS IN PHOSPHORYLATION

The fragmentation approach to the mechanism of oxidative phosphorylation has not produced dramatic results. It has been possible by certain treatments to obtain sub-mitochondrial fragments that will carry out oxidative phosphorylation. Lehninger *et al.* have used digitonin treatment of rat liver mitochondria for this purpose while Green and co-workers have used high phosphate concentration or sonic treatment to disrupt mitochondria and yet retain phosphorylative capacity.

Certain factors have also been extracted from mitochondria that are required to recouple phosphorylation. Linnane and Pullman *et al.* have both obtained a protein fraction that is necessary to couple phosphorylation to electron transport. The identity of the coupling factor remains unknown. McMurray and Lardy have shown a requirement for coenzyme-*A* for phosphorylation in liver mitochondria. On the other side of the picture, Remmert and Lehninger have reported a "loosening" factor that releases respiration from dependence on ADP in otherwise tightly coupled mitochondria without acting as an uncoupling agent when ADP is present.

Greater success with fragmentation studies is seen in the demonstration by Brodie and co-workers that vitamin K_1 is required for oxidative phosphorylation in a bacterial system. Light-inactivation of oxidative phosphorylation in animal mitochondria also has been reversed by addition of vitamin K_1, but there is no evidence as yet that vitamin K_1 is destroyed in mitochondria by the treatment or that vitamin K_1 is even present in mitochondria. Light-inactivation of adenosine triphosphatase activity of mitochondria has also been reported by Low *et al.* and Beyer. The effect of light in this instance has been attributed to destruction of flavin. It would seem that the time has come to find out exactly which compounds in mitochondria are destroyed by light and which are involved in each of the inhibition effects.

OTHER ENZYMES OF MITOCHONDRIA

A large group of enzymes have been reported in mitochondria, especially from liver (1) : the fatty acid oxidizing system, choline oxidase, α-glycerophosphate dehydrogenase, and the enzymes involved in lecithin synthesis and special protein synthesis. Previous reviews have discussed animal (1) and plant mitochondria with respect to various activities. Further systematic study of the enzymes of plant mitochondria continues. Pierpoint has continued his studies of the enzymes of tobacco leaf mitochondria. Cheniae and Evans have also found a nitrate reductase associated with mitochondrial material from soybean nodules. Bone has described ornithine and citmulline synthesizing systems in mung bean seedling mitochondria as well as glutamic dehydrogenase. Spencer was able to show ethylene production in mitochondria from ripening tomatoes. The activity reported in the mitochondria was no greater than that in intact fruit so evidence of localization of the ethylene producing system in mitochondria is not conclusive. Some of these functions are obviously important in the oxidation of respiratory substrates whereas others may be involved in the synthesis and breakdown of mitochondrial structure as shown by incorporation of choline into mitochondrial phospholipids.

COMPOSITION OF MITOCHONDRIA

Our understanding of mitochondrial functions increases as we learn more about the molecular composition of mitochondria. The overall composition indicates a basic lipoprotein unit with most mitochondria consisting of 30 to 40 per cent lipid and 60 to 70 per cent protein. Trace elements and vitamins are also present [primarily for their catalyic potential] but perhaps, as in the case of nonheme iron, they are structurally important also. Composition of mitochondrial lipids needs more study. At least 70 to 80 per cent of the lipid fraction is phospholipid. These phospholipids are rather unique in their high level of unsaturated fatty acids (especially polyethenoid) and their large proportion of plasmalogen. The bulk of the remaining lipids in animal mitochondria are neutral fats and cholesterol. No full recovery of mitochondrial lipid has been achieved yet, so an uncertainty remains about possible unknown material. It is also intriguing to consider whether certain plant sterols are specifically associated with mitochondria.

Analysis of lipid composition in fragments of animal mitochondria has shown that lipids are not qualitatively nor quantitatively the same in all parts of the mitochondrion. This uneven distribution has led to the proposal that mitochondrial lipids are contained in discrete packets or lipoprotein units in the mitochondria and that each lipoprotein unit is associated with a certain catalytic function. Further evidence for discrete functional lipoprotein units comes from detergent-induced fragmentation of beef heart mitochondria and the isolation of lipoproteins or lipid packets. Cytochromes-*c* and -c_1, diaphorase, and coenzyme *Q* have been isolated in functional lipoprotein or lipid units from heart mitochondria. It can be assumed that other members of the electron transport chain will be associated with other lipoproteins or phospholipid packets.

Certain other lipid materials are present in smaller amounts appropriate to a catalytic function. These include coenzyme *Q* tocopherol, and possibly carotenoids. Some reports state that vitamin K is a mitochondrial material whereas others do not indicate significant amounts. In both plant and animal mitochondria other materials exist that show absorption spectra not relatable to known mitochondrial constituents. Since any compound with absorption in the ultraviolet or visible region must be considered of potential interest for the electron transport function, it is obvious that these unknown materials represent a large gap in our understanding of mitochondria.

Nucleic acid has also been reported in mitochondria although the amount is variable (Table 4.2). The presence of nucleic acid is consistent with the demonstrated synthesis of proteins in mitochondria. It would perhaps be easier to conceive of a synthesis of mitochondrial proteins in an ordered sequence *in situ* to account for the consistent and ordered structure of mitochondria rather than a localization of proteins, after synthesis in the cytoplasm, onto lipid surfaces. Interpretation of nucleic acid function in this way would indicate a higher proportionate nucleic acid content in young mitochondria than in old. Indication of a role for nucleic acid in mitochondrial function has also been presented by Pinchot's studies of phosphorylation in bacteria and Hansen's studies on the effects of ribonuclease.

Analysis for discrete and unique molecules may be extremely useful in studies on the distribution of mitochondria or in the

Table 4.2. Ribonucleic Acid Content of Plant Mitochondria.

Source	*RNA Content* ($\mu gP/mg$, *protein N*)
Pea seedlings	29
Wheat roots	9
Silver beet	8
Castor bean endosperm	26
Wheat roots	25*
Peanut cotyledons	[(0.164 (E_{260})/mg. N)]
Corn scutellum	19 μg RNA/mg. dry wt.
Rat liver	17 μg/mg protein

*Estimated from dry weight value.

localization of enzyme functions. More of this is needed in correlation with enzyme studies because of the problems of inhibition and activation in enzyme assays. In other words, mitochondria may not always oxidize succinate especially if an inhibitor like malonate or catechol is around, but we hope they will always contain coenzyme *Q*. Likewise the distribution of copper in the cell should be consistent with assays for ascorbic acid oxidase, phenol oxidase, and cytochrome oxidase activities. If not, certain facets of the enzymatic activites in a given tissue remain to be explored, or other copper-containing proteins must be considered.

PURIFICATION

A technical problem that deserves much further study is the improvement of methods for purifying plant mitochondria. Plant cells seem to be blessed with a multitude of particulate or lipoprotein structures primarily defined by cytological studies. The diverse particles are also obvious in electron micrographs of plant mitochondrial preparations. Some of the problems encountered in the determination of the cytochromes of plant mitochondria may derive from the presence of other particulates. Lysosomes, described as a possible part of the electron-dense granule fraction from liver homogenates by Novikoff, Beaufay and deDuve, have been separated from liver as well as carotenoid granulcs from beef heart. The presence of both plastoquinone and coenzyme *Q* in a mitochondrial type of preparation from cauliflower buds points to a mixture of mitochondria and proplastids in the preparation. Precise determination of the chemistry and enzymological properties of plant mitochondria will be greatly advanced by better methods of separation and purification.

The methods of electrophoresis and gradient centrifugation applied by James and Das to leaf mitochondria should be useful in other material. New techniques such as special filtration systems and the use of non-aqueous media will provide approaches to the purification problem. Even techniques that do not yield physiologically intact mitochondria may be useful if they provide better definition in the essay or isolation of mitochondrial components.

Precise sedimentation studies as carried out with liver particles by Beaufay *et al.* should give better definition of enzymatic functions of the particles in plant cells. Forti *et al.* have reported separation of particles in pea stems by differential centrifugation.

5

Plastids

The plastids are characteristic large cytoplasmic bodies of the cell intimately related to the metabolic processes and which impart colour to the tissues. They are found mostly in plants with possible exception of the very lowest group of plants. However, plastids are also found in few animals, especially the protozoan flagellates. They are not found in bacteria, certain algae, myxomycetes and fungi. On the basis of presence of the plastids, the living world has been distinguished into plant and animal kingdoms. Old literature indicates that *Pringshetn* first investigated the plastids. The term plastid was used in 1885 by *A.F.W. Schimper* to include the cell organelles which are primarily involved in the formation and storage of soluble and insoluble carbohydrates. *A. Meyer* in 1883, *F. Schmitz* in 1884 and *A.F.W. Schimper* in 1885 made detailed cytological studies of these cell organelles and showed that green plastids known as chloroplasts could not be formed from cytoplasm, but only from existing chloroplasts. The plastids vary in size, number, shape and chemical organization with the particular tissue and organism concerned. In contrast to the smaller cytoplasmic granules such as the mitochondria, important physical and chemical differences in the plastids can be easily recognized, such as changes in size or shape, the presence or absence of the chloroplastid, carotenoids and starch. Mature plastids can be shown to arise from more or less undifferentiated plastid primordia which occur in the meristematic cells, or those of the developing embryo.

The plastids are self-duplicating bodies and their life history, especially of the chloroplast, may be divided into three phases. In the first phase, self duplication and enzymatic activities with some concern to RNA takes place. The second phase is mainly concerned with differentiation to form the enzymes required for photosynthesis. In the third phase the activities of the enzymes that predominates are those concerned with photosynthesis.

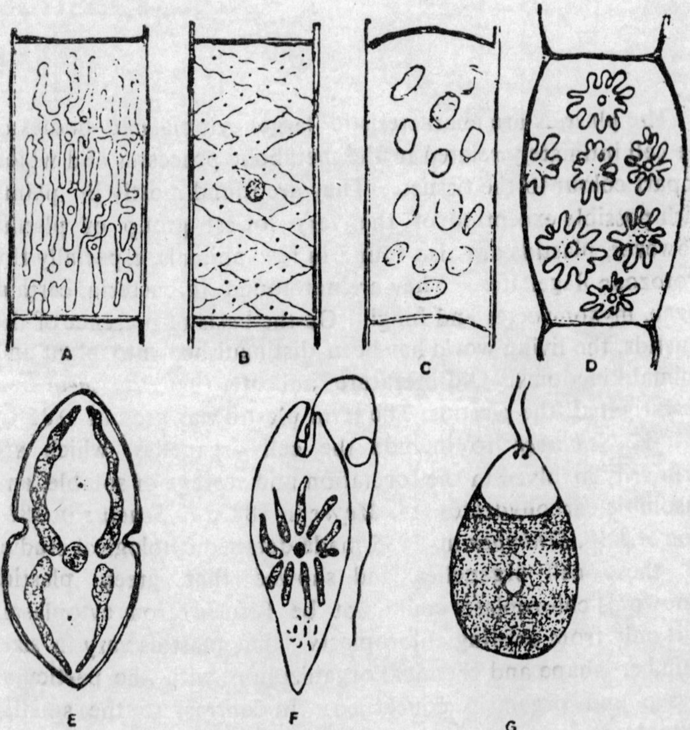

Fig 5.1. Various types of chloroplasts. A-Oedogonium, B-Spirogyra, C-Phyloiella, D-Rhodochlorton, E-Euastrun, F-Euglena, G-Chlamydomons.

VARIOUS TYPES OF PLASTIDS

Plastids may be divided into two general groups ; the *chromoplasts* (coloured plastids) ; and the *leucoplasts* (colourless plastids). The chromoplasts may be further divided into two subgroups ; those that are photosynthetically active and those that are not.

(A) Photosynthetically Active Chromoplasts

The photosynthetically active chromoplasts are chloroplasts, phacoplasts, rhodoplasts, blue-green chromoplasts and chromatophores of photosynthetic bacteria.

(1) *Chloroplasts.* These are green plastids of green algae and higher plants and contain chlorophyll A and B in the ratio of 2.3 : 1. These have different appearance and varying sizes.

(2) *Pheoplasts* (*Brown or yellow plastids.*) This type of plastid is commonly present in the brown algae, diatoms and dinoflagellates. In these pheoplasts, brown carotenoids mask the chlorophyll.

As reported by *Strain* (1951) the yellow or brown colour of the pheoplasts appears to result from the physical state or condition of the pigment from their geometrical arrangement in the plastids. Brown algae contain fucoxanthin which can absorb light and transfer energy to chlorophyll.

(3) *Rhodoplasts.* Various members of Rhodophyceae contain red coloured plastids. The red colour is due to a pigment phycobilin and protein phycoerythrin. The colour of plastids may be the result of light absorption. There may be phycocyanin too in traces. *Haxo, Blink* and *Young* (1952) found that light is absorbed by phycobilin and is transmitted to chlorophyll *a*.

(4) *Blue-green chromatophore.* Two accessory protein phycocyanin and phycoerythrin impart blue-green colour or algae. These pigments and chlorophyll *a* alongwith carotenoid form lamellar structure, the *chromatophore.*

Wilson and *Morrison* (1967) mention that all the organisms capable of performing photosynthesis do not necessarily contain plastid structures ; for example, blue-green algae and photosynthetic bacteria. In blue-greens, the light-sensitive pigments are chlorophyll and phycobilin pigments. These pigments are associated with a system of membranes and lamellae in the cytoplasm of the cell. In photosynthetic bacteria also wide diversity of structures are associated with the photosynthesis. These structures are as follows :

(1) Lamellae in some purple bacteria ; for example, *Rhodomicrobium*, (2) Vesicular membrane-bound structures, known as chromatophores found in non-sulphuric and purple sulphur bacteria ; for example, *Rhodospirillium*, *Chromatium*, and (3) Ribosome like granules found in green sulphur bacterium, *Chlorobiumthio sultatophilum*. All these photosynthetic bacteria possess a pigment system consisting of a specialized form of chlorophyll, known as bacteriochlorophyll and various carotenoid pigments.

(B) Chromoplasts without photosynthetic activity :

These types of chromoplasts contain carotenoids only. They are red to yellow in colour and are much more variable in shape. They impart specific colour to various plant tissues, fruits and flowers. The new colour that arise often due to the formation of special carotenoids, for example, the *leucopene* of tomato fruits, *prolycopene* of berries, and *capsanthin* of red pepper. Xanthophyll is also present in adequate proportions. Coloured carotenoids are also important constituents of the coloured parts of plant. These may be found in fungi and bacteria. Acidic carotenoids of fungi are similar to astaxanthin. *Haxo* (1949) isolated a mixture of carotenoids from the *Neurospora crassa* which includes lycopene, carotene, and spiriloxanthines as the main components. The growth of carotenoid becomes double in presence of light.

These chromoplasts are supposed to develop from the leucoplasts that contain starch. As the carotene increase in concentration the starch disappears. The older chromoplasts appear thin plates as needles with typical crystalline appearance. They are direfringen and dichroic.

Leucoplasts. The noncoloured plastids are called leucoplasts. They do not contain pigments and the membranous structures are absent. They are found mainly in the embryonic and sexual cells. They are also found in meristematic cells and in those regions of the plants which do not receive light. They serve mainly for the storage of starch, oils and proteins. Plastids located in cotyledons and the primordium of stem are first colourless but eventually become filled with pigments and become coloured plastids. Leucoplasts are further categorized into following three types :

(1) *Amyloplasts.* These are mature plastids with predominant starch contents and generally found in storage tubers, cotyldons and endosprem. Starch is characteristic of the coloured and the colourless plastids of green algae and the higher plants. The deposition of other carbohydrates may not necessarily be with the plastids. For example, in red algae starch is formed at the surface of the plastids. In *Euglena*, paramylum grains are considered to be produced outside the plastids. The starch grains may be of two types : assimilative starch and reserve starch. The former is characteristic of actively photosynthesizing chloroplasts and the latter is most familiar in storage organs. The starch grains are made of a series of concentric layers successively deposited about a centre, the *hilum*. As starch continues to be deposited, the walls of the amyloplast become greatly distended. *Frey Wyssling* (1953) observed that the formation of a new concentric layer begins as a dense region of high refractive index, then it becomes a dense layer due to more hydration. Chemically starch contains two components : mainly α-amylose (which stain red with iodine) and minorly β-amylose (only 20-30 per cent forming helical chain structure and stains blue with iodine).

(2) *Elaioplasts.* The plastids which stone oils are known as elaioplasts. In most monocotyledons, the chloroplasts after losing their chlophyll start storing oil in epidermal cells (exp. Orchidaceae and Liliaceae). The disorganised plastids fuse to form oil droplets. In brown algae, the reserve storage is oil and this oil appears to be formed in the pheoplasts.

(3) *Aleuroneplasts.* These are the plastids that contain much protein and often termed as the *proteinoplasts.* Protein crystals

and granules are produced in the seeds of numerous plants especially in those seeds that also form large amount of oil, *e.g.* the Brazi Neof and seeds of *Ricinus*. *Mottier* (1921) referred that protein and oil formation involves the activity of permanent plastid primordia. Protein crystals may develop independently in the cytoplasm without any aid of the aleuroneplasts, as for example, in peripheral layers of potato the cells are poor in starch. Protein crystals in epidermal cells may also represent virus material or may be the result of virus action.

Amyloplasts are the most important of all the different types of leucoplasts. The amyloplasts resemble proplastids and differ from them only in size. They contain fewer lamellae, but in presence of light they can build up a thylacoid structure found in chloroplasts. However, chloroplasts can also synthesize starch and store it in stromal region. The starch in chloroplasts disappears quickly and is, therefore, known as transitory starch. The starch present in leucoplasts, on the other hand, can be stored for longer periods and is, therefore, known as reserve starch.

The starch development begins with the formation of a particle in the stroma. This particle has several tubuli and thylcoids and is surrounded by additional rings of starch. It grows till the amyloplasts is filled with starch. As a result, the tubuli are pressed against the wall and ulimately starch granule exceeds the size of amyloplast. The wall of the plastid then ruptures and finally withers away so that only the starch granules can be seen.

Chloroplasts : Chloroplasts are the plastids that are most common and of greatest biological importance, since by *photosynthesis* they produce most of the chemical energy used by living organisms. Without chloroplasts there would be no animals, because animals feed on the food stuffs produced by plants.

Distribution and Orientation : Chloroplasts are sometimes distributed homogeneously within the cytoplasm, but are frequently packed near the nucleus, or close to the cell wall (Fig. 5.1). The distribution and orientation of the chloroplasts within the cell may vary with the amount of light energy. Chloroplasts apparently multiply by division—elongation of the plastid and constriction of the central portion. They have a higher density

Table 5.1. Showing occurence, functions and pigments of various plastids.

Kinds of Plastids	*Occurrence*	*Functions*	*Pigment*
I Chromoplasts.			
A. Photosynthetically active chromoplasts.			
1. Chloroplasts	Higher plants and green algae	Photosynthesis	Chlorophyll-a and Chlorophyll-b
2. Pheoplasts	Brown algae, diatoms and dinoflagellates	Light absorption	Fucoxanthin
3. Rhodoplasts	Red algae (Rhodophyceae)	Light absorption	Phycoerythrin
4. Blue-green	Blue-green algae	Photosynthesis	Phycocyanin Phycoerythrin Chlorophyll-a and Carotenoids
5. Chromoplasts of Photosynthetic bacteria	Purple and non-purple sulphur bacteria	Absorption of infra red portion of light	Bacterio-chlorophyll
B. Chromoplasts devoid of photosynthetic activity			
6. Carotenoids	Tomato, red peppar, flower parts, fungi, ornamental bacteria and fruits		Lycopene and Capsanthin
II Leucoplasts			
7. Amyloplasts	Storage tubers, cotyledons and endosperm	Starch-storage	None
8. Elaioplasts	Epidermal cells of Orchidaceae and Liliaceae	Oil-storage	None
9. Aleuronplasts	Epidermal cells of *Helleborus* and seeds of *Ricinus* and Brazil nut	Protein-storage	None

than the cytoplasm and migrate to the centrifugal pole of the cell when submitted to the action of centrifugal force.

MORPHOLOGY

Shape. Higher plant chloroplasts are generally biconvex or planoconvex. However, in different plants cells, chloroplast may have various shapes, *viz.*, *filamentous, saucer-shaped, spheroid, ovoid* or *discoid shaped.* They are vesicular and have a colourless centre. Some chloroplasts are *club shaped*, having a thin middle zone and bulging ends filled with chlorophyll. Algae often possess a single huge chloroplast that appears as a network, a *spiral band*, or a *stellate plate*.

Size. The size of chloroplats varies from species to species but is constant for a given cell type. The average size of chloroplasts in higher plants measure 4-6μ in diameter and 1-3μ in thickness. The chloroplasts of polyploid plant cells are larger than those in the corresponding diploid cells. In general, chloroplasts of plants grown in the shade are larger and contain more chlorophyll than of those plants grown in sunlight.

Number. The number of chloroplasts is relatively constant in the different plants. In higher plants there are 20 to 40 chloroplasts per cell It has been calculated that the leaf of *Ricinus communis* contains about 400,000 chloroplasts per square millimeter of surface area. When the number of chloroplasts is insufficient, it is increased by division ; when excessive, it is reduced by degeneration.

Conformational changes and ion movements

Chloroplasts undergo striking conformational changes which are dependent on the action of light. In the dark they undergo swelling and separation of the membranes of the thylakoid. This phenomenon is reversible ; in the presence of light the membranes again return to close apposition. Mitochondira eject H^+ ions during the electron transport and that the interior becomes more alkaline. In the case of chloroplasts the opposite reactions occur. Under the action of light H^+ ions which are absorbed, the interior becomes acidified. At the same time K^+ and Mg^{++} are ejected into the medium. In the dark, H^+ is ejected and K^+ and Mg^{++} are reabsorbed by the chloroplasts. Thus, the chloroplast membrane has a vectorial organization that is the reverse of the mitochondiral membrane. In physical terms it is understood that dark reactions, being of a thermochemical nature, do not need an ordered structural arrangement (*i.e.* they are carried out by soluble enzymes). On the contrary, photophosphorylation requires the separation of positive and negative charges into specific pathways of electron flow.

Chemical composition. Chloroplasts of spinach leaf cells contain about 56% protein, 32% lipid and 8% chlorophyll. The lipids include xanthophylls, carotenoids, triglycerides, steroids, waxes and phospholipids. Chloroplasts of higher plants contain approximately 35-55% proteins, 20-30% lipids, 9% chlorophyll, 4-5% carotenoids and 2-3% nucleic acids. In cytoplasm the protein-lipid ratio is generally 90 : 1. In plastids this ratio is about 40 : 30. Thus the grana have a high concentration of lipid. The nucleic acid DNA and RNA have been demonstrated in the chloroplasts of higher plants.

Chlorophyll has a tetrapyrrole structure and consists of a hydrophilic head and a hydrophobic tail. The head is made up of a magnesium atom surrounded by four pyrrolic nuclei (tetrapyrrole structure). The long tail consists of a phytol china. Green plants have two types of chlorophyll, *chlorophyll a* and *chlorophyll b.* In chlorophyll *b* one of the methyl ($-CH_3$) groups of chlorophyll *a* is replaced by a formyl (-CHO) group.

Structure. Plastids are dynamic organelles whose structure changes in response to changes in environment. A mature plastid is having the following parts :

(1) Limiting membrane
(2) Stroma
(3) Grana

(1) *Limiting membrane.* Each mature plastid is covered by a double membrane as in mitochondira, each unit is about 40 to 60Å in thickness. The limiting membrane functions as a semipermeable membrane as well as a boundary.

(2) *Stroma.* The inner space which is formed by a limiting membrane is filled with a proteinaceous matrix called the 'stroma'. A number of starch grains and sometimes osmiophilic droplets are embedded in the stroma. When the plastid becomes inactive or ages, the number of osmiophilic droplets may increase.

(3) *Grana or lamellar system.* These are flat bodies shaped like platelets or discs, which in a lateral view appear as dense bands perpendicular to the limiting membrane of plastid. A number of grana are embedded in the stroma or matrix of plastid. The size of the grana varies between 0.3-1.7μ depending on the species. Each granum consists of a stack of double membrane or lamellae. The lamellar membranes are composed of lipids and proteins. Protein contents are about 45% while lipids contents are 55% in amount. Each lamella is 30-35A° thick and the space between two such lamellae is of 65—70A°. The lamellae can be placed under two groups :

(*i*) Stroma lamellae

(*ii*) Grana lamellae

Those lamellae, extending through the entire length of the plastids are called the stroma lamellae. These are not directly connected with the cell wall. The grana lamellae are always densely packed. Any way, the system of lamellae represents a closed unit, and thus both stroma and grana lamellae are closely related to one another, *i.e.*, each granum lamella represents a local adhesion of two or more stroma lamellae. Both stroma and grana lamellae may stay coupled for a long distance and then separate again from each other. Another term *thylakoid* (*Greek* ; sac like) has been used for the double membranous lamellae already explained. The stroma lamellae are larger thylakoids while the smaller, stacked thylakoids are the grana lamellae. One more term has been added by *Weier* and others, which is *fret* to discribe stroma lamellae as a flattened tubular channel connecting adjacent grana. Recent studies have provided evidence that the membranes within the chloroplasts form a continum and the cavities of the thylakoids (loculi) communicate via fret work connection.

The total number of grana in plastids of certain plants have been calculated. There is a considerable variation in the number of grana in a single plastid, depending upon the species. In *Euglena* only one granum is present in one plastid while in Spinach, where about 40-60 grana are present in each plastid.

In *Chlamydomonas,* the plastid envelope is sometimes continuous with elements of the endoplasmic reticulum. In *Ochromonas* as well as in *Chlamydomonas,* connection with other cytoplasmic membranes with single plastid have also been traced.

QUANTOSOME CONCEPT

The membranes of thylakoids are composed of many paracrystalline spheroid particles known as the quantosomes. The quantosomes are about 160A° by 180A° and about 100A° thick. *Sauer* and *Calvin* (1962) have shown that 3 to 6 quantosomes many aggregate together to form a large particle of 500×100A°. *Park* and *Beggins* (1964) have reported that each quantosome is composed of four subunits. According to *Howelland Moundrianakis* (1967) these quantosomes are the smallest units capable of carrying out the photochemical reaction and each of them cotains about 200 chlorophyll molecules.

According to the size, nature and arrangement of quantosomes in the membrane of thylakoids and stromal lamellae, *Branton* and *Park* (1967) have recognised three types of membranes which are as follows :

1. Membrane with quantosomes.
2. Membrane with smaller particles (110A°).
3. Membrane with rough surface and with a few or no particles.

The chloroplasts, thus, act as elaborate biochemical machines and transform the energy at the molecular level.

FUNCTIONS OF THE CHLOROPLAST

Photosynthesis : Photosynthesis involves two major sets of reactions, each set consisting of many steps. One set, the *light reactions,* transform ("transduce") energy acquired through the absorption of light into usable chemical forms, namely ATP and the reduced form (NADPH) of the coenzyme NADP. This is accomplished via oxidation-reduction sequences that include the removal of electrons from water, releasing oxygen and protons (H^+). These protons and others transferred across the thylakoid membranes during the oxidation-reduction reactions contribute to an ATP-producing proton gradient like the one described for mitochondria.

The other reactions in photosynthesis, the *dark reactions,* occur both in the light and in the dark. They use the ATP and NADPH to build carbohydrates from CO_2. The most common pathway by which this is done is the "carbon reduction cycle" (the *Calvin cycle*), which starts with the reaction of CO_2 with the 5-carbon sugar ribulose 1,5-diphosphate. This generates two 3-carbon molecules that undergo the cycle schematized. This initial CO_2 fixation is catalyzed by the enzyme ribulose 1,5-diphosphate ("bisphosphate") carboxylase, often called RuDP carboxylase or carboxydismutase.

The Light Reactions. The green colour of chlorophyll is due to its selective absorption of light with wavelenghts of 650 to 700nm. (Light of these wave-lengths appears to the eye as red ; when red light is removed from white light, as by absorption, what is left appears green.) Chlorophyll also absorbs light at shorter wavelenghts (400 to 475nm, corresponding to the violet-blue region of the spectrum). Two major chlorophyll types are present in eukaryotic plastids : Chlorophyll *a* shows absorption maxima (peak absorption) at about 670 nm and at about 490 nm, Chlorophyll *b* absorbs maximally at about 650 and 470 nm. Wavelength is an indicator of the energy carried by packets (*photons*) of light. The shorter the wavelength, the greater the energy. The absorption maxima of the chlorophylls and other photosynthetic pigments indicate the portion of the spectrum of sunlight that can be captured for energizing photosynthesis. When a chlorophyll molecule absorbs a photon, an electron of the chlorophyll's porphyrin ring acquires the photon's energy and thereby is excited. This is the basis of the chloroplast's ability to transduce light energy to chemical energy. The excited electron can enter a series of electron transport reactions involving iron-containing proteins (ferrodexins, cytochromes), a copper-containing protein (plastocyanin), lipid-soluble molecules (plastoquinones) resembling coenzyme Q and other participants, some yet to be identified. Two distinguishable "photochemical units", *photosystems*, cooperate in photosynthesis. This cooperation as currently envisaged by a majority of investigators. The functional heart of each unit is a reaction center (action center) that carries out the essential processes by which energy derived from light is made available to the electron transport sequences and other phenomena of the light reactions. Each photosystem is characterized by a particular form of pigment molecule associated with the reaction center; the pigments almost certainly are chlorophylls. In photosystem I (PSI), the reaction center pigment absorbs light maximally at 700 nm and hence, is called P 700. Excited electrons from P 700 pass from the PSI reaction center to the electron transport sequence that reduced NADP. The electrons "lost" from P 700 in this process are replaced through electron transport from photosystem II (PSII); the cyclic pathway diagrammed represents an alternative replenishment mechanism, which some investigators believe is actually the predominant one, at least under some physiological circumstances. The reaction center pigment of PSII, P 680, absorbs light preferentially at 680 nm. The PSII reaction center transfers electrons from this pigment to an electron transport chain that links PSII to PSI. The "loss" of electrons from P 680 is made up by the removal of electrons from water molecules; it is this PSII-mediated oxidation ("splitting") of water that generates O_2.

In net effect the two photosystems collaborate via electron transport to transfer electrons from water to NADPH. Accompanying this transfer is the establishment of an electrochemical H^+ gradient across the thylakoid membranes: The thylakoid interior is maintained at low pH relative to the stroma. The H^+ ions freed in the "splitting" of water also contribute to this gradient. The proton gradient is used to generate ATP (photosynthetic phosphorylation). As with mitochondira, vesicles formed from fragmented thylakoids are capable of both electron transport and phosphorylation but can no longer phosphorylate if broken open or otherwise treated to disrupt the permeability barrier inherent in the membrane. Moreover, chloroplast vesicles can produce ATP for a brief period if simply soaked for a time in a medium of low pH, until their interior equilibrates with the medium, and then transferred into a medium of high pH, establishing a transient H^+ gradient across the vesicles' membranes. Experiments of this sort were crucial in verifying the plausibility, in general of chemiosmotic mechanisms for production of ATP.

ORIGIN OF THE PLASTIDS

Plastids are generated only by other preexisting plastids or proplastid body. They are self reproducing cell organelles like the chromosomes and the centrioles. The proplastid may be regarded as the "stem" plastid, giving rise to metabolically active leucoplasts or chloroplasts which may later develop into other types of chromoplasts through a kind of degeneration. Plastids are transmitted directly to daughter cells during division, providing a genetic continuity in plastid lineage. Although the origin of initials that develop into proplastids is not entirely clear, some observations in fern suggest the nuclear envelope as a possible source.

In Lower Plants. In algae and lower plants, the plastids are quite large and often of different shapes and the process of their division is easy to follow. Here the division begins with an infolding of the inner membrane of the limiting chloroplast envelope. Later on the outer membrane invaginates until the plastid is divided into two. The daughter plastids thus formed, may or may not be equal in size.

In Higher or superior Plants. In higher plants the process of formation of plastids is more complicated. The process can be effected even by sun light, as the formation of plastids in sun light and in dark, show vast differences in either case (Fig. 5.2).

In sun light. During the development the proplastids increase in size. Protoplastids are limited by a double membrane. In the presence of light the inner limiting boundary invaginates at several places, producing small vesicle like structures. These vesicles later on pinch off as free units, which then aggregate in several arrangements to form the lemellae of the mature plastid.

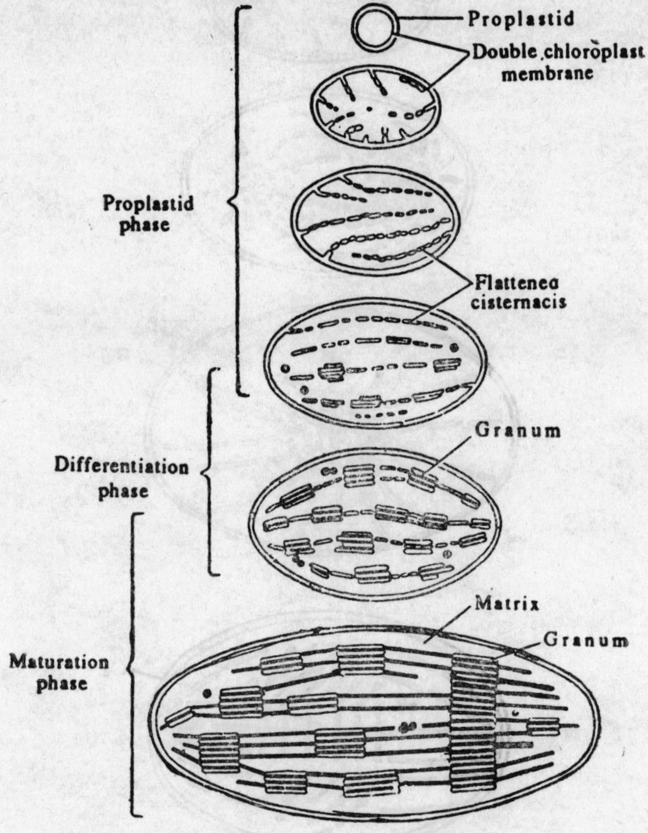

Fig. 5.2. Development of a chloroplast in sun light.

In dark. If the proplastids are left in dark, the process of formation upto vesicle stage is the same, as in presence of light. But the arrangement of vesicles is different. In dark the clusters of vesicles or discs constitute the lamellar system or a series of concentric rings are developed (Fig. 5.3).

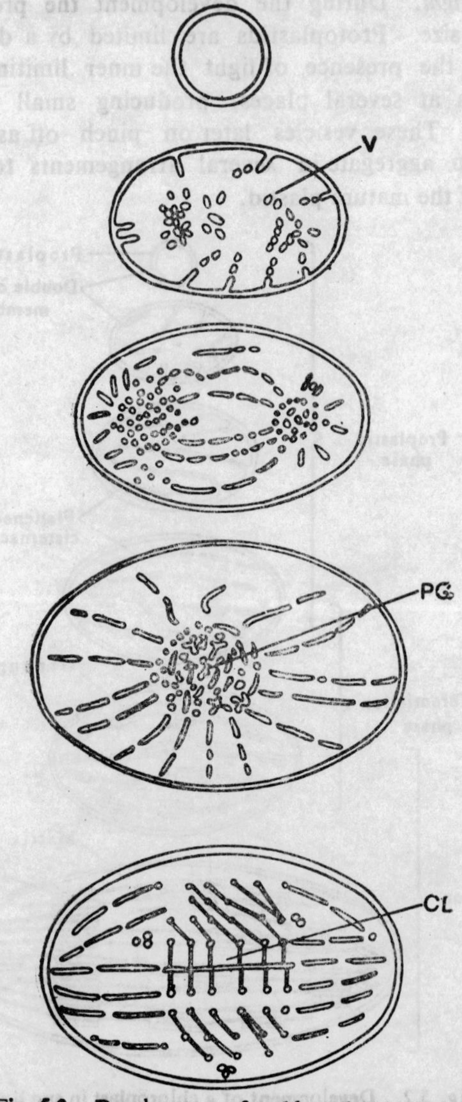

Fig. 5.3. Development of a chloroplast in dark.

Formation of leucoplasts, chromoplasts and chloroplasts from proplastids. The non-lamellated proplastids can also transform themselves into leucoplasts. These, in turn, can develop into immature lamellar plastids or by a reversible process, into starch containing amyloplasts. The mature proplastids then grow to become either mature chloroplasts or other chromoplasts or even, by storing proteins, aleurone plastids. Mature chloroplasts can become chlorophyll free chromoplasts or, by absorption of many lipids elaioplasts.

Nuclear origin : A large number of initials around the nucleus have been observed under special conditions, where the cell organelles are breaking down. This suggested that the initials originated due to evaginatin of the double nuclear membrane and gave rise to mitochondria and proplastids.

Symbiotic Origin. To much surprise, resemblance has been observed between mitochondria and bacteria on the one hand and chloroplast and blue-green algae on the other hand. They are not only some what similar in shape and structure but also in function. All possess DNA, much smaller than nuclear DNA and circular, not organised in chromosomes; their ribosomes are all about the same size, smaller than the cytoplasmic matrix ribosomes of animal and plant cells. In bacteria the respiratory chain is localised in the membrane comparable to inner membrane of mitochondria. Further comparable to cristae of mitochondria, infoldings of plasma membrane (mesosomes) have been reported in certain bacteria. In mitochondria and bacteria both, tbe protein synthesis is inhibited by chloramphenicol, whereas the exracellular protein synthesis of higher cells is not effected. These similarities have led some to consider the mitochondria and chloroplasts of today as generally modified descendents of an ancestral free living bacteria and blue green-algae respectively (autotrophic microorganisms) which at one time came together with the ancestral eukaryotic cell (heterotrophic host cell) to form a symbiotic relationship. This hypothesis is no doubt highly attractive, but still it is evident that the electron transport system in chloroplasts and the enzymes required for making the photosynthetic pigments are controlled by the nuclear genes.

6

Physiology of Vacuoles

Several aspects of vacuoles have been dealt with in connection with other topics such as cellular compartmentation of intermediary metabolites fluxces and compartmentation of ions, phosphate pools and osmoregulation. Vacuoles are generally believed to represent storage compartments for inorganic ions and a variety of organic compounds. However, most of the corresponding information stems from indirect approaches to vacuolar compartmentation. The analysis of tracer kinetics, the assessment of electrical potentials in giant algal cells, and other techniques have led to views and conclusions about inactive pools of compounds which are remote from the centers of metabolism, or about the fluxes of ions in and out of vacuoles. In the past few years methods which allow a direct analysis of vacuoles have been developed and used in a number of laboratories. As a consequence, the present review covers these direct approaches to the biochemistry and function of vacuoles. Work with isolated vacuoles and attempts at selective extraction of vacuolar compounds will be stressed.

Not only ions and small molecules are deposited in vacuoles but also certain macromolecules. Protein bodies represent a type of vacuole specialized in the storage and mobilization of seed proteins. As this aspect of vacuoles has recently been reviewed, it will not be considered here. One feature which appears to be typical for aleurone grains, the presence of certain hydrolytic enzymes, has also been noticed when other types of vacuoles were analyzed for their enzyme content. Hence, intracellular digestion, which may be associated with vacuolar hydrolases, is another important functional aspect of vacuoles which will be discussed here.

The term vacuole designates an (optically) empty space within the cytoplasm. This space is encircled by the vacuolar membrane, the tonoplast, which faces the cell sap with its inner surface and the cytoplasmic matrix with its outer surface. It is obvious that the tonoplast must play a prominent role in metabolic processes associated with vacuoles. Some of its properties have been elucidated recently and views on the interactions between the cytoplasm and the vacuolar fluid are gradually emerging.

It should be mentioned that the term vacuole is subsequently used in a purely descriptive sense. In the era of ultrastructural cytology the term has become somewhat vague as vacuoles turned out to be but comparatively voluminous components of an elaborate system of endomembranes.

ANALYSIS OF VACUOLES

Isolation

The main difficulty in preparing vacuoles is caused by the rigidity of cell walls. Large vacuoles are inevitably disrupted if drastic mechanical means are used for disintegrating cells or tissues. At best only small vacuoles may be available for isolation in homogenates prepared by grinding young tissues with sand using a pestle and mortar. In order to avoid difficulties resulting from the cell wall, naked protoplasts have been used successfully for isolating large vacuoles. A variety of procedures for disrupting yeast protoplasts gently are now available. Mechanical homogenization metabolic lysis in the presence of glucose and a chelating agent as well as osmotic lysis of yeast protoplasts have been used for liberating intact vacuoles. In contrast to the techniques which allow the maintenance of isotonic conditions, osmotic lysis of yeast protoplasts has the disadvantage that the tonoplast is rendered permeable for small molecules if the vacuoles are exposed to hypotonic media, yet if the medium is readjusted to isotonic conditions immediately after the osmotic bursting of yeast protoplasts, micromolecules are retained in the liberated vacuoles. Perhaps the most elegant method for isotonic lysis of yeast protoplasts is based on the irreversible binding to the plasma membrane of the polycation DEAE dextran which causes the disrupture of protoplasts. This method may be useful also for preparing vacuoles from protoplasts of vascular plants. Instability of the plasma membrane at pH 8 in conjunction with slightly hypotonic conditions result in the liberation of intact vacuoles from protoplasts prepared from a variety of plant tissues. Protoplasts have the disadvantage that the properties of vacuoles may be changed upon prolonged exposure of tissues to wall-degrading enzymes. In addition, the preparation of protoplasts depends on a number of parameters and may not be possible in all cases of strain and age of culture of fungi or of plant tissues. A method to overcome these disadvantages is based on the slicing of strongly plasmolyzed beetroot storage tissue; the slicing apparatus has been so designed that organelles (including vacuoles) from cells cut open are released into an undisturbed reservoir of osmoticum. Similarly, but in a more primitive fashion, vacuoles have been liberated from plasmolyzed root tips by gentle chopping with a razor blade. In the case of highly vacuolated parenchyma cells one would expect the slicing of the plasmolyzed tissue to result in the liberation of protoplasts rather than of free vacuoles. In fact the same technique has been used for preparing protoplasts mechanically as an alternative to the conventional enzymatic method. Properties of the medium, particularly a slightly alkaline pH, may cause the instability of the plasmalemma and, hence, allow the mechanical isolation of vacuoles.

A good source of vacuoles is the latex released upon tapping plants with articulated laticifers. It contains large numbers of small vacuoles which can easily be separated from the cytoplasm (serum) by centrifugation.

Stepwise Extraction of Solutes from Cells

A procedure for the stepwise extraction permitting an estimate of the distribution of soluble substances between cytoplasm and vacuole is based on the observation that the retentiveness of the plasmalemma is rapidly lost if yeast cells are exposed to macromolecular polycations at low ionic strength. In a first step of extraction, yeast cells are treated with a polycation such as cytochrome *c* in the presence of an osmotic stabilizer which prevents the bursting of vacuoles. Cytoplasmic pools of solutes are extracted by this treatment. In a second step the vacuoles are disrupted by hypotonic treatment of the cells with water so that the vacuolar pools are released. This procedure appears to be very useful for the study of interactions between cytoplasm and vacuole. Its reliability depends, however, on whether or not the cells of a given population are equally sensitive to the polycation. Excellent effects in inducing permeability of the plasmalemma in *Candida utilis* and *Saccharomyces cerevisiae* strains were recently obtained with DEAE dextran. Although this compound has a molecular weight of ca 5×10^5 dalton and is, therefore, unlikely to penetrate the cell walls and to bind directly to the plasmalemma, it efficiently renders the cytoplasm accessible for the fluorescent dye primulin which was used for probing the effect quantitatively. As plasma membranes of higher plant cells are sensitive to macromolecular polycations, the application of the procedure of stepwise extraction is not necessarily restricted to unicellular fungi

VACUOLES AS STORAGE COMPARTMENTS

Constituents of Cell Saps

It is not the aim of this section to compile the information available on constituents of vacuoles. The idea is rather to illustrate the uneven distribution of substances between vacuole and cytoplasm with a number of selected examples, and to prepare the reader for the presumably more attractive section on transport across the tonoplast and mechanisms of accumulation. In this context it is necessary to make a brief statement about the assessment of intracellular concentrations of substances.

Water is certainly the most abundant among the constituents of cell saps. In fact, it is the solvent for a majority of vacuolar substances, and quantitative data about the composition of cell saps may be given in terms of concentrations calculated on the basis of vacuolar volume. However, concentrations are meaningful only if they can be compared with concentrations in the cytoplasmic matrix. It is obvious that corresponding calculations cannot be based on estimates of the cytoplasmic volume because the fraction of that volume which is occupied by free water (available as solvent) is unknown. The cytoplasm may be conceived as an integrated system of macromolecules, micromolecules, inorganic ions, and structured water. The actual concentration of a compound in the cytoplasm may therefore be much higher than the concentration based on cytoplasmic volume. Hence, it is not justifiable to draw conclusions about the nature of transport processes across the tonoplast (energy dependency) merely from the uneven distribution of substances between vacuoles and cytoplasm. Yet in some cases compounds are accumulated in vacuoles to such an extent that the concentration in the cell sap appears almost certainly to be higher than in the cytoplasm. In *Neurospora crassa* more than 90% of the total arginine and ornithine are contained in vesicles most probably in the form of tiny vacuoles which occupy only a small fraction of the cytoplasm of exponentially growing hyphae. In isotonically isolated yeast vacuoles a concentration of *a*-amino nitrogen as high as about 1 molar has been measured As only 0.6-0.7 M sorbitol is necessary for stabilizing isolated yeast vacuoles, it may be assumed that not all of the vacuolar amino acids are present in an osmotically active form. In yeast the vacuoles occupy about 25% of the cell volume. The basic amino acids account for a large proportion of the total amino acid pool of yeast cells. These amino acids, rich in nitrogen and particularly arginine but also glutamine, are accumulated preferentially in the vacuoles. On a volume basis, the concentration of arginine in *Saccharomyces cerevisiae* grown on argining is about 20 times higher in vacuoles as compared with the concentration in the cytoplasm. In contrast, glutamic acid, which accounts for about one third of the total amino acid pool, is largely absent from the vacuoles.

The differential distribution of different classes of amino acids within yeast cells strongly suggests the existence of specific transporting capacities in the tonoplast. The same conclusion must be drawn from the differential distribution of sulfonium compounds which, under certain conditions, can be induced to accumulate in cells of bakers' yeast : *S*-adenosyl-methionine is deposited largely in vacuoles, dimethyl-3-propiothetin is accumulated predominantly in the cytoplasm, while *S*-methyl-*L*-methionine is about equally distributed.

Organic acids have been assessed in the small vacuoles (so-called lutoids) and in the cytoplasm (serum) of latex from *Hevea brasiliensis*. On a volume basis citric acid is enriched considerably in the lutoids (10-15% of total latex volume), whereas malic acid is evenly distributed. A recent study with protoplasts from mesophyll cells of *Bryophyllum daigremontianum* has yielded evidence that the isolated vacuoles contain practically all of the malic acid.

Large storage pools of sugars present in organs such as internodes of sugarcane have commonly been thought to be located in vacuoles. In the case of red beet storage tissue, a corresponding compartmentation has now been demonstrated : as much at 90% of the total sucrose of this tissue is present in the vacuoles.

The accumulation of phenolics in vacuoles is obvious, especially in the case of colored compounds. These glycosides are retained in vacuoles isolated from protoplasts prepared from petals. Betacyanin has been used as a marker substance for the isolation of vacuoles from red beet storage tissue. In this sections electron dense iron complexes of tannins appear to be localized exclusively in the vacuolar space. In the case of alkaloids the situation is more complex. On the one hand, almost all of the morphine present in the latex tapped from the capsule of *Papaver somniferum* has been recovered in vesicles, presumably small vacuoles comparable with the *Hevea* lutoids. On the other hand, vacuoles isolated from the latex of *Chelidonium majus* contained the four major isochinoline alkaloids of this species, but the extent of accumulation differed greatly ; sanguinarine and chelerythrine were enriched in the vacuoles, berberine and dihydrocoptisine were distributed evenly between vacuoles and cytoplasm.

Certain inorganic substances are extremely plentiful in vacuoles. An eccentric example is provided by the statoliths of the *Chara* rhizoid. These dense bodies are identical with small vacuoles containing barium-sulfate as the main constituent. Another remarkable example concerns the vacuolar polyphosphate of yeast. A careful analysis of polyphosphate distribution in yeast cells and protoplasts provided strong evidence in favour of an exclusive location of this polyanion in the vacuole. In *Scenedesmus* the accumulation of polyphosphate is associated with the appearance of electron dense granules in small vacuoles.

As to the distribution of inorganic ions in yeast cells, the results are not yet conclusive. A considerable vacuolar accumulation of Mg^{2+} and orthophosphate has been claimed. Quantitative microprobe analysis of thin freeze-dried sections of bakers' yeast cells yielded evidence that several univalent ions are about equally distributed between vacuoles and cytoplasm. In contrast, the cytoplasm appeared to contain the bulk of K^+ and Na^+ when the distribution in protoplasts was calculated on the basis of the ion content of isolated vacuoles. It is possible that these results reflect a changeable distribution of inorganic ions in yeast cells depending on the physiological conditions.

Significant differences between the ion contents in vacuoles and cytoplasm could not be detected when whole protoplasts and intact vacuoles of *Tulipa* petal cells were analyzed. Only in the case of *Hevea* latex has a remarkably uneven distribution of inorganic ions been reported. The lutoids contain a large proportion of the total Mg^{2+} but comparatively small fractions of Ca^{2+}, K^+, and Cu^{2+}. On a volume basis the concentration of Mg^{2+} is about 90 times higher in the vacuoles than in the cytoplasm, whereas in the case of K^+ the concentrations are equal.

Hence, at least some of the data available suggest that the deposition in vacuoles of inorganic substances is associated with a remarkable specificity. This specificity is obvious in the case of vacuolar compartmentation of organic compounds. It is also obvious that the tonoplast must play a decisive role in determining the composition of the vacuolar fluid.

Properties of Tonoplasts

A number of observations in yeast suggest that the properties of the tonoplast and those of the plasmalemma differ considerable. Differences concern, for example, the stability of protoplasts and vacuoles as a function of pH and osmolarity of the medium and the sensitivity of plasmalemma and tonoplast to polyene antibiotics and polycations.

Although the results of chemical analysis are incomplete, they provide some hints as to the chemical background of these differences. The tonoplast of bakers, yeast is comparatively poor in sterols and rich in phospholipids, with a high proportion of unsaturated fatty acids. These findings may explain the resistance of this membrane to the polyene antibiotic nystatin as well as the remarkable elasticity of yeast vacuoles. In contrast, the microvacuoles of *Hevea* latex are sensitive to hypotonic treatment. The tonoplast of this organelle is characterized by a most unusual lipid composition ; phosphatidic acid accounts for more than 80% of the total phospholipids, and almost half of their fatty acids are saturated.

Isolated yeast vacuoles do not bind concanavalin A unless they have been rendered permeable for this lectin. Polysaccharides associated with the yeast tonoplast have also been demonstrated cytochemically. The yeast plasmalemma as well as the tonoplast are asymmetric with regard to binding sites for concanavalin A : the corresponding polysaccharides appear to be absent from the membrane surfaces facing the cytoplasmic matrix.

The presence of ribosomal RNA in *Hevea* lutoids is interesting with regard to the possibility that protein synthesis is associated with vacuoles ; however, the data presented do not enable the conclusive distinction between RNA associated with the cytoplasmic surface of the membrane and RNA localized inside the lutoids. In any case, the question of whether or not ribosomes are associated with tonoplasts deserves special attention with regard to the synthesis of vacuolar enzymes. In isolated and positively contrasted yeast vacuoles, polysome-like structures appear to be present.

A general feature of tonoplasts appears to be the presence of certain oxidoreductases which, however, are also typical for ER membranes. ATPase activity associated with vacuolar membranes deserves particular attention as it may be involved in transport processes requiring metabolic energy. Membrane bound, Mg-dependent ATPase activity has been detected in vacuoles isolated from petal protoplasts of *Tulipa* and *Hippeastrum* from *Hevea* latex and in yeast vacuoles. ATPase seems to be absent from vacuoles of red beet storage tissue. Since sucrose is enriched in these vacuoles it is exciting to learn that glucose specific hexokinase is tightly bound to tonoplasts ; moreover, under certain conditions sucrose synthetase activity and other elements of sugar metabolism appear to be loosely attached to this membrane.

Due to the lack of a specific marker enzyme, it has not yet been possible to isolate tonoplasts from conventional homogenates. Only in the case of bakers' yeast vacuoles does α-mannosidase seem to be specifically located in the tonoplast.

Membrane Transport in Tonoplasts

Using isotanically prepared yeast vacuoles which retain their large amino acid storage pool, the group of Wiemken has discovered the existence of an argininespecific permease in the tonoplast. If vacuoles of *Saccharomyces cerevisiae* are exposed to labeled arginine and reisolated after a period of incubation, uptake at a constant rate over 7-10 min can be assessed. Although an enormous difference in arginine concentration between cell sap (100-200 mM) and medium (*e.g.* 100 μM) suggests the requirement for metabolic energy to pump arginine against the chemical potential, the uptake does not depend on a corresponding supplement to the medium. According to a number of criteria, the transport across the tonoplast is due to a specific permease which facilitates diffusion of arginine between the cell sap and the medium (cytoplasm) which surrounds the vacuoles. In fact, absorption of L-arginine from the medium is associated with a stoichiometric efflux as demonstrated in double labeling experiments with preloaded vacuoles. Concisely, the arginine permease has the following properties which clearly differ from the properties of arginine transport in protoplasts (P). It exhibits saturation kinetics with K_m=30 μM (P : K_m=1,8 μM). It is optimally active at pH 7.0-7.5 (P : in *Candida utilis* transport is practically unchanged between pH 4.5 and 8.0 ; The activation energy, as drawn from the temperature dependency of transport, is 39 kJ/mol (P : 51 kJ/mol for temperatures > 11°C). Metabolic inhibitors such as dinitrophenol and azide are ineffective (P : strong inhibition). As shown by competition experiments, the L-arginine-permease transports L-histidine (K_m=270 μM), L-lysine (K_m=1,3 mM), and L-ornithine (K_m= 3,0 mM). However, a number of other amino acids such as L-leucine, L-phenylalanine, L-glycine, and L-glutamic acid are not transported. Structural analogues such as D-arginine or L-canavanine inhibit the transport of L-arginine competitively. The pattern of inhibition by a variety of structural analogues differs clearly for transport in vacuoles and protoplasts. Lastly, the analysis of arginine transport in vacuoles isolated from a number of mutants of *Saccharomyces cerevisiae* which lack the specific arginine, and/or general amino acid-permease of the plasmalemma, has shown that the properties of the vacuolar arginine-permease are unchanged in these mutants ; this permease is therefore genetically independent from the permeases that catalyze the uptake of arginine from the culture medium.

Preliminary transport studies with vacuoles isolated from the cell wall-deficient "slimy" mutant of *Neurospora crassa* showed the existence of an arginine-specific permease corresponding to the permease of the bakers' yeast tonoplast.

The susceptibility of arginine transport in isolated yeast vacuoles to proteinases has been used for demonstrating that the permease represents a proteinaceous factor of the tonoplast. Thermolysin causes an activation of transport by a factor of 2-3 and even more. Pronase, in turn, increased the rate of transport at low concentrations and caused its decay at higher concentrations. The decay proceeded according to a first order reaction, indicating that a single hydrolysis in the active center of the permease results in complete inactivation. In fact, if vacuoles were incubated with pronase in the presence of L-arginine, the active center thus being occupied, the inactivation was largely prevented. The comparatively smaller protecting effects of other substrates of the permease correspond with its lower affinity for L-histidine, L-lysine, and L-ornithine. The authors tentatively conclude that the action of pronase abolishes the catalytic properties of the permease while thermolysin, as also pronase at low concentrations, alters a proteinaceous factor which hampers the "mobility" (that is, controls the reaction rate) of the permease in the membrane. Vacuoles appear to be ideally suited for this sort of investigation as the cytoplasmic face of the tonoplast is not coated with polysaccharides which may prevent the access of proteinases to membrane proteins; corresponding experiments with protoplasts may be unsuccessful because the outer surface of the plasmalemma is coated with polysaccharides.

In a similar fashion and with similar results a specific transport system for *S*-adenosyl-L-methionine has been characterized in isolated vacuoles of *Saccharomyces cerevisiae*. The uptake of this metabolic intermediate exhibits saturation kinetics ($K_m=68\ \mu M$) and is competitively inhibited by *S*-adenosyl-L-ethionine and *S*-adenosyl-homocysteine. It is not sensitive to metabolic inhibitors and is resistant to the polyene antibiotic nystatin. In all of these properties the transport of *S*-adenosyl-L-methionine in vacuoles differs clearly from transport in the protoplasts. An important difference with respect to the arginine permease concerns the absence of *S*-adenosyl-L-methionine efflux which must be expected if the permease specifically facilitates diffusion across the tonoplast. However, the efflux may concern another vacuolar constituent which escapes analysis if vacuoles preloaded with labeled *S*-adenosyl-L-methionine are used for assessing efflux. It should be mentioned that an earlier attempt to demonstrate transport of *S*-adenosyl-L-methionine, arginine, lysine, and methionine in isolated vacuoles of *Candida utilis* was unsuccessful; it seems that the permease systems were inactivated by the osmotic lysis of the protoplast which caused a transitory stretching of the tonoplast.

It is interesting to note that vacuoles have an extremely poor capacity for transporting *S*-adenosyl-L-methionine if bakers' yeast cells are cultured in the absence of L-methionine, that is, under conditions which do not induce the formation of a large vacuolar pool of *S*-adenosyl-L-methionine. Hence, vacuolar permeases may be inducible proteins. This is also suggested by the finding that in *Saccharomyces uvarum*, glycine has access to vacuoles and is accumulated there only if the cells are first starved of nitrogen and then fed with glycine. Likewise, considerable differences in the arginine-transporting capacity as dependent on the culture conditions suggest that the vacuolar arginine-permease in bakers' yeast cells is subject to N-catabolite repression.

In the latex of *Hevea brasiliensis*, free amino acids are largely located in the cytoplasmic fraction (serum); however, basic amino acids are enriched in the vacuoles. Transport of arginine and lysine in these vacuoles (lutoids) is mediated by a permease as suggested by saturation kinetics, competitive inhibition (lys/arg), and other features. In contrast to the corresponding permease of the yeast vacuole, the affinity for basic amino acids is very low (10^{-2} molar range) and the transport is sensitive to dinitrophenol; moreover, uptake of lysine is not associated with efflux and the rate of transport is enhanced in the presence of ATP. These findings suggest that the mechanisms of transport and particularly the mechanisms of accumulation of amino acids may be different in vacuoles from fungal and vascular plant cells.

As *Hevea* lutoids contain a large pool of citrate, they were also used for the study of citrate transport. These vacuoles take up considerable proportions of exogeneously supplied citrate against a concentration gradient (*e.g.* 50 mM citrate in cell sap, 7.5 mM in medium). The rate of uptake is constant for at least 30 min. Saturation kinetics ($K_m = 5$ mM at pH 7.0) suggest the existence of a specific citrate permease. Efflux of citrate was not assessed. ATP enhances the rate of uptake considerably, dinitrophenol is a potent inhibitor of citrate transport.

In conclusion, there is little doubt that tonoplasts are equipped with specific permeases. The discovery of further permeases seems to be a matter of experimental know-how in isolating and handling vacuoles. The existence of such permeases is not surprising, as the distinct distribution of individual compounds between the cell sap and the cytoplasm points to a control of fluxes in and out of the vacuolar compartment. This control is obviously exercised by specific permeases of the tonoplast. However, the existence of permeases does not explain the phenomenon of accumulation of compounds in the cell sap. In particular, the action of permeases, such as the arginine permease of yeast vacuoles which catalyze diffusion, cannot account for the enrichment of arginine in vacuoles.

Accumulation

A common feature of vacuolar permeases, which emerges from studies with isolated vacuoles, appears to be the absence of a direct dependency of transport on metabolic energy. If this property reflects the situation in the living cell, the permeases cannot be responsible for the establishment of a chemical potential between the cell sap and the medium or the cytoplasm, respectively. In a few cases of vacuolar accumulation of compounds the factors responsible for the active transport into vacuoles are now being elucidated.

Although in isolated yeast vacuoles arginine is exchanged between cell sap and medium, the large vacuolar pool is retained. This basic observation suggests that in vivo the vacuolar arginine is partially present in a nondiffusible form.

Equilibrium-dialysis studies have yielded substantial evidence that arginine cations are immobilized through binding with the vacuolar polyphosphate anions. Thus the retentiveness of isolated vacuoles for arginine (and other basic compounds) can be interpreted in terms of a Donnan equilibrium. The yeast vacuole can therefore be conceived as an ion exchange system, as shown in the model of transport and accumulation of arginine presented in Figure 6.1. Its principal features include the presence of a nondiffusible anion (polyphosphate) in the cell sap and a selectively permeable tonoplast. Transport mechanisms, whether experimentally demonstrated (Figures 6.1 A and B) or merely postulated (Figures 6.1 C and D), are based on the condition of electroneutrality. As far as a reversible accumulation of arginine is concerned, the model predicts that changes in size of the vacuolar pool must be associated either with changes of the polyphosphate complement or with the replacement of arginine molecules by other basic compounds or inorganic cations. Experimental data on antiport processes to support this aspect of the model have been obtained. The expansion of the vacuolar arginine pool would require the synthesis of polyphosphate. In the hypothesis presented in (Figure 6.1 E) an electrogenic transfer of phosphate groups from cytoplasmic ATP to vacuolar polyphosphate acceptor molecules is postulated. This group transfer across the tonoplast would represent an electrical coupling of net influx of arginine to the formation of the vacuolar polyanion. Nothing is known about this important aspect of arginine accumulation. However, mobilization of arginine as associated with the degradation of polyphosphate has been observed in isolated yeast vacuoles after prolonged incubation. Endo- as well as exo-polyphosphatase activity is present in these organelles. In the context of the model it is important to note that upon the gradual hydrolysis of the vacuolar polyphosphate, the ratio between acid-labile phosphate groups and arginine remains constant; obviously efflux of arginine takes place, and the model (Figure 6.1 D) predicts that symport of this cation and an anion, presumably orthophosphate, takes place.

In the vacuoles isolated from *Hevea* latex, citrate is transported without the addition of an energy source to the medium. As this transport takes place against a gradient of citrate

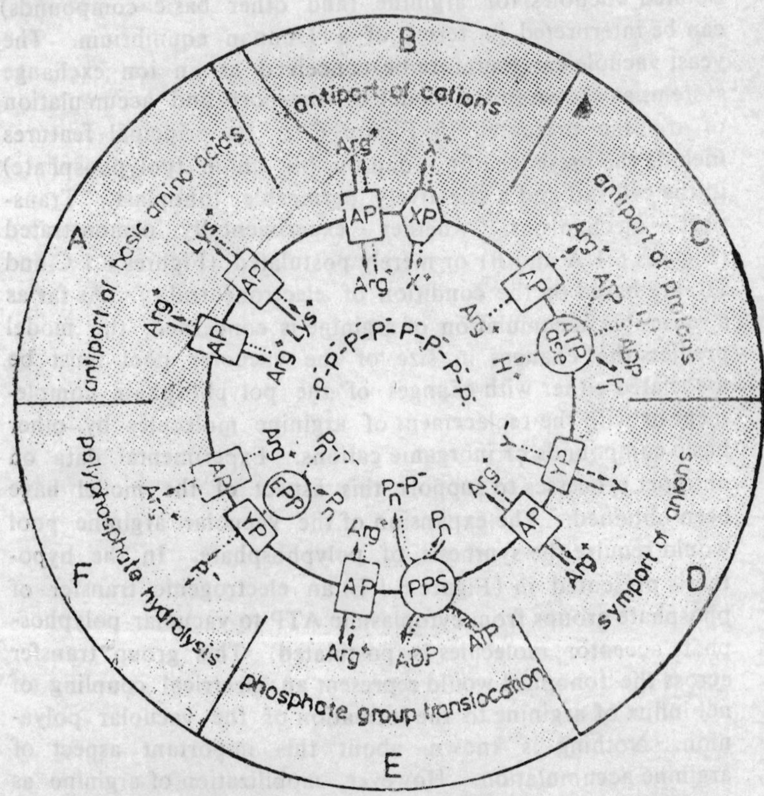

Fig. 6.1. Possible mechanisms of accumulation and mobilization of arginine in the yeast cell vacuole.

Symbols : Stippled area, Cytoplasm ; Central area, Vacuolar space ; AP, Arginine-permease located in tonoplast ; Arg^+, Arginine-cation ; $P^-—P^-—P^-$, Vacuolar polyphosphate ; - - →, Processes resulting in accumulation ; —→, Processes resulting in mobilization.

A : Antiport of basic amino acids. B : Antiport of basic amino acids and other cations. C : Antiport of arginine and protons. D : Symport of arginine and anion Y^-. E : Polyphosphate-synthesis through phosphate-group translocation. F : Hydrolysis of vacuolar polyphosphate.

concentration, it must require energy. The possibility that citrate is chelated by Mg ions accumulated in the vacuoles has practically been ruled out. However, the energy required for citrate accumulation appears to be supplied by the electrical potential between the cell sap (pH 5) and the medium (pH 7). Indeed, dinitrophenol greatly inhibits the uptake of citrate, probably through the abolishment of the proton gradient. Moreover, ATP added to the medium causes an increased rate of transport. It is feasible, therefore, that the ATPase present in the lutoid membrane acts as a pump for protons which neutralize the citrate anions transported by the citrate permease. If this is so, citrate accumulation in lutoids has the character of a symport catalyzed by two independent carriers. A proton pump has also been proposed in connection with transport of basic amino acids in *Hevea* vacuoles. This transport is also considerably enhanced in the presence of ATP. In this case the proton pump is a constituent of an antiport system.

In the light of mechanisms of vacuolar accumulation of charged molecules now being elucidated, it is interesting to consider the mechanism which may be responsible for the accumulation in vacuoles of uncharged compounds such as sugars. In sugarcane, group translocation of sucrose by the action of tonoplast-associated sucrose phosphatase has been proposed as a mechanism of sugar accumulation. Recent results obtained with vacuoles isolated from redbeet indeed suggest that enzymes of sucrose metabolism such as invertase and sucrose synthetase are loosely attached to the tonoplast. As the presence of these enzymes in preparation of tonoplasts depends on the aging of the tissue prior to the isolation of vacuoles, they may be involved in the mobilization of stored sucrose rather than in accumulation. Preliminary results obtained with vacuoles isolated from protoplasts of the pericarp of immature grapes suggest the existence of a sugar transporting system in the tonoplast ; sucrose is transported at a much higher rate than glucose. Thus the accumulation of monosaccharides in this tissue may be brought about by a sucrose-specific permease in the tonoplast in conjunction with the vacuolar acid invertase.

Functional Aspects

A few selected examples will illustrate the functional significance of vacuolar compartmentation of certain compounds.

Storage Pools of Intermediates. The storage pool of basic amino acids localized in the yeast cell vacuole has an important advantage for the maintenance of metabolic activity under changing environmental conditions. This has been demonstrated by subjecting an exponentially growing culture of *Candida utilis* to abrupt changes in the supply of nitrogen. Surprisingly, exponential growth and concomitant protein synthesis are not interrupted if a complete medium is replaced by a medium without a source of nitrogen. It is true that the continuation of growth in the absence of exogeneous nitrogen results in the initial reduction of the small cytoplasmic pool of amino acids, yet this pool is not depleted because it is continuously refilled from the vacuolar storage pool. Whether the mobilization of stored basic amino acids involves the hydrolysis of polyphosphate or the exchange of polyphosphate-bound amino acids with inorganic cations has not yet been investigated. It may be expected that under conditions of nitrogen deficiency the vacuolar polyphosphate is preserved, whereas nitrogen in conjunction with phosphate deficiency causes the utilization of both N and P-reserves. It is interesting to note that yeasts behave in a species-specific fashion when they are subjected to nitrogen starvation. *Candida utilis* continues to grow at the normal growth rate and becomes stationary after about one generation when the vacuolar nitrogen reserves are exhausted. In contrast to this prodigal way of life, *Saccharomyces cerevisiae* turns immediately to a nitrogen-saving reduced growth rate and is able to remain in a growing condition for much longer. Needless to say, starved yeast cells immediately start to refill their vacuolar storage pools when they are supplied with appropriate nutrients.

Polyphosphate and basic amino acids of yeasts can be compared with phytic acid and reserve proteins which are stored in seeds. In both cases, but for different reasons, vacuoles (aleuron grains) are used as storage compartments. The mobilization of raserve proteins upon seed germination requires proteolysis which takes place within living cells. The spatial separation of this process from the cytoplasmic centers of metabolism appears to be meaningful. As reserve proteins of seeds make the young plant independent of exogenous nitrogen, the vacuolar storage pools of amino acids render the fungal nitrogen metabolism substantially independent of the cellular environment. The micromolecular form of nitrogen-storage has the advantage over reserve proteins of being immediately available. The significance of vacuolar compartmentation of arginine is evident when the regulation of arginine metabolism is considered the rapid and precise regulation within the cytoplasm requires a small metabolic pool and hence the spatial separation of the large storage pool from the site of arginine metabolism. Thus the vacuolar pools buffer the cytoplasm against changes in environmental conditions and, therefore, represent an important tool for maintaining homoeostasis within the cytoplasm. It is true that in this respect the plant vacuole can be conceived as a subcellular analogue of the body fluids in complex animal organisms.

Another example that demonstrates the importance of vacuolar compartmentation of metabolites in maintaining homoeostasis is provided by Crassulacean acid metabolism. Malic acid is the most prominent of metabolites, with a diurnally oscillating level in leaf cells of plants exercising this type of photosynthesis. With regard to the corresponding fluctuations of pH and also in consideration of the property of malate as an effector of key enzymes of malate metabolism, the spatial separation of storage and metabolic pools of malic acid appears to be crucially important. In fact, isolated vacuoles contain the bulk of malate present in protoplast from leaf cells of *Bryophyllum daigremontianum*. The maintenance of homoeostasis within the cytoplasm is likely to involve the removal of excessive malate by virtue of a transport system residing in the tonoplast. A fascinating problem of regulation of membrane transport is posed by the phenomenon of circadian rhythmicity of malate accumulation and mobilization in vacuoles. A corresponding hypothesis based on changes of the permeability of the tonoplast dependent on turgor has been proposed. Whether or not this hypothesis will be supported by future experimental work, it is obvious that the compartmentation of malate in space and in time provides plants with Crassulacean acid metabolism a prodigious possibility of coping with severe environmental conditions.

Turgor. In *Avena* coleoptiles, auxin-induced ion exchange reactions in the plasmalemma are associated with the accumulation (in vacuoles ?) of malate and K^+ ions. One aspect of these processes certainly concerns the regulation of turgor in the expanding cells. The budding process in yeast provides another example of involvement of vacuolar micromolecules in turgor production. The extrusion of buds is associated with the inflation of vacuoles. In *Candida utilis* grown on arginine it has been demonstrated that this inflation is due to the sudden doubling of the rates of arginine uptake from the medium and arginine catabolism, as well as to the doubling of rates of amino acid synthesis at the onset of bud formation. As in yeast cells amino acids account for the bulk of osmotically active constituents, this timing of their synthesis is undoubtedly significant with regard to the adjustment of turgor in the budding cell.

Detoxification. A single example will illustrate the role of vacuoles in the removal of toxic secondary products of metabolism from the cytoplasm. Two major alkaloids present in the latex of *Chelidonium majus*, sanguinarine and chelerithrine, are toxic in the sense that they abolish the semipermeability of membranes. In contrast, the two other major alkaloids, berberine and dihydrocoptisine, do not affect membranes. It is interesting to note that the toxic compounds have a higher affinity for the vacuolar tannins and are therefore enriched and trapped in the vacuoles. Hence the vacuolar compartmentation of tannins appears to be a prerequisite for storing harmful alkaloids within the living laticifer.

VACUOLES AS LYSOSOMES

Vacuolar Hydrolases

The presence in vacuoles of certain hydrolases which are typical for lysosomal organelles has been demonstrated in a considerable number of plant species. This aspect of vacuolar compartmentation has recently been reviewed. The most complete analysis of subcellular hydrolase distribution has undoubtedly been achieved in the model organism yeast. This section shall therefore be limited to the presentation of findings in this fungus and to the discussion of some recent reports on vacuolar hydrolases in other objects.

Vacuoles isolated from protoplasts of *Saccharomyces cerevisiae* contain a proteolytic system comprising an acid proteinase (A), a serine proteinase (B), and a serine carboxypeptidase (Y, formerly protease C). The vacuoles represent the principal compartment of this system. However, other localizations of proteases cannot yet be excluded on the basis of data available. In whole extracts from yeast cells or protoplasts the assessment of proteolytic activities is complicated by the presence of several specific inhibitor proteins ; the activation of the proteolytic system requires the preincubation of extracts under conditions which favour the autocatalytic degradation of inhibitor proteins by the action of proteinases A and B. It would be tempting to conclude that in yeast cells the proteases are present in the form of zymogens. However, it has been demonstrated that the inhibitor proteins are located outside the vacuoles, most probably in the cytoplasmic matrix. This differential location of protease and their specific inhibitors suggests that the proteolytic system is fully active in the living cells.

Other vacuolar hydrolases of yeast include ribonuclease, alkaline phosphatase, and α-mannosidase. Some hydrolases which are present in the yeast vacuole—invertase, β-glucosidase, aminopeptidase, and acid phosphatase—are also secreted into the extracytoplasmic space.

Incidently, not all the yeast hydrolases are located in the vacuole ; α-glucosidase is an enzyme of the cytoplasmic matrix. Tentatively it can be stated that specific and harmless hydrolases may be located in the cytoplasm, while unspecific and harmful enzymes such as phosphatases, proteinases, and nucleases are located in the "lytic compartment" of which the vacuoles are a principal intracellular component. Protection of the cytoplasmic constituents from an uncontrolled attack by the digestive enzymes is the suggested reason for this intracellular compartmentation.

Almost nothing is known about the site of synthesis of vacuolar enzymes. Hydrolases of the yeast cell vacuole that have been obtained in pure form turned out to be glycoproteins. This property may be important for the stability of these enzymes in the cell sap, in terms of a possible inaccessibility of their protein moiety for proteinases. Carboxypeptidase Y contains 14% mannose. Its formation is inhibited by tunicamycin as is the formation of the secreted mannanprotein invertase or the vacuolar proteinase A. In the presense of tunicamycin, nonglycosilated carboxypeptidase Y is not accumulated, suggesting that synthesis of the polypeptide is linked with glycosylation. As mannan synthesis involves a lipid intermediate and is associated with membranes, it is tempting to assume that the synthesis of hydrolase-polypeptides and the subsequent glycosylation are integrated in the tonoplast or in components of the endomembrane system which are in communication with vacuoles. A short-lived precursor form of carboxypeptidase Y has recently been detected in a membrane preparation from baker's yeast. Its conversion into the final form is associated with a reduction in molecular size. Hence, this vacuolar glycoprotein may be released from the site of its synthesis into the vacuolar compartment upon the cleavage of an extra extension of its polypeptide chain.

A part from the isolation of vacuoles, cytochemistry has been widely used for localizing hydrolases, mainly phosphatases, in plant cells. The corresponding data are consistent with the vacuolar localization of acid unspecific phosphatase; however, the reaction product of this enzyme has also been observed within the entire endomembrane system (ER, Golgi) as well as in the cell wall space. Hence, vacuoles appear to be only the most voluminous components of an elaborate membrane system. A similar distribution as demonstrated cytochemically for acid phosphatase may be expected for other hydrolases. Indeed, results obtained from tissue fractionation studies reflect a complex subcellular distribution of hydrolases. It is important to note that cytochemists have observed the absence of acid phosphatase from the cytoplasmic matrix. As basic metabolism depends so much on intermediary phosphate esters, the confinement of this unspecific hydrolase to vacuoles, endomembrane system, and extracellular space convincingly illustrates the significance of lysosomal compartmentation.

Depending on the conditions of homogenization, and the tissue concerned, a variable but usually large proportion of total hydrolase activities is recovered in the soluble fraction. It is impossible to draw conclusions about the intracellular compartmentation of hydrolases from this kind of result. The enzymes concerned may be localized either in vacuoles which were disrupted upon homogenization, or in the mural space, or they may be localized in the cytoplasmic matrix as α-glucosidase in yeast cells. It is to be expected that the novel techniques for gently releasing subcellular structures, including large vacuoles, from higher plant protoplasts will be useful for elucidating the compartmentation of hydroglases. Corresponding results about the complement of hydrolases present in central vacuoles of mature plant cells are not yet conclusive. Acid phosphatase and nucleases are present in these vacuoles whereas glycosidases appear to be localized outside the vacuoles. It will be interesting to learn whether the glycosidases or the substantial fractions of phosphatase which are not recovered in the large vacuoles are localized in other components of the endomembrane system.

From *Nitella* and *Chara* cell saps have been obtained upon perfusion of cut internodes ; they contain the bulk of acid phosphatase and carboxypeptidase, whereas only small proportions of oxido-reductases are extracted by this procedure. A corresponding result has been reported on hydrolase distribution in stratified subcelles of stalk of *Acetabularia.* RNase has been localized, by means of immunocytochemistry, using specific fluorescing antibodies, in vacuoles of morning glory flower tissue.

Intracellular Digestion

The role of compartmentation. Although the information available on the localization of hydrolases within plant cells is far from being complete, it is sufficient to justify the consideration of compartmentation in studies on the metabolic function of hydrolase action. The necessity of complementing data from biochemical approaches with a morphological, structural frame appears convincingly from recent work on proteolytic regulation of enzyme activities in yeast. Principles which in homogenates caused the inactivation of tryptophan synthetase or the activation of chitin synthetase turned out to be identical with those of known yeast proteinases. As the proteinases in question are located in the vacuole, tryptophan synthetase, in the cytoplasmic matrix, and chitin synthetase in the plasmalemma, the discovery of proteolytic inactivation of activation of these enzymes obviously has no direct significance for the understanding of the regulation of their activity in vivo. A considerable number of similar proteolytic artifacts have been compiled, and an antiartifact methodology for preventing the proteolytic attack on native proteins upon the abolishment of compartmentation in homogenates has been proposed. If considered from the viewpoint of possible proteolytic artifacts, it is almost a miracle that so many enzymes have been purified in the past from autolyzed yeast under conditions which undoubtedly favor the occurrence of proteolytic inactivation !

If subcellular compartmentation of proteolytic enzymes and their potential endogeneous substrates is not considered in the first instance, one is tempted to assume that the proteolytic action bears some specificity with regard to the inactivation of distinct protein species. In fact, if cell-free extracts from yeast cells are incubated in the presence of the yeast proteinases A or B, or in the presence of carboxypeptidase Y, differential inactivations of various enzymes of the intermediary metabolism occur and suggest a selective action of proteolytic enzymes. However, this surprising selectivity of protease action is unlikely to reflect a selective degradation of enzymes in the living yeast cell. Rather does it seem that the selectivity observed is caused by the differential vulnerability of catalytic centers due to the conformation of individual enzyme proteins. Little is known about the mechanism of intracellular protein degradation. This process undoubtedly represents a major factor in the determination of cellular concentration of a protein. Its significance is particularly obvious with regard to turnover reactions which lead to the biochemical differentiation of cells. Considering the compartmentation of unspecific hydrolases which may be responsible for the degradation of biopolymers, it must be concluded that catabolism takes place in the lytic compartment. In rat liver the products of intracellular proteolysis have in fact been localized in lysosomes. Corresponding experiments have not yet been performed with plant cells. Yet the similarities between the compartmentation of hydrolases in animal and plant cells suggest that protein degradation in plant cells also takes place in a lysosomal cell space. In any case, the site of protein degradation must not necessarily be identical with the site from which a species of protein disappears under certain conditions. The conclusion that the loss of carboxydismutase during senescence of detached leaves is due to proteolysis in chloroplasts may not be drawn unless the corresponding localization of the protease(s) involved has been demonstrated.

Autophagic vacuoles. The autophagic nature of vacuoles is suggested by numerous ultrastructural observations concerning the presence within vacuoles of structured cytoplasmic material. The bad preservation of such vacuolar inclusions is likely to be due to the action of digestive enzymes. Membrace processes seem to be responsible for the sequestration of cytoplasmic material in vacuoles. On the one hand the engulfment by the invaginating tonoplast results in the formation of intravacuolar vesicles containing a portion of cytoplasm. On the other hand the transfer of cytoplasm into the digestive compartment coincides with the formation of vacuoles portions of cytoplasm are encircled by components of the endomembrane system which ultimately fuse together and form an autophagic vacuole. These observations suggest that intracellular digestion proceeds in such a fashion that the compartmentation of vacuolar hydrolases is never abolished. Yet, autophagy cannot be understood unless the morphological observations are complemented with corresponding knowledge about the biochemical events. At the moment we are dealing with a mosaic of somewhat isolated ultrastructural observations of autophagic vacuoles, assessments of turnover reactions, localizations of enzymes, and correlations between catabolic processes and hydrolase activities. What is needed is an integrated study of one organism which covers all the relevant aspects.

Hypothetical evaluations. The principal problem concerns the mechanism by which a cytoplasmic macromolecule or a structure is specifically subjected to degradation. Yeast is undoubtedly one of the favorite organisms for corresponding studies. The vacuolar compartmentation of the proteolytic system and the presence in the cytoplasm of a specific inhibitor system strongly suggest that the degradation of protein takes place within the vacuoles. Furthermore, bakers' yeast is ideally suited, because biochemical differentiations which are caused by exogenous glucose are associated with the inactivation and degradation of catabolite-repressed enzymes. It is therefore not surprising that an attractive hypothesis for the selective degradation of proteins deals with yeast. It is termed "catabolite inactivation" because its principal feature concerns the allosteric effects and interconversions produced in enzymes by catabolites. These structural changes convert the enzymes into inactive and vulnerable proteins which are recognized by the autophagic system. In this concept it is thought that autophagy and proteolysis in vacuoles unspecifically comprise the inactivated enzymes which are no longer integrated in metabolism. Specificity would, in turn, be provided by the interactions between enzyme molecules and effective catabolities. A similar concept attributes the specificity of degradation to the action of group-specific proteases. Such enzymes have been isolated from several animal tissues. They bring on a limited proteolysis in susceptible forms of enzymes, that is, in the absence of coenzymes, substrates, or effector molecules. This irreversible step taking place in the cytoplasm is followed by random digestion of inactivated enzymes in lysosome. Corresponding group-specific proteases have so far not been detected in the plant kingdom.

These concepts may be most valuable guides for investigating the mechanism of protein degradation. It would certainly be interesting to test them with isolated vacuoles. However, the autophagic activity is likely to be dependent on the living cytoplasm. In fact, isolated lysosomes of animal cells do not take up protein. This is not surprising as the morphological observations suggest that autophagy is associated with pinocytosis-like membrane processs which depend on metabolism and on metabolic energy.

7

Cell Division

There are two types of organisms uni-cellular and multicellular. The growth and development of an individual depends exclusively on the growth and multiplication of the cells. It was *Virchow* who first of all adequately stated the cell division. In animal cell, the cell division was studied in the form of segmentation division or cleavage by *Prevost* and *umos* in 1824. The mechanism of cell division was not precisely investigated until long afterward but *Remak* and *Kolliker* showed that the process involves a division of both the nucleus and the cytoplasm. The term *Karyokinesis* was introduced by *Schleicher* (1878) to designate the changes of nucleus during division, and the term *cytokinesis* was introduced by *Whiterman* (1887) to designate the associated changes taking place in the cytoplasm.

Cell division is necessarily the avoidance of aging, and secondly for the segregation of an individual into semi-independent units which leads to efficiency. Thus we see that cell division is a widespread phenomenon that is essential not only for the maintenance of life but also for the development of the organism itself.

Cell division can be conveniently described as :

(*i*) *Direct division.* Where the nucleus and cell body undergoes a simple mass division into two parts. It is also called *amitosis*.

(*ii*) *Indirect division.* Here the nucleus undergoes campli-cated changes before it is divided into two daughter nuclei.

AMITOSIS

The amitosis or direct cell division is the means of asexual reproduction in unicellular organisms like bacteria and pro-tozoans and also a method of multiplication or growth in foetal membranes of some vertebrates. In amitosis type of cell division the splitting of nucleus is followed by cytoplasmic constriction (Fig. 7.1).

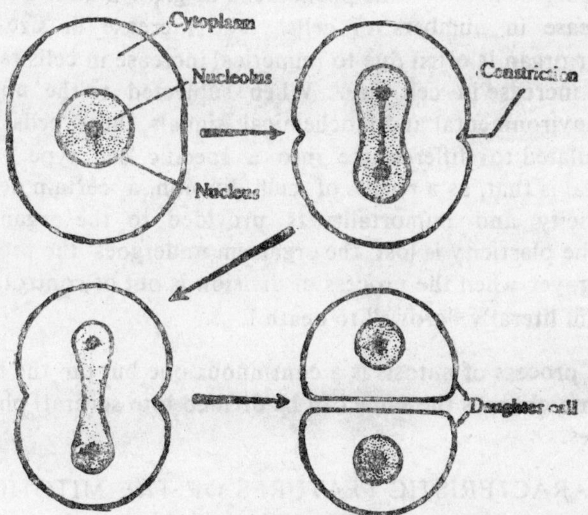

Fig. 7.1. Different stages showing amitosis.

During amitosis the nucleus elongates first and then assumes dumbbell-shaped appearance. The depression or constriction increases in size and ultimately divides the nucleus into two nuclei. The division of nucleus is followed by the constriction of cytoplasm which divides the cell into two equal or approxi-mately similar halves. Therefore, without the occurrence of any nuclear event two daughter cells are formed.

MITOSIS

In mitosis, one cell divides into two, both of which are genetically identical to each other and to the parental cell. In other words, both the chromosomes and the genes are the same in all of the cells. This type of cell division is necessary if the organism and/or cell is to persist and survive. There are many facts to the need for cell division, and they vary depending on the particular biological function. For example, in liver tissue when some cells die or are damaged, others divide and provide new cell to replenish those that are lost. Other cells in the organism actually grow (increase in size), and it is possible that when they reach a point where there is too much cytoplasm for a given amount of nuclear material, they divide, and the whole process begins again. The phenomena of growth also involves an increase in numbers of cells. The increase in size of a tissue or organ is often due to numerical increase in cells rather than an increase in cell size. When subjected to the appropriate environmental and biochemical signals, these cells may be stimulated to differentiate into a specific cell type. The sum total is that, as a result of cell division, a certain degree of plasticity and immortality is provided to the organism. When the plasticity is lost, the organism undergoes the process of aging, yet when the process of division is out of control, the organism literally "grows" to death !

The process of mitosis is a continuous one but on the basis of certain changes the same can be divided into several phases or stages.

CHARACTERISTIC FEATURES OF THE MITOTIC PHASES

In cell division, the first visible changes occur in the nucleus, when the chromonemata condense into *chromosomes* (Gr ; coloured bodies). This stage is called the *Prophase* (Gr ; early figure).

Then the nuclear boundary disappears and the chromosomes line up in or near one plane in the cell. This is *Metaphase* (Gr ; middle figure).

Following this each chromosome separates into two parts, and these two parts migrate away from each other to the ends of the cells. This is *Anaphase* (Gr ; up and down figure).

Next the chromosomes at each end of the cell reconstitute a nucleus. This is *Telophase* (Gr ; final figure).

Then the cytoplasm divides into two daughter cells, bringing about the *Interphase* again.

Interphase

The period of metabolic activity during which cell division is not in a process, has been called the 'interphase.' This is frequently referred to as the 'resting phase'. But this term is not appropriate because the cell is metabolically most active at this stage that is why *Berril* and *Huskins* (1936) referred it as the *energy phases*. The interphase is the period between the telophase of one division and the prophase of the new cell division. During this phase the cell does every thing except division. It is during interphase that the genes self-duplicate and carry on their function of supervising synthesis.

The chromatin granules in the nucleus are not readily distinguishable in the living cell, but may be brought out by treatment with chemicals which kill, fix and stain them. They appear at first glance to be scattered throughout the nucleus, but careful study has produced evidence that they are arranged in a definite pattern as long coiled strands and appear as a thread like net work in ordinary stained preparation.

"Nuclear sap" or, "Karyoplasm" fills the interstics between the chromosomes. One of more rounded bodies, the *nucleoli* are usually present. Between the nucleus and the surrounding cytoplasm, is the *nuclear membrane*. In the cytoplasm adjacent to the nucleus, there is a body, the *central body*, which consists of two granules or, after each granule has replicated, of two pairs of granules, the *centrioles*.

A typical cell cycle, including interphase lasts from 20-24 hrs. Interphase is the longest period in the cell cycle, and may last for several days in cells.

Interphase can be further divided into form sub phases :

(1) G_1-phase

(2) S-phase

(3) G_2-phase
(4) M-phase

G_1-phase includes the synthesis and organization of the substrate and enzyme necessary for DNA synthesis. Therefore, G_1 is marked by the synthesis of RNA and protein.

G_1 phase is followed by the S phase where the synthesis of DNA occurs.

During G_2 phase, all the metabolic activities are performed. M phase is the period of chromosomal division.

The relative lengths of these phases differ in different organisms. A human cell in culture at 37°C, completes the mitotic cycle in about 20 hours and the M-phase lasts for only one hour. Temperature and cell environment plays an important role in determining the rate of cell division. Even the nonmeristematic cells can sometimes be made to divide by changing the environmental conditions. Those cells which are not going to divide any more, have the mitotic cycle at the G_1-phase and start differentiating.

The cell shows following changes :

(1) The cell, as a whole, attains the maximum growth and possesses synthesized proteins for energy for various divisions and processes.

(2) The nuclear membrane is intact and the chromosomes are found in the form of more or less loosely coiled threads, some-what closely appressed to the membrane. In this condition of chromosomes, most of the cytologists regard them to be duplicated, while some workers are of the opinion that they are multipartite.

(3) The two centrioles, which are found at right angles to each other replicate into two each. *Mazia* (1961) has described that, if the replication is checked, then division will not take place.

(4) For the future spindle condensation of protoplasm into a coherent area of jelly-like consistency also takes place. The spindle also starts growing and pushes the centrioles apart.

(5) DNA synthesis occurs during autosynthetic interphase, when the chromosomes are dispersed.

(6) Chromocentre are also conspicuous during the interphase.

Prophase

The prophase is the longest in M-phase and may take about one to several hours. In the neuroblast cells of grasshopper it takes about 102 minutes. Various important changes in this stage are as follows :

(1) The cell tends to become spherical and increases its refractivity and turgidity by increasing its surface tension by removal of water.

(2) The nucleus imbibes water from cytoplasm, and the chromosomes get themselves organized differently. Each chromosome shows its peculiar structure with marked isolation of chromatids, each of which undergoes a regular cycle of coiling. At the beginning of division the chromosomes start to contract, get thickened and undergo their coiling. Partially this change appears to be associated with the development, at right angles to old gyres of the coils. *Sparrow* (1941) referred that total end to end contraction is approximately one fifth of the initial length. Physiologically, the coils are formed due to continuous condensation. The coils are of two types : smaller minor coils and larger somatic coils. The mode of twisting during the formation of coils and gyres is also categorized in two distinct varieties :

(*i*) *Plectonemic*. Twisted in such a way that it is not easy to isolate them, and

(*ii*) *Paranemic*. Coiled chromatids can be separated laterally.

(3) The chromatids are connected with centromere. The chromosomes are separately distributed in the nuclear cavity. The RNA and phospholipid contents gradually increase.

(4) The nuclear boundary becomes disrupted, the spindle apparatus begins to form, and the nucleolus and the centromere generally disappear. The spindle formation takes place in two ways, *viz.*,

(*i*) Single centriole divides into two daughter centrioles and on separation, astral rays appear as delicate filaments, called the spindle. The centrioles migrate along with asters until they become located at antipodal positions. This type of spindle is referred as to the *central spindle*.

(*ii*) The two centrioles are already polarized before the beginning of the division and the spindle formation takes place at metaphase. This type of spindle is known as the *metaphasic spindle*.

The mitosis in which achromatic figure and spindle are formed by centers is called as the *amphiastral mitosis* and whence centers are absent, the mitosis is called as '*anastral*'. The anastral mitosis takes place in plants.

Prometaphase

This stage follows the complete diappearance of the nuclear membrane, the chromosomes tend to aggregate in a central position in the cell, near the equator. The term prometaphase is described by *Coin* (1964). *White* (1963) defined it as the period during which the spindle is being formed and during which the chromosomes give the impression of struggling and jostling one another in an attempt to reach the equator of developing spindle. At least, in plants, this stage corresponds to the first appearance of an organised spindle. *Wilson* and *Hyppio* (1955) considered this positioning of chromosome to play an important role, both in the development and functional organization. The spindle fibers are tubular, elastic, fiberous and proteinous in nature.

Metaphase

The metaphase chromosomes are sharply defined and discrete bodies and are tightly coiled. In this stage the chromosome number can be easily counted and it is possible to recognise the different chromosomes from their sizes, shape and gross structure. In early metaphase characteristic spindle-shaped figure appears in the clear nuclear region. This consists of fibrous radiations which extend from the central broad-portion called the *equator* and converge at two points on the opposite ends of the nuclear region, called the *poles*. The chromosomes which were till now scattered in a random manner in the central nuclear portion of the cell begin to show peculiar movements and arrange themselves in a single plane midway between the two poles of the spindle, forming an equatorial plate. To be more precise it is the centromeres of the chromosomes that are lined up at the equatorial plate. Usually, the smaller chromosome occur near the center of the *equatorial plate* and the larger ones near the outer ends. It is, however, not necessary that the two homologus should be near each other on the equatorial plate. The position of each chromosome is independent of other on the equatorial plate. The metaphase chromosome is a visibly double structure and is clearly seen to be split longitudinally into two exactly identical chromatids.

The centromere plays an essential role in the development of the spindle and the separation of the daughter chromosomes. The chromosomes are so arranged at the equator that one chromatid of each chromosome faces one pole and the other faces the opposite pole. Two spindle fibres are attached to the centromere of each chromosome, one on either side of it. These connect the chromosome to the two opposite poles of the spindle and are called *chromosomal* or *tactile fibres.* Other fibres of the spindle extend from pole to pole and are not attached to the chromosomes. These are called the *continuous spindle fibres*. The spindle fibres are composed mainly of protein, some ribonucleic acid (RNA) and lipids.

Form and arrangement of the chromosomes

On the basis of their morphological study, one can divide them into three categories :

(*i*) *Straight rods or threads.* These chromosomes arise directly by shortening of the spireme thread.

(*ii*) *Loops, V's or hook forms.* These chromosomes are formed by a flexture at the middle point or near one end.

(*iii*) *Ovoidal or Spheroidal forms.* These forms arise by the extreme shortening of threads.

All the three forms mentioned above are double at metaphase due to their longitudinal splitting. The longitudinal splitting among rod shaped and thread shaped chromosomes, can be traced clearly but in spheroidal forms, it often appears as an apparently transverse constriction due to the extreme shortening of chromosomes.

The Spindle attachment of the chromosomes

The mode of attachment of the chromosomes to the spindle fibres based on the structure of chromosomes. But even though, the arrangement and the way of attachment is constant for every particular chromosome and that it is inherited from generation to generation. Chromosomes are attached to spindle fibres by centromeres. The chromosomes lacking centromere fails to attach with spindle fibres. The way of their attachment may be of two types :

(*i*) Terminal or telocentric
(*ii*) Nonterminal or atelocentric

In terminal attachment the chromosomes may have their connection with spindle fibres at the free end. Nonterminal attachment may be at the middle point (*median*) or at an intermediate point *submedian* or sub *terminal* (Fig. 7.2).

Anaphase

In anaphase the pairs of centromeres move apart along the spindle and carry one daughter chromosome of each pair to opposite poles. The spindle eventually grows longer. The movement of chromatids is complex. First separation of chromatid from centromere takes place and then a flow of current along the spindle governs their poleward movement. This stage lasts for very short duration, varying between 6 to 12 minutes. In the late anaphase the zone between the two sets of chromosomes or the equatorial region gradually increases. It seems that the fibers stretch and are named *interzonal fibres*. The pushing body looks like a gel which pushes the chromosomal sets towards respective poles. During the poleward movements the chromosomes assume peculiar 'J'

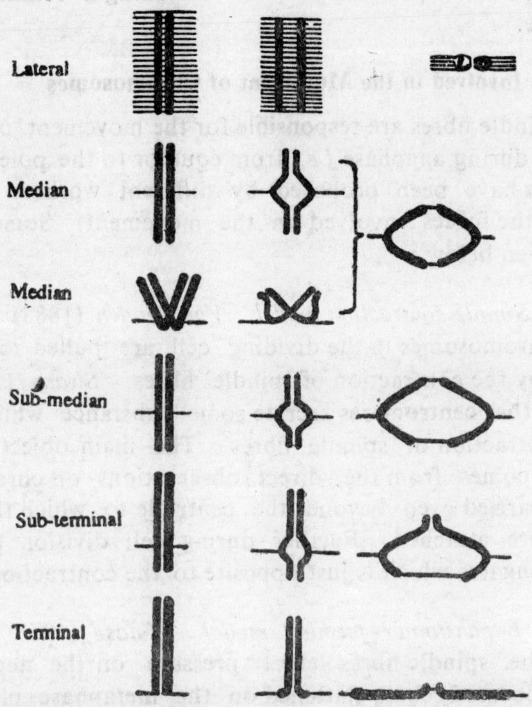

Fig. 7.2. Showing the movement of chromosomes according to their mode of attachment, during anaphase.

or 'V' shapes, depending upon the position of centromere (Fig. 7.2). At this stage, 'J' are called *heterobranchials* and the 'V' chromosomes as *isobranchials* (Fig. 7.3).

Movement of Chromosomes

The movement of the chromosomes are controlled by spindle fibres. Actually two processes may be at work; a continuous enlargement and elongation of the spindle and the shortening of the chromosomal fibres. Since the fibres shorten without becoming thicker, the process probably involves the removal of water or other molecules from fibres. A band of 'interzonal fibres' is often seen for a time after the separation has been accomplished, connecting the chromosomes which have pulled apart, and often including a remnant of the spindle.

Forces Involved in the Movement of Chromosomes

Spindle fibres are responsible for the movement of chromosomes during anaphase *i.e.*, from equator to the pole. Several models have been proposed by different workers to explain about the forces involved in the movement. Some of them are given below :

(*i*) *Simple contraction model. Van Benden* (1883) suggested that chromosomes in the dividing cell are pulled towards the poles by the contraction of spindle fibres. *Swann* (1962) described that centromeres secrete some substance which causes the contraction of spindle fibres. The main objection to the theory comes from the direct observations on chromosomes being carried even beyond the centriole to which the spindle fibres are attached. Further during cell division the whole cell elongates which is just opposite to the contraction model.

(*ii*) *Expansion or pushing model. Watase* (1981) suggested that the spindle fibres exert pressure on the nucleus and chromosome become flattened on the metaphase plate. The chromosomal fibres become attached to the chromosomes, which are now pushed towards the opposite poles.

(*iii*) *Contraction and expansion model* (*Traction model*). *Belar* proposed that initial separation of daughter chromosomes is an autonomous process but further separation is the result of both contraction or expansion of various parts of spindle. The chromosomal fibres which extend from the centromeres of chromosomes to the poles of spindle contract and pull the attached chromosomes to the poles. The interzonal fibres present between the separating daughter chromosomes expand and push the daughter chromosomes towards opposite poles.

(*iv*) *Equilibrium dynamic model.* This provides the most convincing explanation about the possible mechanism of chromosome movement. *Inove* and *Sato* (1967) have described the occurrence of an equilibrium between a large pool of monomeres which constitute the proteins of microtubules. During polymerization some hydrophobic interactions occur in between the nonpolar groups of protein monomeres and some structured water is dissociated.

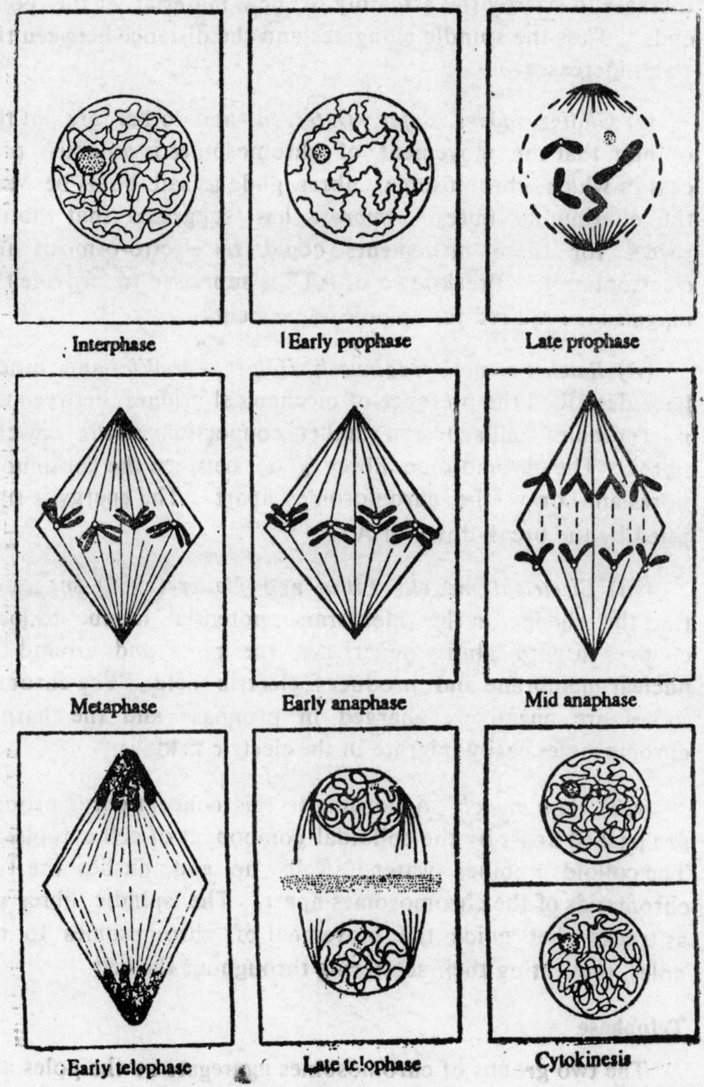

Fig. 7.3. Mitosis in a plant cell.

During chromosomes movement both contraction and elongation of spindle fibres occurs by the substraction or addition of new monomeres to the fibres. During anaphase chromosomal fibres contract by the delation of monomere from the polar end of the fibres and the continuous fibres increase in size by the assembly of new material at the polar ends. Thus the spindle elongates and the distance between the poles increases.

(*v*) *Gliding model. Bajer, Ostergen* and other are of the opinion that the movement of chromosomes is an active process in which chromosomal fibres glide as sail boats between the continuous fibres. *Ambrose* has suggested that motive forces for these movements could be electro-osmosis and electrophoresis. Breakdown of ATP is supposed to provide the magnitude required for spindle movements.

(*vi*) *Ratchet model. McInuosh, Helper, Van Wie* and others have described the presence of mechanical bridges between the microtubules similar to the ratchet connections in the muscles fibres. The chromosome fibres glide between the continuous fibres and draw the chromosomes apart. The energy is supplied by the break-down of ATP.

(*vii*) *Electrical model. Lillie* and *Coller* (1936) suggested that the change in the membrane potential is due to local changes in permeability occurs near the poles and around the nuclear membrane and produces electric field. The chromosomes are negatively charged in prophase and the charged chromosomes easily migrate in the electric field.

(*viii*) *Push model.* According to this concept chromosomes are pushed apart by the colloidal component of cell cytoplasm. The colloid imbibes water, swells up and pushes the two chromatids of the chromosomes apart. The spindle fibres act as tracks that guide the movement of chromosomes to the poles, preventing their scattering throughout the cell.

Telophase

The two groups of chromosomes aggregate at the poles and on arrival they progressively lose their chromaticity. Decondensation takes place. Reformation of the nuclear membrane by certain unknown process takes place. Perhaps new meterial is synthesized by RNA during telophase or it may be possible that the endoplasmic reticular system may be giving rise to a new membrane round the nucleus. All chromosomes become uncoiled. The spindle fibers are absorbed in the cytoplasm. The nucleolar organizer or SAT zone again forms the nucleolus. Finally, these changes result into two daughter nuclei which are equivalent to the parent nucleus in all respects.

Cytokinesis

Nuclear division or mitosis, as it is called, is followed by the division of the cytoplasm. While the daughter nuclei are being organised at the poles, the mitotic spindle disappears except at the equator, where the continuous spindle fibres become more dense. This region is now called the *phragmoplast*. According to *Porter* and *Machade* (1960) the formation of cell plate is initiated by migration of tubular elements of the endoplasmic reticulum toward the interzonal region of the spindle where they spread out to form a close lattice along the mid line. The droplets appear in phragmoplast and contain pectic substances which fuse to form in the middle of dividing cell a partition known as *cell plate*. More droplets are added to the cell plate to form the middle lamella which now begins to extend outwards until it reaches the outer walls of the original cell. Primary cell wall material is now deposited on either side of the middle lamella. This divides the cytoplasm into two roughly equal parts. Cell division is now complete and two cells are formed. The division of the cytoplasm is called *cytokinesis*.

The cytokinesis varies in plants and animals. In animal cells, the separation takes place by a process of invagination of the cell membrane in the equatorial region. The cytoplasmic division takes place at the late anaphase stage. The invagination furrow originates as an indentation, gradually deepens to cleave the cells into two daughter cells which may be of equal size or may be quite unequal. *Gray* (1931) reported about two types of cleavages and named them as *disjunctive* and *astral cleavages.* The disjunctive cleavage occurs in those cells where the daughter cells proliferate, *e.g.*, leucocytes or in cells of the tissue culture. Astral cleavage is characteristic of the marine eggs where the zygote cells go on divinding to result in the formation of large number of bastomeres intact.

Role of Centriole During Mitosis

The centriole acts as an epicentre for the development of the central spindle. Whether the centriole produces the spindle fibres or serves to direct their formation or on the other hand is entirely passive in the process, can not be unquestionable deduced from all the informations at hand. According to *Cleveland* (1957) the spindle fibres and astral and chromosomal fibres are recognizable first in the close vicinity of the centrioles. The completely developed achromatic apparatus then functions to separate daughter chromatids towards respective poles. With relation to the mitosis, *Cleveland* concluded two important points : (*i*) relative temporary independence of the achromatic and chromatic apparatus as regards apparent duplication and function ; (*ii*) final dependence of the cell upon the integrity of these two systems and their united function.

Duration of Mitosis

The time required for mitosis differs with species and environment. Temperature and nutrition, in particular, are important factors. The entire sequence of phases may be completed in 6 minutes to many hours. Normally the entire cycle of cell division takes approximately 18 hours ; about 17 hours for the interphase. Different phases of mitosis are of different duration. Anaphase is the shortest, the prophase and telophase the most prolonged, and the metaphase of intermediate duration (Table 7.1).

SIGNIFICANCE OF MITOSIS

Equal Distribution of Chromosomes

The essential feature of mitosis is that chromosomes are distributed equally among the two daughter cells. With every

Table 7.1. Duration of Mitotic Stages (in minutes).

Example	*Temp °C*	*Pro-phase*	*Meta-phase*	*Ana-phase*	*Telo-phase*
Mouse spleen	38	21	13	5	4
Chicken (mesen-chime cells in tissue culture)	39	30-60	2-10	3-7	2-10
Frog (Fibroblasts in culture)	20-24	32	20-29	6-11	
Grosshopper (Neuroblasts)	38	102	13	9	57
Sea Urchin Embryo	12	19	17	2	18
Onion (root tip)	20	71	6.5	2.4	3.8
Pea (root tip)	20	78	14.4	4.2	13.2

cell division there is a division of chromosomes. The constant number of chromosomes in all the cells of the body is the result of mitosis.

Surface-volume Ratio

Mitosis restores the surface-volume ratio of the cell. A small cell has a greater amount of surface available in relation to volume that a large cell. As the cell increases in size the available surface area in relation to increased volume becomes less. By undergoing division the cell becomes smaller in size and the surface-volume ratio is restored.

Nucleoplasmic Index

The growth of a multicellular organism is due to mitosis. A cell cannot grow in size to a large extent without disturbing the ratio between the nucleus and the cytoplasm. After a particular size has been reached the cell divides to restore the nucleoplasmic index. Thus growth takes place by an increase in the number of cells, rather than by increase in the size of the cells.

Repair

The repair of the body takes place because of the addition of cells by mitosis. Dead cells of the upper layer of the epider-

mis, cells of the lining of the gut, and red blood corpuscles are constantly being replaced. It is estimated that in the human body about 500,000,000,000 **cells are lost daily.**

ABNORMALITIES IN MITOSIS

Under unfavoriable conditions the mitotic divisions become defective and this various abnormalities are formed when exposed to physical and chemical agents like temperature, radiations, nacrotics and enzyme inhibitors, etc. Some abnormal mitotic divisions are given under the following heads :

C-Mitosis (abnormal spindle formation)

Brachet (1975) has described that colchicine inhibits mitosis by disorganizing spindle formation. The results are the formation of polyploid cells, and reduplication of chromosomes. Besides, during mitosis, some time cell nucleus is either deformed into a single spheroid compact mass (pycnosis) or becomes broken down (karyorrhexis).

Cytasteral Mitosis

Wilson (1901) noticed the formation of numerous asters in the cytoplasm of unfertilized eggs called cytasters by placing them in hypertonic sea water. In the eggs of marine invertebrates, production of many small cytoplasmic asters is of frequent occurrence. The intermixing of nucleoplasm and hyaloplasm seems to be necessary condition for the aster formation. *Costello* (1940) showed that asters can be produced only if germinal vesicle has broken down.

Multipolar and Catenar Mitosis

Mitotic division showing several spindles and centrosomes are common in many protozoans and fish eggs. Multipolarity is usually caused by uneven division of centrosomes as well as irregular distribution of chromatids of the different spindles. It results in the formation of cells as aneuploidy (uneven chromosome number). Catenar mitosis have been described by *Dalcq* and *Simon* (1932) in amphibian eggs. During this, dividing cell forms a large number of asters without spindles.

for chromosomes to link them. It suggests that asters are capable of dividing in the absence of spindle or nucleus in an autonomous way.

Achrosomal Motosis

Sometimes, mitosis is induced by various agents without use of nucleus or chromosomes. *Briggs et al* (1951), in fertilized eggs with heavily X-rayed sperms, removed the egg's maturation spindle by pricking and sucking. In this way, he obtained very nice non-nucleate blastulae. Similarly, *Stauffer* (1945) obtained normal non-nucleated Axolotl blastula.

Anastra mitosis

In plant cells and many oocytes, generally asters around the centrosomes are absent, thus producing anastral mitosis. Batallion and Tchou Su (1933) described in details the anastral mitosis in amphibian interspecific hybrids.

ACHROMATIC FIGURE

Achromatic figure consists of spindle fibers and astral rays (Fig. 7.4). Astral rays are developed just before the disintegration of nuclear envelope during late prophase, while the formation of spindle rays can be traced only after the disappearance of nuclear membrane.

The Spindle

In fixed preparations of dividing cells, fibers are seen forming a spindle shaped body lying between the centrioles. It is generally believed that spindle fibers are not visible in normal living cells. They can be produced by adding acid to the medium, and, on removing the acid, the fibres disappear. As most fixatives contain acid, it is stated that the fibers are artifacts. There has been considerable controversy as to the reality of fibers and various theories have been put forward to explain their presence in fixed material. *Schrader* in (1944) observed the spindle fibers in living cleavage cells of mite and certain flagellates.

The spindle is superficially similar to the spindle shaped area found between the poles in the magnetic field. The

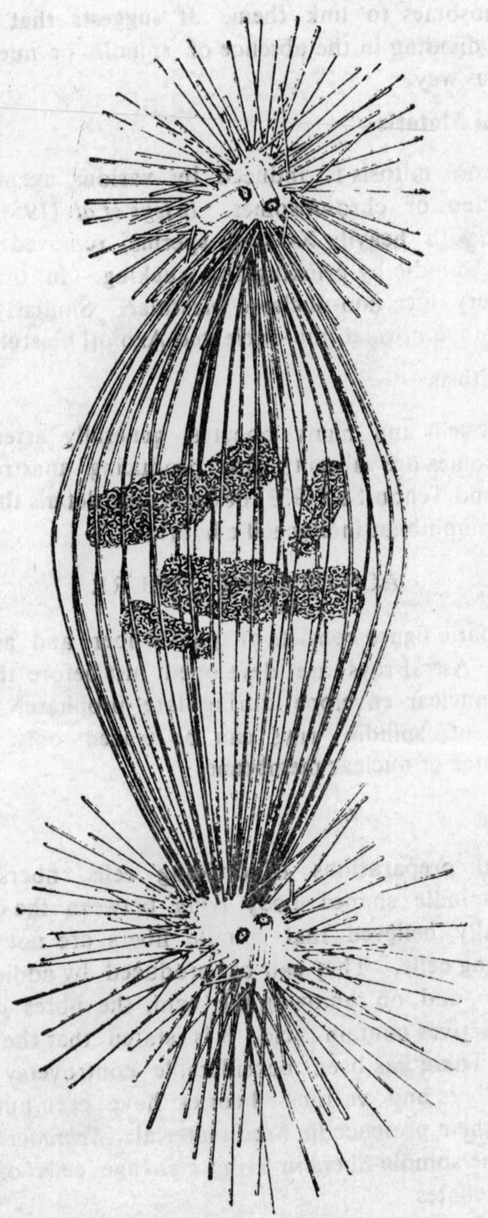

Fig. 7.4. Showing achromatic figure.

spindle consists of two kinds of fibers which have been supposed to differ widely in functional significance.

Those spindle fibers extending from the poles to the chromosomes are called 'half spindle', incomplete fibers or discontinuous fibers or chromosomal fibers. The remaining spindle fibers extend without interruption from pole to pole and are called the continuous fibers or direct or complete fibers.

It is believed that the spindle is usually formed in two parts. A part arises between the centrosomes before the disappearance of the nuclear membrane ; it is cytoplasmic in origin and is called the central spindle. As the nuclear membrane disappears the central spindle moves into nuclear region. Then the nuclear material undergoes gelation around the central spindle to from the 1/2 spindle fibers or chromosomal fibers. In the metaphase the chromosomes are attached to the 1/2 spindle fibers at the equator. *Schrader* suggested that the spindle may consist of protein molecules arranged in parallel lines.

It appears that the spindle is organized from a pre-existing pool of protein. It is postulated that protein monomers aggregate to from microtubules of the spindle fibres. The spindle fibres are thus oriented polymers. It has been proposed that the continuous fibres are elongated by the insertion of subunits within the fibre. This leads to the elongation of the spindle. Conversely, the shortening of chromosomal fibres takes place by loss of subunits. This depolymerization of fibres takes place in the region of the spindle poles. Thus the fibres do not contract like a rubber band, but shorten by loss of material at one end. This is in agreement with observations that no change in the thickness of fibres takes place during shortening and elongation.

There is evidence that the microtubules of the chromosomal fibres originate at the kinetochore. Microtubules are also found in cells which have no centrioles. *e.g.*, those of higher plants. Here the microtubules end in the pole region and are not attached anywhere.

THE ASTERS

Astral rays are developed by the centrosome present in the visicinity of nuclear region. Astral rays together with the spindle from the *amphiaster*. In early stages the astral rays are straight, simple and few in number. But in latter stages they rapidly elongate in all directions, and increase in number. In certain cases the astral rays of both asters cross one another in the equatorial region outside the spindle. Sooner or later, the crossing of the rays disappear and rays adjust themselves in such a way that their free ends move towards the opposite asters and meet in the equatorial plate thus giving the impression of continuous fibres. These continuous fibres which are formed by the fusion of astral rays are lying in the region outside the spindle. In has been suggested that the long protein molecules of the protoplasm become arranged in rows parallel to the lines of protoplasmic flow and thus constitute the astral rays.

MEIOSIS

Mitosis is a device of increasing the number of cells. During asexual reproduction it is the means of multiplying the individual. Therefore, in asexually reproducing organisms, set of characters are transferred unchanged from parents to offspring, except of course accidents. There is uniparental inheritance due to mitosis. But in sexually reproducing organisms, gametes are produced as a result of meiosis. Two gametes unite to give rise to a zygote which divides mitotically to form the individual. The gametes from different parents unite to form a zygote, thus ensuring biparental inheritance. This results into a blending of characters from two parents into their offsprings and ensures variation which is important for evolution. Gametes contain half the number of chromosomes, as compared to somatic cells. Union of two gamates restores the somatic number of chromosomes in the zygote. Gametic union results into the mixing of cytoplasms of two cells which is followed by the fusion of two nuclei. During meiosis or at the time of gamete formation, therefore, halving of the number of chromosomes or the amount of DNA is achieved by two successive divisions of cytoplasm and nuclei (meiosis I and meiosis II) accompanied by only one replication of chromosomes. For example, each somatic cells of man has 46 chromosomes. Somatic cells are derived from the zygote due to mitotic divisions. Zygote is the result of a fusion of two gametes, each of which contain only 23 chromosomes. Both the male and female gametes contain the same amount of DNA and equal number of morphologically similar chromosomes. The zygote and somatic cells derived therefrom, therefore, contain 23 pairs of homologous chromosomes, one of each pair derived from the male parent through the male gamete and the other from the female through the female gamete. The members of each pair are homologous to each other and nonhomologous to the other chromosomes. In some organisms, the chromosome complements of male and female cells are slightly different and this forms the basis of sex differentiation. Those cells which possess two sets of chromosomes, are called *diploid* and they have 2n or dipoid or unreduced or zygotic number of chromosomes. Gamets, on the other have n, or *haploid* or reduced or gametic number of chromosomes. In sexual reproducing organisms the diploid and haploid phases of the life cycle alternate and this phenomenon in called *alternation* of *generation*. In plants, there is an alternation of sporophytic (2n) and gametophytic (n) generation. Reduction division heralds the beginning *of* gametophytic generation and fusion of gametes that of the sporophytic phase. Thus, meiosis can be studied at the time of gamete formation which occurs in specialised organs *i.e.*, the embryo sac and the pollen sac. A single cell undergoes meiosis in the embryo but in the pollen sac a large number of spore mother cells divide meiotically, almost synchronously, to produce haploid pollen grains. It is, therefore, easier to study meiosis in anthers, similarly in testes in animals.

Meiosis consists of two divisions which take place one after the other, during which the number of chromosomes is halved. The two divisions are known as the first meiotic division or heterotypic division and the second meiotic division or homotypic division. The first meiotic division has the same stages as mitosis. These stages are known as prophase I, prometaphase I, metaphase I, anaphase I and telophase I. Except for prophase, the other stages are similar to those of mitosis. Following telophase I is a period of interkinesis or rest. The second meiotic division consists of prophase II, metaphase II, anaphase II and telophase II.

HETEROTYPIC OR FIRST MEIOTIC DIVISION

This is the first half of meiosis and is true reductional in nature. Here the diploid cell divides into two haploid daughter cell. As stated already that the phases of this division are called prophase I, Metaphase I, Anaphase I and Telophase I.

Prophase I

It is the longest stage and its duration makes from a few hours to a few days. The nucleus starts increasing in volume and this increase is much then the mitotic increase to some extent. This is due to an increase in hydration.

Nobel and *Ruttle* (1936) have called this stage as the *premeiotic spiral prophase*. Prophase I is further divided into six substages.

(*i*) Pre-leptotene
(*ii*) Leptotene
(*iii*) Zygotene
(*iv*) Pachytene
(*v*) Diplotene
(*vi*) Diakinesis

(*i*) *Preleptotene*. As noted previously, replication of DNA has already taken place during interphase. The chromosomes at the beginning of meiosis are, therefore, not single but double strands. The beginning of meiotic prophase is marked by an increase in nuclear volume. During preleptotene the chromosomes are very thin and cannot be easily seen, except sometimes for the sex chromosomes.

(*ii*) *Leptotene* ("*slender thread*") *or Leptonema*. During this stage the chromosomes become more distinct, and their double nature is seen in many organisms. Leptotene extends from the time single unpaired chromosomes are first seen to the time when they begin to undergo homologous pairing. Under the light microscope the chromosomes appear as slender threads bearing a series of granule-like structures called *chromomeres.* Whole mount electron microscope preparations show that leptotene chromosomes have an axial filament to which chromatin fibres are attached as a series of lateral loops. In certain regions these loops appear in bunches, and form the chromomeres of light microscopy. The axial filaments are

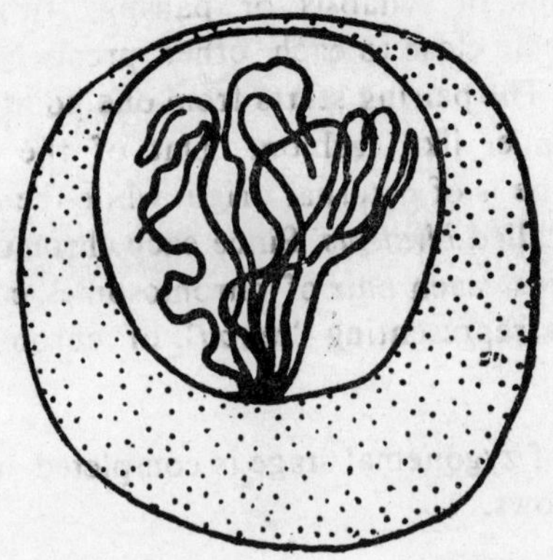

Fig. 7.5. Leptotene chromosomes arranged in a 'bouquet'.

apparently composed of two independent protein fibrils, one for each chromatid. The later become transformed into the lateral elements of the synaptonemal complex. In the leptotene cells of the quail there is evidence of premeiotic pairing. Two homologous chromosomes become apposed to each other, chromomere for chromomore, even though no synaptonemal complex is formed. In leptotene cells there is a tendency for the ends of homologous chromosomes to be associated with each other at the nuclear membrane, where they are attached.

Leptotene chromosomes may be irregularly arranged, or may be polarized toward the centrioles forming a '*bouquet*' (Fig. 7.5). Electron microscopic studies have shown that bouquet formation results when a group of chromosomes is attached close together on the nuclear membrane.

During leptotene the cytoplasm has many polyribosomes, but endoplasmic vesicles are few. In maize (*Zea mays*) the nucleus increases in size during leptotene, reaching its maximum size during pachytene. This increase in size is correlated with synthesis of RNA and protein.

(*iii*) *Zygotene* ("*mating thread*") *or Zygonema*. The chromosomes becomes much shorter and thick and number of coils is also increased. Homologous chromosomes begin to pair by force of attraction. In synapsis or pairing, two homologous chromosomes come close to each other precisely chromomere to chromomere. The pairing starts from one point and proceeds onward in a zipper like fashion. Out of the two synaptic chromosomes, one is of paternal origin while the other is maternal. These are called *bivalents*. Since each chromosome is made of two chromatids, each pair of chromosomes consists of four chromatids thus representing '*tetrad*' of chromatids at this stage.

The pairing of zygonemal stage is completed in three different ways, as follows.

(*i*) *Proterminal synapsis*. The two homologous chromosomes start pairing at the terminals, which gradually then progresses towards the centromere.

(*ii*) *Procentric synapsis*. The pairing starts near the centromere and progresses onwards on either side.

(*iii*) *Random synapsis*. The pairing may be at random at many points which then finally fuse together.

Two views have been advocated to explain the mode of initiation of synapsis. *Darlington* (1937) proposed precocity theory which explains that the chromosomes undergo synapsis due to singleness, but this theory does not explain the extra synthesis of DNA and duplication at leptotene stage. Another view is proposed by *Sax* (1932) which is referred as the retardation theory. It explains that pairing takes place due to the retardation or cessation of metabolic activities of the cell. During zygonema, the nucleus remains intact and the centriole moves apart initiating the aster formation.

(*iv*) *Pachytene* "*thick thread*" *or pachiynema*. It generally begins when the pairing of chromosomes ends. It represents one of the longer substages of prophase I and is characterised by the following important events :

When the pairing has completed, the opposition of homologues becomes so tight that it is difficult to identify two separate chromosomes. Within the major coils, and at right angles to them smaller minor coils appear. At the height of pachytene, the pairing chromosomes twist about one another as relational coils and each soon longitudinally splits if no splitting had occurred. This coiling is further complicated by a coiling of two chromatids of each chromosome around each other. This coiling put the chromosomes under considerable strain. This means that at this stage, each pachytene element consists of four chromatids. The chromatids of each homologue are called *sister chromatids*. Formerly this element of four chromatids was called a *tetrad*-a name which has now been replaced by *bivalent*. When the chromosomes have longitudinally splitted, transverse breaks occur at the same level in two of the adjacent homologous chromatids. Soon both segments of the chromatids interchange and fuse together with those of homologous. In this process which is called *crossing over*, portions of the two chromatids with their genes are interchanged and two remain intact. The new chromatids thus interchanged will be mixed.

After this interchange, the parts begin to separate, repalling each other. However, this separation is not at all complete since the homologous chromosomes remain joined at their point of interchange. This point where the two are joined is called the *chiasma*. This chiasma is an immediate consequence of crossing over and are found in all plants and animal with few exceptions (*e.g.*, males and females silk-worm). The number of chiasmata is variable but for one bivalent atleast one chiasma is found. Bivalents with 2, 3 and 4 chiasmata are rare. The highest number of chiasmata has been observed in the long chromosome of the broad bean, *Vicia faba*, where there are probably 12 chiasmata in a bivalent. The average number of chiasmata is known as *chiasma frequency*. This number seems to depend upon the length of the chromosome. Short acrocentric chromosomes have a greater number of chiasmata than metacentric chromosomes. In small chromosomes, chiasmata are localized at the distal ends. Nucleus remains as such, being somewhat larger than the proceeding substage.

Cytomixis. It is a botanical term which refers to the passage of chromatin from one microsporocyte (pollen mother cell) to another during early prophase. This phenomenon is so widespread in the normal species of hybrids that one wonders if it is a general phenomenon.

(*v*) *Diplotene or Diplonema.* The separations of homologous chromosomes initiates diplotene. The synaptic forces of attraction between them lapse due to breakage at one more point, so that the homologous chromosomes uncoil and start separating. But the separation is nonetheless incomplete since the homologues are in, contact at one or more points where the crossing over has already taken place. These points of contact are known as chiasmata (sing-chiasma ; meaning cross). The chiasma is the morphological equivalent of genetic crossing over. Their number and position varies in the same pair of chromosomes and in different cells of the same individual.

By the end of diplotene, the chiasmata begin to move along the length of chromosomes from the centromere towards the end. This displacement of chiasmata is termed as terminalization. When the terminalization of chiasmata is complete the homologous chromosomes are kept in contact by terminal chiasmata. The degree of terminalization is generally expressed as a coefficient of terminalization (T)

$$T = \frac{\text{Number of terminal chiasmata}}{\text{Total number of chiasmata}}$$

The average number of chiasmata in bivalents is known as frequency of chiasmata (fq).

$$\text{Frequency (fq)} = \frac{\text{Total number of chiasmata}}{\text{Total number of bivalents}}$$

According to *Darlingtion* two types of repelling forces operate on the chromosomes at diplotene. One of the forces is electronegatively charged and operates on the surface of the chromosome throughout its length and the other with electropositive charge is localized on the centromere. The former controls the repulsion of the chromosomes and the later causes distal movement of the chiasmata.

Diakinesis

At this stage the chromosomes are best seen as very short and thick bivalents. The contraction is nearly maximum and the chromosomal pairs are well spread throughout the cell as though by mutual repulsion. The chromatids of each chromosome are very close to each other and their identity is not distinguishable. Terminalization may be completed in the substage. The bivalents migrate to the periphery of the nucleus. The nucleus generally disappears during this stage but may persist in reduced size until anaphase. The nuclear membrane disrupts and eventually disappears thus all the chromosomes are set free in the cytoplasm. The spindle fibres are organized with an establishment of the pole.

Prometaphase

Spiralization reaches its virtual maximum in this stage ; the nuclear membrane disappears, and the chromosomes become arranged at the equator of the cell to begin the metaphase. Nuclear membrane disappears and spindle and astral rays are formed during this stage.

Metaphase I

The bivalent chromosomes arranged themselves in the equatorial plane so that the two members of each homologous pair are found with their centromeres directed toward opposite poles. Whereas in mitosis there is a single centromere which unites the chromosomes, in meiosis there are two centromeres. Each homologous chromosome has a centromere to which the daughter chromatids are united. The repulsion of the centromere is accenuated, and the chromosome is found on the verge of dividing.

Anaphase I

In this the chromosomes are pulled apart by the movement of their centromeres which remain undivided, with the result

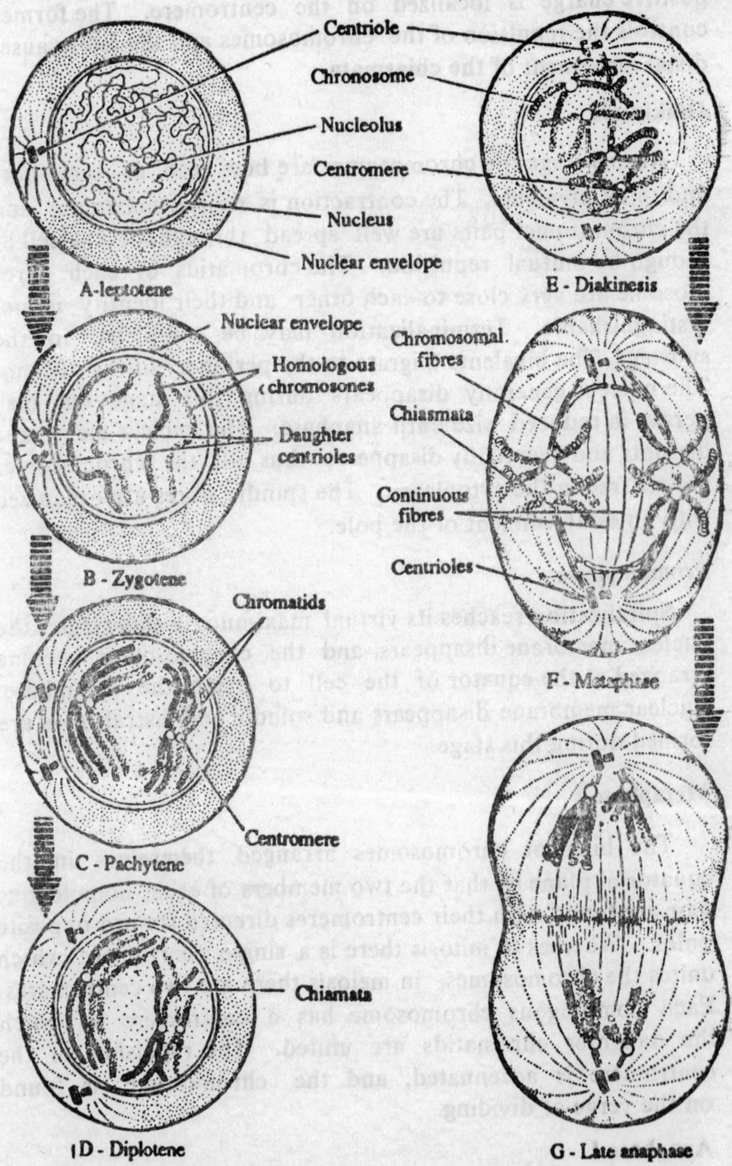

Fig. 7.6. Different stages of first meiotic division.

that each tetrad breaks up into two component pairs of chromatids or dyads. The dyads separate and move along the tactile spindle fibres to the opposite poles. **This is the true reduction division.**

Telophase I

The group of chromosomes at each pole forms a nucleus. Each daughter nucleus has half the number of chromosomes as compared to the parent nucleus. Unlike the mitotic telophase each chromosome consists of two distinctly separated chromatids united only at their centromeres. This is followed by the formation of a cell plate between the separated sets of chromosomes, dividing the mother cell into two daughter cells. Sometimes, cytokinesis fails to occur after the first meiotic division. In such cases the second meiotic division takes place in the common cytoplasm of the mother cell (Fig. 7.6).

Interkinesis

After telophase, two haploid daughter nuclei or cells sometimes undego typical resting stage as in mitosis. This intervening stage between first telophase and the beginning of second prophase is termed as interphase or interkinesis. This stage is either very short or may be entirely absent, if present, the coiling of the chromosomes persists. No increase in DNA takes place between the I and II meiotic divisions. The occurrence of an interphase between the two division suggests that additional RNA and protein sythesis is required to carry meiosis through to completion.

HOMOTYPIC OR SECOND MEIOTIC DIVISION

Before the start of this division, most of the activities of heterotypic division cease for some time. This resting stage is termed as the interkinesis. Second Meiotic division involves following steps :

Prophase II

This stage superficially resembles that of the mitosis prophase with following exceptions :

At the early prophase II, the two chromatids of each dyad looks like X, since they are conjoined by a common centromere.

The four arms are widely separated. There is no rational coiling. The X-shaped dyads are quite longer than at telophase. The chromonemata are still not completely coiled. The genetic constitution of the two chromatids of each dyad depends upon the kind and number of cross overs which took place in the phase I. If there is no crossing over, the dyads would consist of two identical sister chromatids of either paternal or maternal origin. At the end of the prophase II the nucleolus disappears. The nuclear membrane disappears and the acromatic figure is developed.

Metaphase II

As usual, all the chromosomes arrange themselves on the equator for a short duration. The centromeres touch the equator but their arms radiate out in different directions. Later on the centromere in each dyad divides into two sister centromeres.

Anaphase II

Sister centromeres separate to the poles, pulling with them the chromatids to which they are attached. The chromatids, however, become much thick and stout.

Telophase II

The chromosomes at each pole uncoil and thin out to form the nuclear net. Each group gets surrounded by a nuclear membrane. Nucleolus reappears. Thus two nuclei are recognised in each cell. This is soon followed by cytokinesis and two cells are formed from each haploid daughter cell. Thus as a result of meiosis four cells are produced, each with a haploid set of chromosomes, *i.e.*, each contains just one members of each homologous pair (Fig. 7.7).

Cytokinesis. Sometimes meiosis I is followed by cytokinesis and sometimes it is deffered until the end of meiosis II. The details of the process are same as during mitosis. Cell plate formation results into a tetrad of cells with reduced chromosome numbers. The arrangement of cells in a tetrad is

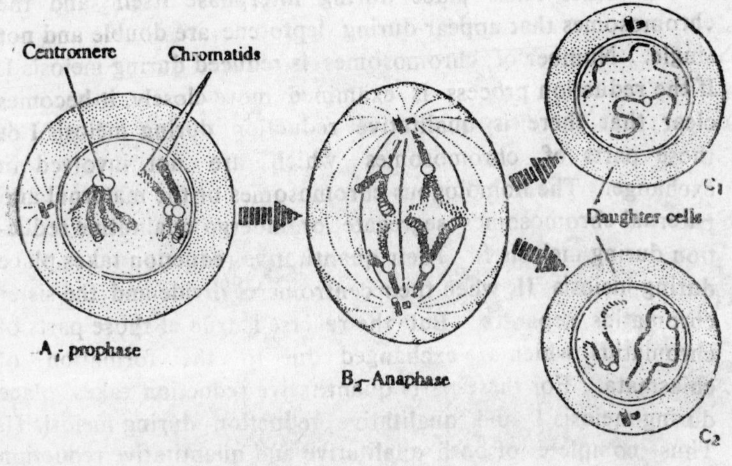

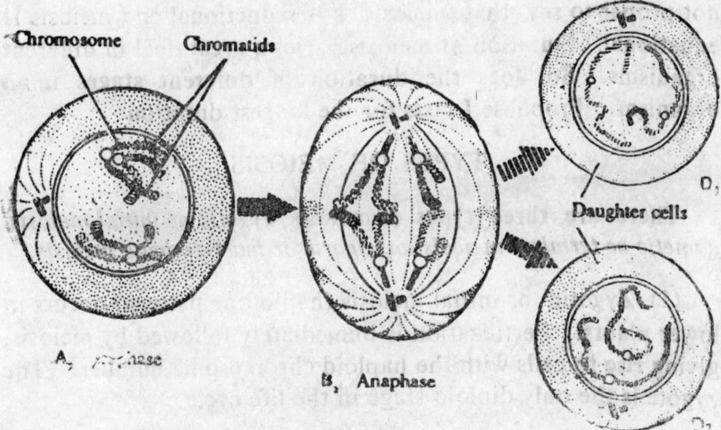

Fig. 7.7. Different stages of second meiotic division.

different in different organisms, but is characteristic of the species.

Like mitosis, meiosis too is a dynamic process and its different stages merge into subsequent ones. There is no sharp demarcation between different stages. DNA synthesis takes place during interphase, prior to prophase I. Duplication of chromosomes takes place during interphase itself and the chromosomes that appear during leptotene are double and not single. Number of chromosomes is reduced during meiosis I. If the reduction process is examined more closely, it becomes clear that there is qualitative reduction during meiosis I of those parts of chromosomes which are not involved in exchange. The homologous chromosomes or the maternal and paternal chromosomes segregate or undergo qualitative reduction during meiosis I. Their quantitative reduction takes place during meiosis II, when their centromeres divide and the sister chromatids separate. But the reverse is true of those parts of chromatids which are exchanged due to the formation of chiasmata. For these parts quantitative reduction takes place during meiosis I and qualitative reduction during meiosis II. Thus, complete of both qualitative and quantitative reduction takes place as a result of meiosis I and II and none of them alone can bring about complete reduction. It is, therefore, not proper to say that meiosis I is reductional and meiosis II equational. Duration of meiosis varies a great deal in different organisms. So does the duration of different stages in an organism. Prophase I lasts for the longest duration.

TYPES OF MEIOSIS

There are three types of meiosis, *zygotic or initial meiosis, gametic or terminal meiosis and sporic or indeterminate meiosis.*

(1) Zygotic or initial meiosis (haplotone pattern) occurs in lower plants. Fertilization is immediately followed by meiosis, giving rise to cells with the haploid chromosome number. The zygote is the only diploid stage in the life cycle.

(2) Gametic or terminal meiosis (diplotonic pattern) occurs in animals. The cells resulting from the two meiotic divisions are transformed into sperms or eggs without further cell division. The process is as follows. Primordial germ cells divide repeatedly to give rise to many generations of daughter cells, spermatogonia (in males) or oogonia (in females). The spermatogonia and oogonia are transformed into primary spermatocytes and primary oocytes, respectively. Each primary

spermatocyte undergoes the first maturation (metiotic) division and froms two secondary spermatocytes. Each secondary spermatocyte undergoes the second maturation (meiotic) division to give rise to two spermatids, which are then transformed into spermatozoa. In the female during the first meiotic division the primary oocyte, into a large secondary oocyte and a smaller cell, the first polar body. The secondary oocyte undergoes the second meiotic division and divides into an ootid and a second polar body. The first polar body may or may not divide into two. The ootid is the ovum proper.

(3) Sporic or indeterminate meiosis (diplo-haplotonic pattern) is characteristic of higher plants and some thallophytes, but is not known in animals. In the diploid sporophyte, meiosis takes place at some indeterminate point between the zygote and gamete formation. As a result of meiosis haploid megaspores and microspores are formed, The megaspores and microspores undergo repeated mitotic divisions and form the male and female gametophyte, respectively. Certain cells of the gametophyte differentiate and form the gametes. In plants the cells that undergo meiosis are the microspore mother cell and the megaspore mother cell.

SIGNIFICANCE OF MEIOSIS

The process of meiosis keeps the chromosome number constant from one generation to other. This also involves reshuffling the hereditary characteristic which arises from ressortment of the parental chromosomes through crossing over. *Carothers* (1913) and others mentioned that in the metaphase I of meiosis, it is usually a matter of chance whether the centromere of the paternal or the bivalent lies towards the particular pole. This results in random origin of new set on separation. There may be many new combinations. In fact, since each chromosome can go to either pole, the total possible number of arrangements will be 2n where n is the haploid number of chromosomes.

Darlington (1931) has suggested that the meiosis can be regarded as a precocious mitotic, and that the force which retains daughter chromatids together in mitotic prophase also retains the homologous chromosomes pairs in the earlier stages.

ROLE OF CENTRIOLE DURING MEIOSIS

The centrioles perform precisely controlled synchronous function during meiosis. Where meiosis occurs in two steps, that is, two nuclear divisions take place in sequence, the centriole reproduces and functions to form two achromatic figures. In the first division of meiosis, the centromeres do not duplicate but rather homologous chromosomes pair together and move to opposite poles, while in the second meiotic division the centromeres of chromatids are separated. A scheme for relating various key events is presented briefly as under :

(*i*) Nonduplication of centromeres in meiosis-I.

(*ii*) Nonduplication of chromosomes in meiosis-II, and

(*iii*) Duplication and function of centrioles in both divisions.

By a complete repression of the duplication, simultaneous function of the achromatic apparatus brings about the reduction. The cetriolar function is super-imposed upon the duplication process of the chromatic apparatus in a sequence which leads to the loss of a generation of chromatic apparatus. When meiosis is gametic and fertilization follows after, the immediate formation of zygote as a cell might contain the maternal as well as paternal centrioles, having endowed as it were, with the extra centriole generation. According to *Daniel Mazia* the kinetochore is an essential part of the chromosomes so far as the mitosis is concerned. In gametogenesis there is a gain in total chromatic material, since these types of cell behaviour lead to duplication and subsequent reamassing of chromosomes. The extra set of centriole is rejected normally in fertilization or autogamy. Thus a cell results in a chromatic apparatus. This is accomplished by meiosis : one division meiosis gives the centriole a chance to 'catch up' in a single duplication.

Table 7.2. Showing Differences between Mitosis and Meiosis

Mitosis	*Meiosis*
1. Prolonged resting stage.	1. This stage is very short.
2. The prophase is short without any substages.	2. The prophase is subdivided into six stages, viz., Proleptotene Leptonema, Zygonema, Pachynema, Diplonema and Diakinesis.
3. The chromosomes split longitudinally into sister chromatids during the preceding resting stage. From early prophase they appear as double tetrads.	3. The chromosomes do not split longitudinally.
4. The split homologues are not attracted towards each other.	4. At zygonema or succeeding stages, in prophase of first division, the homologous chromosomes are attracted towards each other and they form bivalents.
5. Synapsis and tetrad stages do not occur.	5. The chromosomes split after synapsis to form sister chromatids which are arranged in tetrads.
6. No chromosomal crossing over takes place.	6. The crossing over is a prominent feature which results in exchange of genic material.
7. The chromatids are coiled in plectonemic manner.	7. Coiling is paranemic type, and much evident.
8. Doubling of chromosomes takes place early in prophase.	8. Doubling of chromosomes occurs in pachytene stage of prophase.
9. Nucleolus and nuclear membrane disappear later on and the formation of nuclear spindle is completed with the commencement of metaphase.	9. The nuclear membrane and nucleolus disappear and the formation of acromatic figure is completed with diakinesis simultaneously.
10. The nucleus enlarges to some extent.	10. The nucleus is greatly enlarged.

(Contd.)

<table>
<tr><td>11. At metaphase, the centrosome of each chromosome divides and the sister chromatides move towards the opposite poles. The somatic number of chromosomes is maintained without any reduction.</td><td>11. At metaphase I, the centrosome does not divide but the homologous chromosomes move towards opposite poles. The chromosome number is reduced to half by reduction division.</td></tr>
<tr><td>12. The chromosomes are seen in dyad consisting of two chromatids.</td><td>12. The chromosomes are in the form of tetrads having four chromatids.</td></tr>
<tr><td>13. Each chromosome of paternal and maternal origin is longitudinally divided and the separation to opposite side results in identical cells at the end of mitosis.</td><td>13. At metaphase I, the orientation of chromosomes at the equatorial plate is in such pattern that either paternal or maternal chromosome goes to either pole. The orientation of one pair is independent of the other.</td></tr>
<tr><td>14. During orientation of dyads the centromeres face towards the equator and the arms directed towards the poles.</td><td>14. During the orientation of tetrads, the centromeres are placed equidistant and face the pole while arms are directed towards the equator.</td></tr>
<tr><td>15. The centromere of each bivalent undergoes division at metaphase.</td><td>15. No division of centromere occurs, but homologous chromosomes simply separate to form two dyads.</td></tr>
<tr><td>16. The initial movement of chromosomes at late metaphase is due to repulsion between the centromeric parts.</td><td>16. The main repulsive force lies between the centromeres of homologous chromosomes.</td></tr>
<tr><td>17. At anaphase the chromosomes are less thick and short.</td><td>17. The chromosomes are very thick and short.</td></tr>
<tr><td>18. At anaphase further separation of chromosomes is due to the formation of stem body.</td><td>18. Most of the changes are similar to mitosis.</td></tr>
<tr><td>19. At telophase two daughter nuclei, each with diploid number of chromosomes, are reconstituted.</td><td>19. At this phase two daughter nuclei are haploid.</td></tr>
</table>

20 There is no evidence of second division in mitosis.	20. There are two division for each meiotic division. The second division follows after a short gap, interkinesis. The second division is homotypic by nature, and results into four cells, in the form of tetrads.
21. In the mitosis neither the chromosomal number nor the quality is changed and the daughter cells carry 2n chromosomes.	21. At the end of meiosis each of tertrads is haploid and quality is changed.
22. This division takes place in growing somatic cells.	22. This division takes place only in gonadial cells and result in the fomation of gametes.

8

Physiology of Cell Division

Mitosis is usually defined as the division of a nucleus during which chromosomes become visible, and separate in such a fashion that two genetically equivalent daughter nuclei are produced. Meiosis is a modified mitosis or sequence of two mitoses that occurs only at one stage during the life cycle. Both types of division are frequently correlated with a cytoplasmic division, cytokinesis, and various other cellular changes so that a discussion of mitosis usually involves the whole cycle of cell division. During the last few decades evidence has accumulated which demonstrates that the events during the cycle between the visible division stages, *i.e.*, during interphase, are in many respects more significant for understanding mitosis than the visible manifestations of prophase, metaphase, anaphase and telophase.

If we adhere to the definition of mitosis as given, then there are many cells that divide without mitosis. To relegate all of these cases to the category of amitosis would be a mistake since this term often implies an unorganized division and, as such, only occurs in pathological conditions or in nuclei with a limited genetic future. As a result, the term mitosis may evolve to mean any division in which genetic evidence shows that there is a regular distribution of genetic material. Likewise, the term chromosome is being used to refer to a linear arrangement of genetic units whether or not it ever appears as a visible structure. For example, the bacteriophage chromosome and the chromosome of *Escherichia coli* are useful concepts in discussing their genetics. Although the chromosome of *E. coli* apparently never condenses, it duplicates and separates in a regular fashion as demonstrated by genetic evidence. Indeed, unless an organism has more than one chromosome, mitosis, *i.e.*, chromosome condensation and spindle formation, would not appear to be necessary. Perhaps mitosis only became necessary in evolution when the organism acquired more than one chromosome or genetic linkage group.

In this chapter no attempt will be made to discuss mitosis and meiosis in separate categories. Although many obvious differences exist between the two types of divisions, the evidence to date reveals no striking physiological differences. The principal difference, of course, is in the length of the prophase of the first meiotic division in which pairing of chromosomes and formation of chiasmata occur. These events, correlated with a unitary behaviour of centromeres at the first division, result in the reduction and segregation of chromosomes typical of meiosis. The timing and sequence of DNA replication, and the synthesis of nuclear proteins in relation to the premeiotic interphase and prophase of the first division are not significantly different from other divisions. A number of reports have indicated differences or unusual phenomena in the behaviour of nucleic acids, but closer examination has revealed these reports to be in error. Likewise, Darlington has consistently maintained that the meiotic nucleus enters prophase precociously compared to nuclei in regular divisions. Again there is little or no biochemical evidence for this view ; the DNA (deoxyribonucleic acid) is usually replicated in premeiotic interphase as in any other division. One unusual behaviour of the chromosomes is the failure of their two chromatids to become visibly separate early in prophase, as they do in most mitoses. The factors determining the separation of chromatids are unknown, however, and the timing and degree of separation are variable when different tissues are compared at mitosis. The variations do not appear to be conditioned by the time of reproduction of the major chromosome components.

THE DIVISION CYCLE

Interphase—The duplication of the chromosomes occurs in interphase. The first measurements of the amount of DNA per nucleus that were made in tissues where divisions were occurring indicated that the synthesis of DNA occupied only a portion of the interphase stage, probably the middle part. Later, Howard and Pelc used autoradiography and P^{32} to chart the events of the cell cycle as related to the incorporation of P^{32} into DNA. With the introduction of tritium-labeled thymidine for autoradiographic studies of DNA synthesis a better method became available for studying the kinetics of the cell cycle. Friedkin, Tilson and Roberts had shown by the use of C^{14}-thymidine that this nucleoside was a highly selective label for DNA, but the production of H^3-thymidine of high specific activity added a convenient and less expensive tool for studies of DNA replication at the cellular level. Its use in studying the events of the cell division cycle is illustrated by several papers.

The technique is well suited for an asynchronous population of cells with a uniform cycle. The cells are exposed to the thymidine-H^3 for a few minutes, then washed and changed to a medium with an excess of non-radioactive thymidine. Those cells in the process of DNA replication will become labeled. If the labeling time is short compared to the whole cycle, the ratio of labeled to unlabeled cells is proportional to the synthesis time divided by the time for the remainder of the cycle. The time required for labeled cells to reach division is a measure of the post-DNA synthesis period of interphase. If the division stages are of sufficient duration, a rough measure of their length can also be obtained from the interval required for labeled cells to arrive at the given stage less the post-DNA synthesis period. The pre-DNA synthesis period is obtained by the difference between the length of the whole cycle and the sum of the other time intervals. The whole cycle is measured by the interval between the arrival of a population of labeled cells at the first division and their arrival at a second division. In practice, populations of cells sometimes vary enough in the length of the cycle and the length of the different phases that such data may not reveal discrete time intervals. The length of the whole cycle may be obscured because a well defined second peak of labeled division figures fails to appear.

Only one report of the cycle in a plant tissue has utilized this method. Wimber studied the cycle in the root tip of *Tradescantia*. The second peak was not discrete enough for determining the length of the cycle ; therefore, he attempted to determine the length of the interval of DNA replication by other means. He plotted the frequency of division figures against time on the abscissa of a graph, as in all such studies, but then used the duration of the first peak, *i.e.*, the time required for the population of labeled cells to pass through mitosis, as a measure of the time of synthesis. He determined the slope of the ascending and descending part of the curve and then took as the synthesis time the interval between the 50 per cent points on the two slopes. From this interval (10.8 hr.) he calculated that the whole cycle was about 20 hr. at 25°C., the post-DNA synthesis period 2.7 hr., mitotic stages 2.5 hr., and the pre-DNA synthesis period 4 hr.

In a study of a population of connective tissue cells isolated from an embryo of Chinese hamster, Taylor used the pulse labeling technique and was able to obtain a distinct second peak in the frequency curve of labeled mitoses. Using the interval between the two peaks, the cycle was estimated as 14 hr. with DNA synthesis occurring in about 6 hr. The post-DNA synthesis period to metaphase was 2 to 3 hr. The division stages were later estimated to require about 30 min. from mid-prophase to mid-telophase. Van't Hof, Wilson and Colon have recently reported some success in measuring the mitotic cycle in roots of *Pisum* by the use of colchicine. They induced polyploidy in a fraction of the cell population by a 30-minute treatment of the roots with the drug. A part of the population of cells was thus marked and synchronized in the division cycle. The roots were removed from the colchicine solution, allowed to recover, and at intervals the frequency of polyploid cells at mitosis was determined in a sample of the roots. When the frequency of polyploid mitotic figures was plotted against time, three peaks in the curve appeared at intervals of about 12 hr. Although the third peak was broader than the first two and therefore indicated increasing asynchrony, the intervals between the appearance of the polyploid cells at division should be a valid measure of their cycle of division. For measurements of similar parameters, Hejnowicz used a technique of sequential photographic records of root growth, plus determination of cell density along the root and frequency of mitotic stages at successive distance intervals (0.05 mm.) from the tip. From these data he calculated that the rate of cell formation per cell was nearly constant throughout the meristem (0.6 to 0.7 mm.) in these roots, possibly excluding a small relatively inactive region near the tip. Clowes had confirmed the existence of such a region just back of the tip in another type of root by means of radioisotopes and autoradiography. Hejnowicz then calculated the duration of the whole cycle (average time between the two successive cell divisions) on the basis that it is given by the reciprocal of the rate of cell formation per cell. The time calculated varied between 20 and 26 hr. for roots grown at 18°C.

All three of the methods mentioned above have possible defects and a comparison of results from all three methods on the same material would be interesting. For example, if too much isotope is incorporated in the labeling method, a mitotic delay could occur that would affect the determination of the cycle. Likewise, the tetraploid cells would perhaps be expected to divide more slowly than the regular diploid ones. The determination of new cell formation per cell is laborious and neither it nor the colchicine method allows the timing of other metabolic events in interphase in the way that the labeling method does.

In most rapidly dividing tissues the synthesis of DNA occurs in the middle of interphase and occupies about onethird of one-half of the cell cycle. Autoradiographic evidence has been presented which indicates that the synthesis of proteins of the nucleus coincides with DNA synthesis in root tip cells and in cells at premeiotic interphase. Later, cytophotometric measurements of basic proteins in individual nuclei indicated that the synthesis of these keeps pace with DNA replication. Rasch and Woodard measured the amount of basic proteins by cytophotometry in a variety of tissues of *Tradescantia*, lily, corn, onion, and broad bean. The fest-green staining technique introduced by Alfert and Geschwind (2) and the extraction of methanol-fixed tissues with 0.01 *N* HCl after removal of DNA were used to characterize and measure the basic proteins (histones). The amounts of nuclear histone fell into approximate multiples of the gametic (egg or sperm) quantity except in meristematic tissues where DNA replication was occurring. Here intermediate amounts of histone occurred in nuclei that were presumably in the process of chromosome duplication.

Amounts of histones per nucleus in root tip nuclei of *Tradescantia*, when plotted against interphase time, showed that the amount of histones was doubled during the middle third of interphase. The basic proteins of sperm nuclei of higher plants were found to be of the histone rather than protamine type ; the latter being found in trout and some other animal sperm cells. Recently, by autoradiographic studies with tritium-labeled arginine, De has confirmed the finding that DNA replication and basic protein (histone) synthesis are simultaneous in root cells. In addition, he has shown by the same technique that the synthesis of histones coincides with DNA replication in *Tradescantia* microsporocytes. Synthesis of both components is completed by early leptotene, one or more days before pairing of meiotic chromosomes can be detected.

By far the largest fraction of nuclear proteins in growing cells are the various non-basic proteins. The histones may represent nearly one-fourth the total in roots, but in many animal cells the ratio is smaller. Probably some proteins are produced continuously in the cytoplasm, and during interphase and prophase in nuclei. Support for this idea comes from the observation that uptake of labeled amino acids is continuous at all stages and from the continuous increase in nuclear volume and non-basic proteins shown by cytophotometric measurements of fast-green bound by staining at pH 2.0. These latter measurements also indicated a decrease in total proteins during the division stages. On the other hand, Grun used cytophotometric techniques based on ultraviolet light absorption and reported that *Tradescantia* root cells did not show a protein increase during interphase. He assumed the increase must occur during division stages ; however, this conclusion is contrary to all other reports and is probably in error.

De reports that during DNA replication in onion roots, accumulation of non-basic proteins by the nucleus is very slow compared to other periods in interphase. In partial agreement with this, cytophotometric measurements by Woodard *et al.* indicates that the most rapid accumulation of non-basic proteins occurs in late interphase and perhaps in early prophase.

Taylor has summarized the evidence that the chromosomes are duplicated, *i.e.* become double as indicated by radiation breakage, simultaneously with the synthesis of their two major components, DNA and basic proteins. In meiosis the chromosomes show their functional doubleness prior to prophase by the types of rearrangements that are produced following irradiation. As in other division cycles, the transition to doubleness occurs simultaneously with DNA replication and synthesis of the major portion of the nuclear proteins.

Although DNA replication frequently occurs during middle interphase, variations have been found. Replication may occur at early or late interphase in microspores and, as mentioned above, synthesis extends into early leptotene of meiotic prophase in *Tradescantia* microsporocytes. However, in the microsporocytes of *Lilium*, synthesis is completed well before the typical leptotene chromosomes become visible. Replication of DNA may sometimes occupy the whole interphase in embryonic tissues, but synthesis has not been observed during the stage when chromosomes are condensed at prophase or other mitotic stages.

Since chromosome reproduction frequently appears to be completed in the middle third of the cell division cycle, the question of what occurs during the remainder of interphase may be asked. In a few instances the synthesis of RNA, which seems restricted to the nucleus in many kinds of cells, does not coincide with the period of DNA replication. In the microspores of *Tulbaghia* the two types of nucleic acids are not produced simultaneously and some evidence was presented to indicate that root tip cells of *Tradescantia* were similar in this respect. This is not a general rule, however, and in many cells the incorporation of isotopes into both types of nucleic acids can be observed simultaneously. Even if DNA serves as the template for all RNA synthesis, the two processes could occur simultaneously in nuclei, for it is now clear that all of the DNA is not replicating at one time. A genetically controlled sequence is indicated in several cell types.

Whether there is a metabolic turnover of RNA in rapidly dividing cells is still an unsettled question and until it can be satisfactorily answered, the related problem of the site of RNA synthesis is difficult if not impossible to solve. In 1954 Hershey presented evidence for the conservation of both DNA and RNA during bacterial growth. Later, Siminovitch and Graham showed that there was no loss of P^{32} incorporated into the RNA of rapidly growing mammalian cells in culture. These experiments show a conservation of nucleotides, but do not rule out the breakdown and efficient re-utilization of the nucleotides. Davern has shown, however, that RNA of *Escherichia coli* does not undergo a detectable turnover. He labeled RNA with the stable isotoped C^{13} and N^{15}, and then transferred the bacteria to a medium with nutrients containing the light isotopes, C^{12} and N^{14}. Samples of RNA taken from the culture at intervals and separated by gradient-density centrifugation did not reveal RNA of intermediate densities as would be expected if the heavy polynucleotides were being broken down and re-utilized. The heavy molecules persisted intact during the subsequent growth and divisions of the cells.

Various experiments indicate that a turnover or breakdown of RNA may occur in some non-growing cells. For example, Watts and Harris found that the macrophage from the rabbit, when placed in culture containing unlabeled adenosine, in 2 hr. lost about one-half of the C^{14}-adenine introduced in the

RNA during the previous 4 hr. Slight loss occurred when inosine was added and no loss if inosine-5-phosphate was used for dilution. Later experiments by Harris indicated a turnover of nuclear RNA in a rapidly growing connective-tissue cell in culture. A turnover of cytoplasmic RNA was not detected in the cells. Harris concluded that the nuclear RNA that disappeared did not appear in the cytoplasm in a stable form because the nucleus continued to lose labeled RNA after an increase in the cytoplasm had already occurred. His argument concerning this point is not entirely convincing, however. His data show some gain in cytoplasmic label during the period of loss. Considering the quantity of labeled RNA in the nucleus as compared to the cytoplasm, a transfer during the latter part of his experimental period might not have caused a significant change in the grains produced in autoradiographic film. A small loss of cytoplasmic label likewise would have compensated for the possible transfer. On the whole, however, his data are at least suggestive of a rapid turnover of a part of the nuclear RNA even in rapidly dividing cells. In addition, the autoradiographic studies of McMaster-Kaye indicate a rapid turnover of the nucleolar RNA in the salivary gland cells of *Drosophila*. In such a situation the transfer of nuclear RNA to the cytoplasm is difficult to prove or disprove. However, there is considerable evidence that most if not all RNA is made in the nucleus. There are, for example, the experiments demonstrating that enucleated cells or cell fragments do not produce RNA ; see Prescott for a review.

In a recent paper Naori, Naori and Brachet have re-evaluated the situation in enucleated fragments of *Acetabularia*. Work from this laboratory had indicated originally that a net increase of RNA occurred in enucleate fragments of this alga. Later measurements by Richter showed no net increase in the RNA of enucleate fragments of *Acetabularia*. Likewise these measurements reported by Naori *et al.* show no net increase, but carbon-14 was incorporated into the RNA when presented as orotic acid-C^{14} or adenine-C^{14}. About 80 per cent of the cytoplasmic RNA in this alga is bound to a large particulate fraction in homogenates that includes the chloroplasts. It was this fraction that became labeled in the enucleate fragments ; the microsomal and soluble RNA remained unlabeled. In fragments with nuclei all fractions became labeled although the amount of radioactivity in the microsomal and soluble RNA was small compared to that in the large particle fraction. Although no net increase in RNA occurs in the absence of the nucleus, the authors concluded that the incorporation of C^{14} into "chloroplastic" RNA indicated synthesis independent of the nucleus. Carbon dioxide-C^{14} was incorporated into purine bases both in nucleated and in enucleated pieces at about the same rate. The failure of a net increase of RNA to occur in enucleated fragments as it did in nucleated fragments is difficult to understand unless cytoplasmic synthesis or incorporation of isotopes utilizes the RNA originally formed in the nucleus. However, if synthesis of RNA independent of the nucleus occurs, the chloroplast, which is known to exhibit independent genetic properties, would be a likely site.

Also, a number of experiments show a long lag period before labeled RNA appears in the cytoplasm as contrasted to its appearance in the nucleus when cells are given a radioactive precursor of RNA. The study of incorporation by the hamster cells in culture indicated that the pool of RNA precursor could be effectively diluted. For example, addition of excess cytidine and uridine (50 times the molar concentration of that in the labeled medium), after removing cells from a medium with tritium-labeled cytidine, immediately stopped the increase in labeled RNA in the nucleus but not in the cytoplasm. Within 4 hr. essentially all of the labeled RNA of the nucleus had disappeared and an equivalent amount had appeared in the cytoplasm. In addition, the labeled nuclear RNA was more quickly dispersed into the cytoplasm when the nuclear membrane broke down at late prophase. However, the question of the sites of synthesis cannot be considered settled without much more data, perhaps with both heavy isotopes and radioactive isotopes as labels.

Division stages. As already mentioned, DNA replication has not been observed during the division stages with the exception that in one species, *Tradescantia paludosa,* the replication extends into the leptotene stage of early meiotic prophase. However, it should be noted that during leptotene in this species the chromosomes are not highly condensed. DNA replication does not occur between the two meiotic divisions even in those species in which the nucleus reverts to a nearly typical interphase condition. As is well known, the chromosomes of many species remain condensed during this interkinetic stage and little if any RNA synthesis occurs.

In mitosis, RNA synthesis stops in mid-prophase and is not resummed until mid-telophase. In meiosis synthesis continues through most of the long prophase until the chromosomes become highly condensed in diakinesis. Protein synthesis continues through the division stages in the cytoplasm of mammalian cells in culture, but in meiotic prophase of *Lilium* the uptake of isotope into proteins is very low compared to that which occurs in premeiotic interphase. On the other hand, many oocytes make their greatest growth during meiotic prophase and these produce large quantities of RNA and protein. The chromosomes at such stages are very much expanded and frequently are not visible.

Measurements of oxygen consumption have shown a fall coincident with division stages in some types of cells. In the ciliate, *Tetrahymena*, measurements of oxygen uptake by cells induced to divide synchronously showed the drop. Erickson noted a fall in the rate of oxygen consumption in whole lily anthers during the meiotic divisions and again at mitosis in the microspores. Stern and Kirk using anthers of *Trillium*, obtained similar results. Nasatir later extended these findings to isolated microspores. Variations in oxygen uptake during the cleavage of frog eggs and eggs of various marine invertebrates have been reported but are so small that their significance has been questioned. Warburg has reported certain changes in the oxidative systems of some tumors and has suggested a linkage with the rapid mitoses in such tissues.

One or more of three possible explanations can be offered for the various results : (*a*) the difference reflects an impairment in the cells' respiratory systems ; (*b*) an energy reservoir is filled during interphase and used during mitosis ; or (*c*) metabolism drops because less energy is required during mitosis.

A high rate of aerobic glycolysis associated with a lower rate of respiration has been reported for cancer cells. The suggestion has been made that an irreversible alteration in the oxidative system induced continued divisions and the resulting cancerous tissue. This idea has provoked much discussion and considerable work, with some contradictions and various explanations of the reported changes. Any primary linkage between the type of respiration and the induction of cell division has not been demonstrated and the chance that one exists appears rather remote. The shift from respiration to glycolysis probably is characteristic only of certain tissues and in view of all of the known steps necessary for completion of the cell division cycle, such a change would seem insufficient to induce mitosis.

Swann's experiments with sea urchin eggs showed that suppressing respiration up to a critical time delayed mitosis for a period equivalent to the interval of inhibition. After a critical stage (probably the beginning of prophase) mitosis proceeded in spite of the inhibition of respiration.

Bullough's idea of an "antephase" build-up of an energy reservoir, which might appear supported by Swann's experiments, can as easily be explained by other means. Any progression of stages or synthesis of materials required for mitosis could be delayed by the inhibitors used. If the inhibition were reversible the delay would be proportional to the period of treatment. Studies by Gelfant on the mouse ear epidermis do not support the concept of an energy reservoir. The synthesis of the nucleic acids and various proteins of chromosomes must be accomplished prior to mitosis. Likewise, evidence now indicates that the spindle proteins are formed before the actual division. The cytoplasmic constituents also approximately double in amount during interphase, although the coupling of these processes with mitosis is perhaps less obligatory than the reproduction of the chromosomes.

Some cells may require less energy during mitosis, for there is no measure of the energy requirements for spindle organization and chromosome movement. The requirement could conceivably be smaller than the interphase growth requirements. There is no DNA synthesis during that period and RNA synthesis appears to stop. In rapidly dividing cells a large part of the protein is nuclear protein. If the synthesis of all of these components stops, even though the cytoplasmic protein synthesis continues, the energy requirements would be smaller unless compensated for by the needs of the division processes.

Stern has reviewed the evidence that sulfhydryl groups are particularly important in the division stages. He points out that the idea is an old one that arose from the observation of (*a*) a strong reaction of a number of proliferating tissues with nitroprusside, (*b*) the rise and fall in concentration of soluble thiols in relation to cleavage in sea urchin eggs, and (*c*) the inhibition of cells at the mitotic stages by thiol poisons and their reversal by cysteine, glutathione, and thioglycollate. Stern examined the soluble thiols in lily anthers during an 11-day interval surrounding mitosis in the microspores. He found a marked increase in total acid soluble—SH and —SS—up to division, and a fall afterward, most of which was glutathione, but no conversion of —SS—to—SH. He suggests a connection with the respiratory changes that likewise are associated with these division stages.

PHYSICAL AND CHEMICAL AGENTS PRIMARILY AFFECTING THE INTERPHASE STAGES

Temperature. Variations in temperature within the physiological range of the organism produce changes in the rate of cell division. In *Lilium longiflorum* the first meiotic prophase requires about 7 days at 23°C and about 10 to 11 days at 20°C. Some species have a rather narrow range of tolerance. In the root tip of *Vicia faba* changes in temperature in either direction from 20°C were soon followed by a reduction in mitotic index. The number of cells in prophase, when taken as a percentage of the total, indicated that the delay was occurring just before the mitotic stages. In the ciliate, *Tetrahymena*, sublethal temperature shocks result in an accumulation of cells at predivision stages. Repetition of such shocks at appropriate times increases the synchrony of cell division until a large percentage of the population divide at nearly the same time. Mammalian cells (strain HeLa) in culture are reported to become synchronized by lowering the temperature periodically to 4-6°C. Presumably the delay is in interphase.

By using repetitive temperature and light cycles, the duration and intensity of which were matched to the synthetic capacity of the cells, James induced division synchrony in *Amoeba proteus*, *Astasia longa*, and *Euglena gracilis*. Similar types of treatments had been shown earlier to induce synchrony in cultures of *Chlorella*.

Radiation. Variations in mitotic frequency are often associated with exposure to visible light, particularly in photosynthetic organisms. In ten species of green algae (Eugleninae) grown in a soil-water medium, mitosis was found to be restricted to a dark period following a day's exposure to daylight. Once initiated, the stimulus to mitosis was not reversed by light even though the cells had not reached prophase ; therefore, the effect is on some interphase stage. Non-green species did not exhibit the response to visible light and one green species, *Euglena gracilis*, did not show light dependence when grown in a rich liquid medium.

Ultraviolet radiations delay mitosis in some cases, especially at wavelengths that are absorbed by nucleic acids. If the dose has not been too large the mitotic index falls slowly, with a gradual recovery. The interphase is the most sensitive stage, although cells at prophase are also affected.

Ionizing radiations affect various stages, but the most sensitive stage in many cells is preprophase. As prophase advances the cells become less sensitive, and after late prophase the division will occur following huge doses of x-rays. When relatively small doses are given there is a rapid drop in the mitotic index followed by an increase above normal as the cells stopped at late stages recover and reach division along with cells that were at earlier stages when irradiated. The processes affected by the radiation are not known. There is some indication that RNA metabolism may be involved since the nucleolus is by far the most sensitive structure. Microbeam irradiation with ultraviolet light for three minutes showed that mitosis was permanently stopped if the nucleolus was in the field of exposure. Irradiating other parts of the nucleus to the same extent produced delay but did not usually prevent cell division. After middle prophase the cell becomes less sensitive and finally at late prophase is insensitive to this amount of radiation applied to the nucleolus. Since the nucleolus is a very active site involved in RNA metabolism and the synthesis stops about the time the sensitivity drops, RNA metabolism is perhaps a good choice for a sensitive process. However, Gaulden has shown that the mitotic delay produced by low doses of x-rays is reversed by hypertonic salt solution. Since the visible change involved is the degree of condensation of the chromosomes, the basic processes affected remain obscure.

Numerous studies have indicated that DNA synthesis is quite sensitive to radiation. Many of these studies have measured effects on masses of cells or whole tissues. In such cases the suppression of mitosis at any stage would result in a drop in DNA synthesis as the cells failed to proceed to the stage where replication begins. The replication process actually is not especially sensitive to radiation damage, for studies on the incorporation of labeled thymidine into DNA show that synthesis continues after relatively large doses of ionizing radiation. Recently some evidence has been obtained that radiation may actually increase the rate of synthesis in actively dividing tissues such as the root meristem. This is perhaps not unreasonable since Kornberg and his collaborators have found that slightly damaged DNA is a better primer for the enzymatic synthesis of DNA on vitro than native DNA.

Chemical agents. Maleuric acid has been shown to produce mitotic abnormalities and depress the mitotic index in both plant and animal cells. Studies with tritium-thymidine to measure progress of cells through the cycle indicated that the interruption occurs both in interphase and at division stages.

The antibiotic, actidione, produces a delay in the division of root cells. Treatment for 15 min. in 2.52×10^{-5} M, solution produced a 5-hr. delay. The mitotic index was rapidly reduced to zero, but after a delay the division rate was restored when the drug was removed. The most sensitive stage appears to be preprophase or some stage just prior to that. After prophase has started, cells will go on through division in the presence of sublethal doses.

Brachet and Ledoux showed that immersion of cleaving eggs at the morula stage in a solution of ribonuclease stops all mitoses in the external layers. Cells already in division at the beginning of treatment showed various abnormalities, but most of the cells were blocked in interphase. Treatment of ascites tumor cells, fibroblasts in culture, and several other types of cells produced a marked reduction in mitotic index. Chevremont and co-workers have also shown that deoxyribonuclease inhibits mitosis and has some peculiar effects in which there is an accumulation of DNA in the mitochondria. McLaren, Jensen and Jacobson have also shown that barley root cells rapidly absorbed ribonuclease, lysozyme, and human hemoglobin when grown in solutions of the proteins.

Halogenated pyrimidines have been used extensively in recent experiments on the inhibition of tumor growth. Karnofsky and Basch applied some of these substances to the sand dollar embryo. Embryos are stopped by 5-fluorodeoxyuridine in early blastula. An excess of thymidine or bromodeoxyuridine protects the embryo, but those treated with bromodeoxyuridine later die. These compounds and other analogs of purines and pyrimidines affect the synthesis of DNA or RNA, or both types of nucleic acid, and therefore would be expected to eventually block mitosis through an effect on the interphase stages. Djordjevic and Szybalski reported that human cells (HeLa) in culture grow and divide indefinitely when bromodeoxyuridine is added to the medium at a concentration of 3×10^{-5} M. Bromodeoxyuridine can replace up to 45 per cent of the thymidine in the DNA when given at higher concentrations. Cells also grow for limited periods in iododeoxyuridine, but prolonged growth results in loss of viability. Cheong, Rich and Eidinoff found that another type of mammalian cell would grow for a time in a medium containing bromodeoxyuridine from 5 to 100 μM. On the average, at 75 μM the cells underwent one division and stopped. They also prepared tritium-labeled bromodeoxyuridine which labeled nuclei sufficiently for autoradiography.

Nutritional and hormonal factors. Wilson and collaborators have studied the mitotic activity in excised pea roots. When maintained in balanced salt solution, mitosis drops as 10 to 15 per cent of that of controls and remains at this level for about 48 hr. A variety of carbon sources, including sucrose, glucose, fructose, dihydroacetone. and glyceraldehyde, restore part or all of the activity for periods of several hours to nearly two days. The effect can be reversed by some respiratory inhibitors.

Das, Patau and Skoog have shown that mitosis and cell division can be induced in excised tobacco pith by the use of kinetin and indoleacetic acid (IAA). In a study of DNA synthesis at the cellular level by cytophotometry, Patau *et al.* showed that DNA doubling was induced first, followed later by mitosis. In control cultures little or no DNA synthesis occurred. The two substances were effective alone, but most stimulation occurred when both were used together. Later, Das *et al.* used thymidine-H^3 to detect the synthesis of DNA in the cultures. In control cultures incorporation occurred in about 20 per cent of the nuclei during the first two days, but few if any were noted thereafter. After two days IAA induced DNA synthesis, but kinetin did not. Continued cell reproduction was maintained only in the presence of both.

In a study of the effects of IAA and kinetin on recovery from the mitotic inhibition produced in irradiated cells of root meristems, Guttman and Brown reported recovery with IAA if the cells were irradiated at or just prior to prophase but no recovery with kinetin. Kinetin and IAA together produced complete recovery under the conditions used by Guttman and Brown regardless of the time in the cell cycle at which irradiation occurred. When the two substances were used together the frequency of chromosomal abberations was also reduced. IAA had no effect when used alone, but kinetin stimulated rejoining of breaks for cells irradiated when close to mitosis.

By the use of radiation and chemical agents Haber and Luipold obtained evidence for different mechanisms initiating cell division and cell expansion. In germinating lettuce seeds cell division and cell expansion usually coincide in time. Gamma-irradiation plus kinetin treatment at 26°C. delayed germination, but recovery and germination occurred several days before cell division began. Growth at 10°C. also produced a similar effect on the beginning of the two processes. At 26°C. in a solution of manitol, cell division preceded germination by many days, while at 30°C. only a small percentage of seeds germinated, but considerably more of the seedlings had mitoses. They concluded that kinetin stimulated cell expansion but not cell division in lettuce seed.

Harding, Donn and Srinivasan have used tritium-labeled thymidine to visualize the wave of premitotic events that radiate out from the point of injury in the lens epithelium of the rabbit. The stimulus for mitosis and the events leading up to it move out from the site of the wound so that an arc or circular group of cells are replicating DNA simultaneously. Indications for a more specific hormonal control of mitosis was obtained by Stich. The effects of sera and homogenates on the mitotic rate in regenerating liver and homologous and heterologous tissues in baby rats were studied to gain information on organ-specificity of the substances. Sera from adult rats inhibited the onset of mitosis and synthesis of DNA, whereas sera taken from rats 24 to 92 hr. after partial hepatectomy, from rats bearing a Walker carcinoma, or from rats with induced hepatomas, stimulated mitosis in regenerating liver. Homogenates of fully grown rat liver decreased mitotic rate, and homogenates of hepatoma and carcinoma increased mitotic rate in regenerating liver. The experiments on young rats suggested that organ-specific inhibitors of mitosis may take part in regulating the mitotic rate.

AGENTS AFFECTING THE DIVISION SEAGES

No new chemical agents primarily affecting the division stages seem to have been reported. However, Mazia and collaborators have used mercaptoethanol for some revealing studies on mitotic inhibition. According to their account the cells of marine eggs can be blocked in metaphase by this sulfhydryl compound. The reaction is reversed when the eggs are removed and division proceeds. If eggs of the sand dollar are blocked for an appropriate time at the first division, a tetrapolar spindle is formed after removal. The next division is abortive since only monopolar spindles form. Chromosome counts show that each of the four cells after division by the tetrapolar spindle have the full complement of chromosomes but measurements of DNA in the four cells indicate that no increase occurred during the delay. If these observations and measurements are all correct, chromosomes can divide in these cells without replication of DNA. Mazia's interpretation is that the mitotic centers divide in the presence of the mercaptoethanol and get out of phase with the chromosomes, but as a result of the monopolar abortive division that follows, the two get back in phase.

THE SPINDLE AND ASSOCIATED STRUCTURES

Improvements in equipment and new biological materials have resulted in advances in the birefringence studies of the spindle in living dividing cells. Bajer has worked out a technique for studying mitosis in endosperm cells in extremely flattened preparations. Not only has he produced elegant cinemicrographic pictures of the process, but excellent pictures with polarized light give a clearer picture of the crystallization patterns in spindle formation and function. Inoue's studies indicate that both kinetochores (centromeres) and centrioles are centers for crystallization. These centers are active in all directions but the activity of the kinetochores is ordinarily polarized. The spindle micelles crystallize into filaments from either pole or betweea two poles and form astral rays and spindle fibers. The aligned micelles are suspended in a pool of randomly oriented particles which are more hydrated. The oriented micelles and the unoriented material are in equilibrium. He believes the oriented micelles to be crosslinked to some degree, but the amount of crosslinking is a sensitive function of pH and ionic strength. The various mitotic movements of chromosomes are explicable in terms of changes in the delicately balanced equilibria which also involves shifts in water balance. Fixation and isolation would readily shift either the orientation equilibrium or crosslinking or both.

Fixed cells of plant root tips with excellent preservation of membranes and various cytoplasmic structures do not show any fibers in the spindle. Tubular structures in association with the spindle are part of the endoplasmic reticulum. Their function perhaps is connected with the reformation of the nuclear membrane rather than the movement of chromosomes.

Sato recently published electron micrographs showing bundles of fibers attached at the centromere of chromosomes and oriented along the axis of the spindle in microsporocytes of *Lilium*. The material was fixed in a mixture of one part 0.1 *M* cadmium chloride and one part absolute alcohol at pH 4.4. It is not clear whether these fibers represent an enhanced aggregation of the oriented micelles that Inoue's results indicate to be present, or a better representation of the actual structure of the spindle. However, one would prefer a fixative that would preserve other parts of the cell in better condition along with the spindle.

CELL AND ORGAN CULTURE

Although animal cells in culture have proven excellent objects for studies of cell division, plant cells in tissue culture are less suitable for various reasons. Their tendency to clump and the rigid cell wall are handicaps. The failure of such cells to adhere to glass surfaces reduces the conditions under which they can be handled and observed. The most interesting developments in this field are the demonstrations that single somatic cells can give rise to whole plants. Muir, Hildebrandt and Riker grew tissues on a shaker in liquid medium. With friable types of tissue cultures of marigold, *Nicotiana tabacum*, and grape they obtained single cells and clumps of cells. Cells, isolated and placed on filter paper that was applied to the freshly cut surface of an established piece of tissue, divided and developed in a few instances into sizable clumps of cells. Braun used their technique to isolate clones from single cells of tobacco crown gall tissue that had been freed of bacteria and grown for several years in culture. The crown gall was the type that produces very abnormal buds and leaves in culture. Braun grafted the abnormal organs formed from these cells onto healthy tobacco plants. After one or two transfers he obtained rather normal branches with flowers, and from these he obtained seeds. These seeds produced normal *Nicotiana tabacum* plants like the ones from which the original cultures were established. He interpreted the change as the loss of a self-duplicating cytoplasmic particle that had maintained the tumor character. Because of the rapid divisions induced in the bud meristems (much faster than the cells ever divide in culture) Braun concluded that the nuclear and cell division proceeded faster than the rate of reproduction of the particle, and that therefore cells eventually divided without distributing the particle to one of the daughter cells.

Stewart *et al.* also obtained isolated cells in liquid cultures from carrots. He observed these divide and form clumps of cells. Whole normal carrot plants were eventually produced from similar clumps.

These experiments indicate that plant cells capable of division usually retain the genetic capacity to undergo differentiation and regenerate the whole plant. Braun interprets his experiments to mean that the crown gall cells have acquired some self-reproducing particle, presumably in the cytoplasm, that causes the change to gall formation. Because of the rapid divisions induced in grafts, some cells of the population lose the particle because its reproduction does not keep pace with mitosis. This, of course, is the most likely explanation, but one must keep in mind that cells can become infected with viruses by means of grafts. Also, perhaps unknown viral-like agents reside in the cytoplasm or become integrated into the nuclear genetic material as has been described for bacteria. Although such a system is unknown in higher forms, indications of a comparable relationship have been reported in mammalian cells by Vogt and Dulbecco. In such a situation cells might either lose or gain the ability for differentiation at either the nuclear or cytoplasmic level of organization.

Studies of meiosis using cell cultures have been only moderately successful. Since cells can not yet be brought through the long prophase of the first meiotic division, isolated cells have little value beyond facilitating studies of the mtiotic spindle and features of the late stages of division. The next best technique is to culture the excised anthers under conditions in which meiosis will be completed in a regular fashion. So far the limitations in this work have not allowed regular survival of meiotic cells in anthers excised earlier than late leptotene. Even when anthers are removed while they have cells at this stage, the microsporocytes often abort or various anomalies appear when the anthers seem to survive. Such cultures are of some use, but since the greater part of the synthesis of materials for development of microsporocytes occurs before leptotene, many interesting physiological studies are excluded. A few papers have appeared indicating improvement in growth with the addition of the growth substances, IAA and kinetin, to basal tissue culture media. In addition, nucleic acids were reported to enhance the survival of anthers in culture. The general applicability of any of these substances is very dubious, however. Lily anthers with sporocytes at late leptotene advance far enough so that the sporocytes often divide on basal media. The addition of the substances mentioned above, combined with various nucleic acid derivatives and amino acids, produced little effect. Metabolism in the sporocytes appears to be very slow in prophase and perhaps the greatest advantage comes from maintaining the growth and differentiation of the tapetal cells. However, the whole problem in this type of organ culture is to maintain differentiation rather than to induce cells to divide, as is the goal in most tissue culture. The maintenance of this differentiation has proven more difficult than keeping proliferating tissues dividing.

9

Plant Microtubules

Current evidence suggests that microtubules constitute a class of morphologically and chemically closely related filamentous elements of general occurrence in plant and animal cells. Microtubules are of special interest at present because they are being implicated in an ever greater variety of important subcellular structures and functions. They compose the fibers of mitotic spindles, the tubular structures in centrioles and basal bodies, and the 9+2 arry of axonemal filaments in cilia and flagella. As further examples, they occur in the sperm manchette, in elongated animal cells including neurons, and as the marginal band in erythrocytes. They are found in numerous structures in the protista, including the long, slender microspikes.

Microtubules probably function in a variety of ways. They participate in a motile system in cilia and flagella, and in movement of a different sort in the mitotic spindle. In many cases they are believed to perform a cytoskeletal role in the production and maintenance of cell asymmetries. In other cases they may aid in the movement of cytoplasmic materials. The uncertainties as to function are well illustrated by Fig. 9.1, which shows microtubules in association with the pellicular ridges in *Euglena gracilis*. These microtubules could have a skeletal function, or could be involved in the body flexures ("metaboly"), in the movement of cytoplasmic materials, or in other ways.

Recognition of the widespread occurrence of microtubules followed upon the introduction of glutaraldehyde as a fixative several years ago. Use of glutaraldehyde followed by osmium tetroxide made the preservation of microtubules possible in many structures in which they had been destroyed previously by fixation either in osmium tetroxide alone, or in permanganate. The wide distribution and importance of microtubules in animal cells were stressed in 1963 by Slautterback. In the same year, Led-better and Porter published their classic paper on microtubules as components of plant cells. Owing to its technical excellence and effective presentation, this contribution did much to direct attention to microtubules as important elements of cells generally.

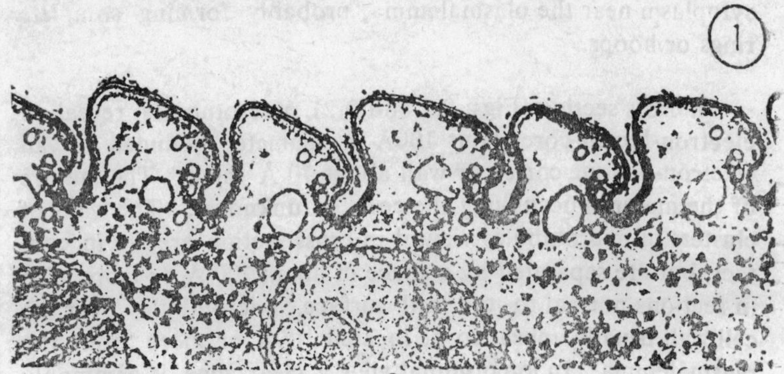

Fig. 9.1. Cross section of microtubules associated with the pellicular ridges in *Euglena gracilis* × 86,000.

The present chapter is devoted largely to microtubules in higher plants although some work on lower plants has been included where particularly relevant to higher plant problems, *e.g.*, microtubules in relation to deposition of wall microfibrils. A brief description of recent work on the protor chemistry of microtubules from animal systems has been included because of its obviously great general significance. Several recent reviews on cilia and flagella are available ; no attempt is made here to revice the botanical literature on the fine structure of these motile organs. Portes has written a general review of cytoplasmic microtubules, while Bracker has covered the occurrence of microtubules in fungi.

MORPHOLOGY AND DISTRIBUTION OF PLANT MICROTUBULES

Microtubules are slender unbranched cylindrical structures frequently encountered in the cytoplasm of both higher and lower plants. Their reported diameters usually fall between 180Å and 300 Å, the average value being about 240Å. Because of technical and biological variations, ordinarily no great significance can be assigned to reported differences in diameter. Although their lengths are difficult to determine, some microtubules can be followed for many microns. In some cases they run circumferentially around the cell in the cytoplasm near the plasmalemma, probably forming complete rings or hoops.

In cross section (Figs. 9.1 and 9.2), microtubules reveal an electron-lucent core about 100Å in diameter bounded by an electron-opaque cortex or wall about 70 Å thick. The nature of the material occupying the core is unknown. The walls of clusters of microtubules viewed in cross section are not in contact, but are separated by a space of about 200 Å or more, also in sections normal to the cell surface, microtubules near the cell wall usually appear to be located 100Å or more from the plasmalemma. In longitudinal view a clear zone is commonly observed along the sides of microtubules. These observations suggest that each tubule may be surrounded by a specialized zone or layer of material. Microtubules appear to be rather stiff and rigid structures, and usually follow straight or gently curving paths through the cytoplasm.

Employing Markham's rotation technique, Ledbetter and Porter showed that the wall of plant microtubules is probably composed of 13 filamentous subunits which have a center-to-center spacing of about 45Å. Subsequent images obtained at improved resolution permitted direct confirmation of these results. Several recent studies of microtubules isolated from various sources, including a plant have shown that the filaments of the walls are composed of a series of globular subunits each about 40 Å in diameter. Microtubutes are known to be composed in large part of protein.

A variety of tubular structures have been reported from plant cells whose morphological characteristics set them apart from typical microtubules as described above. These tubules may be much more sinuous than microtubules, markedly different in size, etc. Among them are the two sizes of tubules derived from the ER in bean leaf glands, the P-proteins of sieve elements, tubular projections on "spiny" vesicles, meandering tubular structures outside the plasmalemma in primary and secondary walls, microtubule-like structures in plastids, tubules in cells of leaf Beltian bodies and plasmodesmatal "desmotubules'. These structures are not discussed further in this review, although it is conceivable that some of them may be closely related to microtubules in substructure and chemical composition. Experiments on the tubules of bean leaf glands, using the agents employed by Behnke and Forer to distinguish four classes of microtubule, have shown clearly that the gland tubutes are not similar to conventional microtubules, however.

Microtubes compose the mitotic and meiotic spindle fibers, and the phragmoplast fibers of dividing cells. They occur in the cytoplasm next to the growing primary wall, and similarly overlie the developing bands of secondary wall. They are occasionally encountered also in small numbers elsewhere in the cytoplasm of interphase cells. Microtubules are also important structural components of centrioles and basal bodies found in many lower plants, the constitute the "9+2" axonemal elements in the flagella of motile plant cells.

Although in animal and protistan cells microtubules are believed to play a prominent cytoskeletal role in controlling cell asymmetries, the cytoskeletal function of microtubules would appear to be much less important in plant cells possessed of rigid, supportive walls. However, microtubules may nevertheless be responsible for cell asymmetries in plants if they control the pattern of deposition of the wall microfibrils.

A case of great interest where microtubules do indeed appear to play a direct cytoskeletal role in controlling the shape of a higher plant cell has been reported recently by Sanger and Jackson. Early in pollen development in *Haemanthus*, the spherical generative cell becomes ellipsoidal, and this change in shape is accompanied by the appearance along the generative cell wall of a sheath of microtubules running parallel to the long axis of the cell. The effects of colchicine and cold are consistent with the hypothesis that the change in shape of the generative cell is controlled by the appearance and orientation of microtubules.

PROTEIN CHEMISTRY OF MICROTUBULES

A brief review of some of the recent work on the comparative protein chemistry of microtubules is included here, for while this work is not based on the study of plant materials, it is clear that it is of general applicability and of the utmost importance for an eventual understanding of microtubule assembly and function. These studies and the closely related biochemical work on the microtubules of cilia, protozoan flagella, and sperm tails now provide considerable evidence that similar actin-like proteins are present in many kinds of microtubules.

In their successful chemical work on microtubules, Taylor and his colleagues have taken advantage of the long-standing observation that colchicine and related compounds are highly active in disrupting the mitotic spindle. They reasoned that if the spindle is a labile structure maintained by a dynamic equilibrium, then colchicine might cause its depolymerization by forming a complex with the subunit protein of which the microtubules are composed.

Borisy and Taylor found that tritiated colchicine, when applied to cultured mammalian cells both *in vivo* and *in vitro*, was bound specifically and non-covalently to a protein with a sedimentation coefficient of 6 S. Two kinds of evidence indicated that the 6S protein to which the colchicine binds is a subunit of the microtubules. First, a survey indicated that the high colchicine-binding activity is characteristic not only of rapidly dividing cells rich in spindle protein, but also of various other materials rich in microtubules, including cilia, sperm tails, and nerve cells. On the other hand, slime molds have few if any microtubules, and bind very little colchicine. Second, Borisy and Taylor found that when the sea urchin mitotic apparatus is extracted, the microtubules disappear and the soluble extract contains the 6S colchicine-binding protein. Furthermore, Shelanski and Taylor found that the 6S colchicine-binding protein is also the major component of the central pair of microtubules of sea urchin sperm flagella.

Thus, although Behnke and Forer have provided evidence that different classes of microtubules can be distinguished on the basis of their sensitivity to colchicine and other treatments, it can be hypothesized that colchicine binding is a property of the subunit protein of microtubules generally. Borisy and Taylor proposed that colchicine acts by binding to subunit protein and preventing its polymerization into microtubules. If the polymerization is an equilibrium process, nucleated perhaps by the centrioles and kinetochores, then its equilibrium can be shifted in favour of the subunits by complexing them with colchicine.

The colchicine-binding proteins from brain cytoplasm, the proteins from the microtubules of sea urchin sperm and the proteins from the outer doublet tubules of *Tetrahymena* cilia, although from three widely different sources, are quite similar, and consist of a dimer with a molecular weight of 110,000 to 120,000 which is convertible into a monomer with a molecular weight of 55,000 to 60,000. This monomer is believed to be the 40 Å globular subunit resolved in the microtubule with the electron microscope. One mole of colchicine is bound per dimer unit. The amino acid compositions of the proteins are quite similar and closely resemble that of muscle actin.

Of great interest is the discovery that, like actin, the microtubule proteins have the ability to associate with a nucleotide. Stephens *et al.* have shown that outer doublet tubules of cilia and flagella contain 1 mole of a guanine nucleotide per 60,000 g of protein, while Adelman *et al.* have reported that either

ATP or GTP is necessary to preserve the colchicine binding site of microtubule protein. Each dimer is believed to have two non-equivalent GTP binding sites.

Adelman *et al.* have pointed out that the data now available indicate the microtubules from sperm tails, cilia, cytoplasm, and mitotic spindles in widely separated animal phyla have subunits which constitute a well-defined group of closely related proteins. These proteins, called "tubulin," all have a molecular weight of *ca.* 120,000, a sedimentation constant of 6S, two binding sites for GTP, and one for colchicine. Furthermore, the ability of colchicine to block mitosis and disrupt microtubule structure generally in plants and animals implies that similar proteins are present in many kinds of microtubules.

Stephens has recently reported the successful reassociation *in vitro* of microtubule subunits into several fibrous forms. Nucleation in the form of tubule fragments was necessary for the polymerization, and it is possible that this is related to the role of orienting centers in the formation of microtubules *in vivo*.

MITOSIS

As is well known from studies with the light microscope, the birefringent fibers of the mitotic apparatus seen in living cells consist of chromosomal fibers running from the kinetochores to the poles, and continuous fibers running from pole to pole. It was shown in the late 1950's that these appear as fine filamentous elements in the electron microscope. Their microtubular nature in root tip cells in anaphase and telophase was clearly described in 1963 by Ledbetter and Porter, who suggested that the fibers which are resolved optically consist of bundles of microtubules. In 1964 Manton published exceptionally fine micrographs of microtubules in the mitotic spindle of a chrysophycean flagellate, *Prymnesium*, and in the mitotic and meiotic spindles of a lower vascular plant, *Equisetum*. It soon became obvious from these and similar observations that the microtubules in the different fibers of the spindle and phragmoplast are quite similar in appearance, and also resemble one another in a wide range of organisms, plant and animal. It was also apparent that as the chromosomes move to the poles in anaphase, the microtubules attached to the kinetochores become shorter without thickening or otherwise changing in appearance, suggesting that they are not contracting but are being digested at the poles.

Lability and dynamics of spindle microtubules. The identification of spindle fibers as bundles of microtubules and the exploration of the protein chemistry of microtubule subunits has allowed thinking in more concrete terms about the many observations and experiments made by optical techniques on the mitotic apparatus in living cells. Some of the most important results and conclusions of this work are given here, since they contribute much to our understanding of microtubule dynamics.

Ultraviolet microbeam experiments and other observations have gone far toward identifying the centrioles and kinetochores as centers of nucleation out of which grow, respectively, the continuous fibers (bundles of microtubules) and chromosomal fibers of the spindle. In plant cells it appears that the cell plate region subsequently nucleates the growth of the phragmoplast fibers. A long series of experiments on birefringence supports the hypothesis developed particularly by Inoue that the fibers are highly labile structures existing in a dynamic equilibrium with a large pool of unassociated subunits available to the nucleation centers.

This equilibrium can be shifted toward depolymerization of the fibers by low temperature or high hydrostatic pressure, and toward polymerization by an increase in temperature or the addition of heavy water. For example, spindle birefringence is destroyed in a matter of seconds by treatment with low temperature, and reappears in minutes by a return to normal temperature, after which chromosome movement can continue. (Related to this is the absence or dearth of microtubules frequently observed in electron micrographs of material fixed in the cold.) At low temperatures the protein subunits are believed to be insulated by shells of bound water molecules which at higher temperatures are dispersed, permitting the subunits to associate more closely and come into alignment to form the polymer.

The effect of high concentrations of D_2O in "freezing" mitosis or over-stabilizing the gel structure of the spindle is of great interest. For example, when 45 per cent D_2O is applied during mitosis, spindle birefringence may increase at least two-fold and the volume occupied by the spindle as much as tenfold. The effect of D_2O is virtually instantaneous and completely reversible. Possible molecular mechanisms that might be involved in the rigidification caused by heavy water have been considered by Gross and Spindel.

The above results illustrating the great lability of the spindle and its ability to recover rapidly from disruption strongly suggest that the subunits from which the fibers are constructed during mitosis have already been synthesized and are readily available from a pool. This view is strengthened by the observation that recovery of the spindle from disruption is not hindered by inhibitors of protein synthesis.

Experiments with low concentrations of colchicine are also instructive. As is well known, a high concentration of this drug completely disrupts the spindle and destroys the spindle microtubules. However, Inoue has shown that a low concentration causes the chromosomes to be pulled poleward, and electron micrographs reveal that the microtubules have simply shortened. It can be assumed that in combining with free subunits the inhibitor in low concentration shifts the equilibrium only sufficiently to cause a slow depolymerization of the microtubules at the poles, resulting in their contraction. This action of colchicine may closely resemble the process taking place in the polar regions during normal anaphasic movement of the chromosomes.

The main hypothesis that has emerged from Inoue's observations is, briefly, that the fluctuation of birefringence during normal mitosis and cytokinesis represents the assembly and disassembly of the same protein subunits sequentially at the different nucleation centers. During anaphase, digestion of the ends of the chromosomal fibers takes place near the poles, accompanied by further growth and elongation of the continuous fibers, both processes being involved in the poleward movement of the chromosomes. A similar hypothesis has been proposed by Pickett-Heaps in his ultrastructural study of mitosis in *Chara*.

Subirana has proposed a molecular model to account for chromosome movements in mitosis based on production of active shearing forces at the surface of the microtubules either by an actin-myosin-like interaction or by a similar mechanism involving a single protein. His model implies a different polarity in the continuous tubules on the two sides of the spindle hence a change in polarity at the midpoint. He also suggests that the chromosome-to-pole microtubules are of opposite polarity to the pole-to-pole microtubules, allowing the latter to act as "pumping units" helping the chromosomal microtubules in their anaphase movement.

A different suggestion is that the continuous microtubules emanating from one pole might be of opposite polarity to those from the other, since there is evidence that some of these microtubules associate as doublets. However, microtubules running from the kinetochores to the poles do not appear to pair with the pole-to-pole microtubules, and it has been suggested that this might be important in allowing the kinetochore microtubules to slide along the pole-to-pole microtubules during anaphase.

The kinetochore. The kinetochore, like the centriole, serves as a point of attachment for microtubules. But more than this, it is believed to be the site from which the microtubules grow by the addition of monomers at their bases during late prophase and metaphase. Harris and Bajer have shown that in *Haemanthus* endosperm a chromosomal fiber consists of a bundle of 50 to 100 microtubules connected to the kinetochore. Some studies suggest that the kinetochore of animal cells contains dense plate-like structures but there is also evidence for a fibrous nature. Electron micrographs of the kinetochore at meiotic metaphase in *Tradescantia* and *Lilium* pollen mother cells, published respectively by Wilson and Dietrich, show that the microtubules arise in a large region that is less electron-opaque than other parts of the chromosome. Wilson believes that the kinetochores in *Tradescantia* are composed of 70 to 80 Å granules rather than fibrils as reported for animal cells, and points out the need for further work.

The structure of the kinetochore is of prime importance for an understanding of the origin of microtubules and of the mitotic events leading to poleward migration of the anaphase chromosomes It is, however, a difficult region to study in the electron microscope. Microtubules emanating from kinetochores are also shown at high magnification in papers by Manton, Harrisand Bajer, Pickett-Heaps and Northeote and Cronshaw and Fsau.

Jokelainen has presented evidence that kinetochore differentiation is affected by low concentrations of colchicine, while Krishan has shown that fibroblast cells arrested in mitosis by vinblastine sulfate have prominent kinetochores showing details of fine structure usually not seen in untreated cells. It should be rewarding to use these drugs in exploring kinetochore structure in plant cells.

The centriole and basal body. The centriole is found only in a few gymnosperms and not at all in the angiosperms, and while it occurs generally among the lower vascular plants it is restricted in these to a brief phase of the life cycle. For this reason, the quite limited objective of this section is to place this organelle in perspective in considering the nature of the orienting centers for microtubules in higher plants. An exhaustive review of the classical literature of light microscopy on the centriole in plants was published by Lepper.

In animal cells and in those of some algae and fungi the centriole serves as an organizing center for the growth of microtubules comprising the continuous fibers of the spindle, and in many cases appears to give rise directly to them. Also, in the form of the basal body the centriole produces the microtubules comprising the familiar "9+2" structure of cilia and flagella. Stubblefield and Brinkley consider that the centriole is a highly specialized organelle for microtubule production and also serves as a temporary scaffolding to anchor the poles toward which the chromosomes move in mitosis. It is a cylindrical body with a remarkably elaborate architecture, which may include both DNA and RNA. Its outer wall consists of a ring of nine evenly spaced triplet microtubules. The microtubules of the mitotic apparatus originating in the centriole appear to grow out from its side wall toward the opposite pole. Presumably this occurs by addition of monomers to the microtubule bases at the centriole.

In lower eukaryotic plants, there are wide variations in the occurrence and manner of participation of a centriole or centriole-like body in the organization of spindle microtubules. In several recently investigated cases in which the nuclear envelope remains intact, so that an intranuclear or "closed" spindle apparatus is formed, a centriole, or a pair of centrioles, lies just outside the envelope and the spindle microtubules do not penetrate the envelope but terminate in a dense region just inside. Ichida and Fuller have studied this type of apparatus in a water mold. Lessie and Lovett have shown that vegetative nuclear division in the water mold *Blastocladiclla emersonii* is similarly intranuclear. In yeast microtubules radiate from the centriole into the cytoplasm, and "centriolar plaques" located at the nuclear envelope serve as polar foci for the spindle apparatus.

The nuclear envelope in the fungus *Armillaria* is discontinuous at the poles where the spindle penetrates the nucleus. The spindle microtubules appear to be attached at the poles to centriole-like bodies that lack the organization of typical centrioles. Johnson and Porter have demonstrated that the nuclear envelope of the alga, *Chlamydomonas reinhardi*, similarly develops discontinuities at the poles. The spindle microtubules arise just beyond these openings in a specialized region of the cytoplasm devoid of organelles. Although centrioles are present in the cell, they are involved not with the mitotic spindle but with formation of cytoplasmic microtubules that participate in cytokinesis.

In surveying the distribution of centrioles in the plant kingdom, basal bodies must be included since they have typical centriole structure and are generally considered to be centrioles that produce the microtubules for the shaft of the cilium or flayellum of motile cells. These microtubules arise as direct extensions of particular microtubules from one end of the centriole, and thus differ in their manner of production from those made by the centriole for the mitotic apparatus.

The origin of basal bodies from centrioles or procentrioles has been observed in a number of studies. In an ultrastructural investigation of the fungus *Allomyces macrogynus*, Renaud and Swin discovered that the basal body in the gamete is derived from a small centriole detectable in the tip of a vegetative hypha before it undergoes gametogenesis.

Centrioles or basal bodies occur not only in many algae and fungi, but also in all plants above these that produce flagellated sperm, including mosses and liverworts, all of the lower vascular plants including the form and finally the cycads and ginkgos among the gymnosperms. They are not known to occur, however, in other gymnosperms or in angiosperms, and in those plants above the algae and fungi in which they do occur, they generally appear only during gametogenesis and their origins are obscure. Mizukami and Gall have studied ultrastructurally the origin of the numerous basal bodies from procentrioles present in the large spherical blepharoplast in the sperm of the fern, *Marsilea,* and the cycad, *Zamia.*

The fine structure ot the very large multiflagellated sperm of *Zamia* has been explored by Norstog. In addition to the flagella and greatly elongated basal bodies, he has described a spiral system of microtubules underlying the plasmalemma in flagellated regions. He suggests that contraction of the microtubules in this system may account for the "euglenoid" movements characteristic of the apical portion of the sperm body.

In two notable series of papers, microtubular structures involved in the developing and mature sperm of a moss, *Polytricum*, have been described by Paolillo and associates and of a liverwort, *Marchantia*, by Carothers and Kreitner.

Origin of pole-to-pole microtubules in plaat cells.—Ultrastructural studies of mitosis in plant cells have made it clear that some of the microtubules of the spindle arise in the polar regions despite the absence of centrioles. If, as suggested the intricate structure of the centriole serves to make it a highly effective machine for the production (and disassembly ?) of microtubules, as well as a scaffolding for chromosome movement to the poles, its loss from most plant cells is difficult to understand. Stubble field and Brinkley stress the role of the centriole as a supportive structure in suggesting that the need for it in plant mitosis is eliminated because the spindle microtubules are embedded in the end walls, which supply the necessary support. Although there is no evidence that spindle microtubules of plant cells are connected in any way to the cell walls, nevertheless the rigidity provided by the walls is conceivably related to the loss of the centriole generally except in those plants and at that point in the life cycle where a motile stage requires formation of a basal body.

If the polar regions of dividing plant cells are examined for clues to the nature of the nucleation centers from which microtubules arise, no unusual structures are encountered. However, Pickett-Heaps and Northcote have observed that elements of endoplasmic reticulum, much of it smooth, become conspicuous at the poles at the onset of mitosis. As they have suggested, these elements may be producing the protein subunits needed for the assembly of microtubules. The occurrence of smooth endoplasmic reticulum at the poles immediately preceding and during mitosis of wheat cells has been reported also by Burgess and Northcote who have suggested that this system may transport microtubule subunits and play a part in their aggregation. An increased amount of finely granular, moderately electrondense material in the matrix surrounding the endoplasmic reticulum has also been observed. Apparently, then, it is to the endoplasmic retionlum and to this material that we must look for the orienting centers replacing the centrioles in the organization of the pole-to-pole microtubules in plant cells. In *Chara*, Pickett Heaps has noted that some spindle microtubules end at small, electron-dense, diffuse bodies.

The discovery of a "preprophase band" of microtubules by Pickett-Heaps and Northcote may shed light on the origin of spindle microtubules in some cells. In ultrastructural studies of mitosis in roots, coleoptile tissue, and young leaves of wheat, these workers found that a dense band of 150 or more microtubules encircling the nucleus appeared in the cells just before they entered prophase. The microtubules were oriented at right angles to the axis of the future mitotic spindle and occurred several units deep in the equatorial region where fusion of the future cell plate would be expected. During prophase the microtubules in the band disappeared and spindle microtubules appeared, suggesting that the band might be contributing microtubules to the developing spindle.

A similar preprophase band has been reported in roots of *Phleum* and in young leaves of tobacco and presumably will prove to be of widespread occurrence.

Burgess and Northcote have also made observations on wheat that suggest that the preprophase band may serve as a source of material for spindle microtubules. Preceding mitosis, numerous elements of smooth endoplasmic reticulum surrounded by an amorphous matrix material containing some microtubules appear in the cytoplasm along the side walls where the band of microtubules arises. Smooth endoplasmic reticulum is also observed here at the start of prophase when microtubules are disaggregating in the band and appearing at the poles. Conceivably, microtubules or portions of them migrate from the band to the poles, where they are reutilized.

The results of Hepler and Jackson with isopropyl N-phenyl carbamate are related to this question of centers of nucleation. When dividing endosperm cells of *Haemanthus* are treated with this chemical, a number of scatt red spindles appear, each with a cluster of microtubules radiating in all directions from the poles. Some of the microtubules can be traced to the kinetochores of nearby chromosomes. Since the microtubules are normal in appearance, as are other cytoplasmic components, it appears that the chemical may be sufficiently specific in its action to prove useful in further study of the nature of polar nucleating centers.

CYTOKINESIS

During mitotic telophase, cell plate formation is commonly initiated by the accumulation of vesicles midway between the daughter nuclei. As these fuse and additional vesicles accumulate and consolidate at the margins, the plate grows laterally until it reaches and joins the side walls. The participation of a fibrous structure, the phragmoplast, lying at the margins of the plate and appearing successively farther out as the plate expands, has been known from light microscopic investigations on fixed and stained cells for a century. In recent years, living cultured cells of the endosperm of the African blood lily, *Haemanthus katherinae*, have been used in light microscopic investigations of cell plate formation by Bajer and associates and Jackson. Observations made on this material with a variety of optical systems (polarized light, phase contrast, and Normarski differential interference contrast) have revealed that the phragmoplast consists of fibrillar elements which arise at the plate and radiate outward more or less normal to its plane, and numerous cytoplasmic droplets which are moved to and incorporated into the plate.

The fibrillar structure of the phragmoplast was recognized in the electron microscope by Sato in 1958. Ledbetter and Porter subsequently observed microtubules among which vesicles were accumulating between telophasic nuclei in an early stage of plate formation in dividing cells of root tips. Detailed ultrastructural studies employing fixatives which preserve the microtubules have now been made on cell plate formation in a variety of plant materials, including meristematic cells of root tips, cambial and procambial cells, meristematic cells of the wheat coleoptile, dividing cells of young leaves, cells giving rise to the stomatal apparatus in leaves, orchid microspores forming the generative cell, and cultured endosperm cells of *Haemanthus*. These studies have confirmed that the birefringent fibrillar elements previously noted in the light microscope are microtubules similar in appearance to those of the spindle.

The microtubules of the phragmoplast first appear in the midregion of the cell in early stages of plate formation when some spindle tubules, continuous through the plate region, are still present. At this time numerous vesicles begin accumulating and fusing in the plane of the future plate. Cinemicrographic studies indicate that in *Haemanthus* endosperm the plate may form in two or more places simultaneously and then grow centripetally until it is consolidated in the center of the cell, after which it grows centrifugally as it does in most dividing plant cells.

According to Inoue and Sato the phragmoplast microtubules arise at the plate, *i.e.*, they are assumed to be polymerized at initiating centers within the incomplete plate and at its margins, from which they grow out in opposite directions into the cytoplasm of the developing daughter cells. This view is supported by ultraviolet microbeam experiments. Irradiation of the phragmoplast specifically at the plate destroys both the birefringent elements and the capability for formation of new elements, while irradiation of the phragmoplast in areas away from the plate destroys the birefringent elements but not the reforming capability.

Inoue and Sato have suggested, from observations with the polarizing light microscope, that near the end of mitosis the spindle microtubules may depolymerize into their protein subunits, which then move to the region of the young cell plate where they are repolymerized into microtubules of the phragmoplast. Similarly, it seems likely that as the plate grows there is successive depolymerization of the older microtubules and incorporation of their subunits into new ones undergoing synthesis at the plate periphery. The lengths of the tubules vary with the stage of plate formation, commonly being shorter at more advanced stages as the side walls are approached.

The base of the phragmoplast microtubule, which may be anchored in the plate region and to which new subunits are assumed to be added as the microtubule grows outward, occasionally appears to be enlarged or bulbous and surrounded by electron-opaque, amorphous material. Hepler and Jackson suggest that this material may contribute to the growth of the microtubules.

As might be expected if they arise at the plate and grow in opposite directions, the microtubules of one developing daughter cell are not continuous through the plate with those of the other. The rare cases in which the microtubules do appear to be continuous are probably due to the contiguity or superimposition of the bases of tubules that arise near one another but grow in opposite directions. In *Haemanthus* the microtubules are interdigitated or overlapping at the plate, since those running into one developing daughter cell arise just beyond the plate in the other daughter cell.

It is generally agreed that the phragmoplast microtubules assist in establishing the zone along which the plate forms, and facilitate the movement of vesicles to this zone. The morphogenetic role of microtubules is best illustrated in certain asymmetric cell divisions, particularly in the formation of the subsidiary cells of the stomatal complex and of the generative cell in pollen where the orientation of the microtubules toward the pole produces a strongly curved plate.

In cinemicrographic analyses, Bajer and Bajer and Allen have observed swellings or droplets sliding along the phragmoplast filaments (clusters of microtubules) and fusing with the plate. It was first postulated by Whaley and Mollenhauer, based on a study of permanganate-fixed material, that the vesicles accumulating at the plate are derived from dictyosomes. Their view was soon affirmed by Frey-Wyssling *et al.* and has been supported by several more recent studies, in which use of glutaraldehyde fixation has permitted observations on the relationship of the microtubules to dictyosomes and to the vesicles arising from them.

On the other hand, several investigators have commented on the abundance of ER elements aligned between the microtubules leading to the plate. Some of these elements may become transformed into plasmodesmata, but the possibility has been recognized that others may contribute material to the plate itself. Hepler and Jackson pointed out that the aligned chains of vesicles near the plate in *Haemanthus* are derived from a cytoplasmic matrix rich in elements of the ER and sparse in dictyosomes. Also Pickett-Heaps, studying incorporation of tritiated glucose into the cell plate, found that there was as much or more label over elements of the ER as over dictyosomes. Cronshaw and Esau and Bajer have also suggested that the ER may be a source of plate material although subordinate to the dictyosomes.

Generally, ribosomes, elements of the ER, and vesicles occur within the phuagmoplast but dictyosomes and other large organelles are excluded, and occur only at its margins and beyond. Dictyosomes at the periphery of the phragmoplast commonly have an associated cloud of small vesicles, some of which are surrounded by a reticulate coat and closely resemble in size and surface investment the vesicles previously described from growing root hairs. Smooth-surfaced and coated vesicles occur abundantly in the plate region, and both types can be observed fusing with the plate. However, there are ordinarily relatively few vesicles in the intervening region of the phragmoplast between the peripheral dictyosomes and the plate. Presumably, undulatory or vibratory motions of the microtubules convey the vesicles rapidly plateward from the dictyosomes. Toward their bases, if the microtubules are anchored in the plate region, the undulations would diminish and the vesicles would come to rest. Undulate phragmoplast microtubules have been reported, but the possibility that they may represent fixation artifacts must be recognized The periodically spaced interconnections observed by Hepler and Jackson between the bases of closely adjacent microtubules are interpreted as cross-bridges which may serve to stabilize the microtubules and thus help to define the plane of the plate.

Schmitt has borrowed the sliding-filament theory of muscle action to suggest how microtubules might function in moving vesicles. The microtubule is assumed to contain subunits bearing the energy source (ATP or GTP) while the vesicle walls contain a binding site for the microtubule subunit and a nearby site with ATPase. With the vesicle and microtubule in contact, the ATP is hydrolyzed, energy is coupled, and conformational changes occur which provide a thrust, moving the vesicle so that a different site is in apposition with a different subunit of the microtubule. As the thrusting process is repeated, the vesicle moves or jerks along the microtubule in a direction dependent on the sense of the microtubule helix.

Hepler and Newcomb searched for evidence to an association between microtubules and vesicles in cross sections of the phragmoplast, but could detect no consistent grouping of microtubules around the vesicles or structural association between the two. Bajer, however, has observed an association in *Haemanthus*.

Considering the well-known effects of colchicine on microtubules, it is not surprising to find that its application inhibits cell plate formation. The spindle and phragmoplast microtubules disappear, and the vesicles fail to accumulate properly in the plate region. If colchicine acts on a cell in which cytokinesis is well advanced, a poorly positioned plate with swollen ends is commonly obtained. The daughter nuclei may be close together, suggesting that the phragmoplast microtubules normally play a cyto-sketelal role in keeping the chromosome masses apart.

Caffeine is of particular interest because it appears to be a specific inhibitor of cytokinesis. It apparently has little effect on mitosis, and does not destroy the microtubules, yet inhibits cell plate formation. Several groups have shown that the vesicles collect in the plate region in caffeine-treated cells, yet fail to coalesce into a plate. This curious effect of caffeine certainly merits further study, as it might lead to a deeper analysis of events in plate formation. Mesquita has found that in dividing cells of *Allium* root tips acenaphthene causes the formation of scattered, wandering sheets of cell plate material even though microtubules are present during their formation. Vincaleucoblastine also causes formation of highly irregular plates, and again microtubules are present, so that this drug appears to be more closely related to acenaphthene than to colchicine in its mode of action.

Finally, attention should be called to the distinctive type of cytokinesis described by Johnson and Porter in a superbly illustrated study of *Chlamydomonas reinhardi*. Cytokinesis is accomplished by appearance of an array of microtubules oriented along the cleavage plane, and then passage of a furrow between them. The authors suggest that the microtubules may facilitate cytokinesis by providing a rigid framework on the opposing sides of the furrow against which cytoplasmic furrowing can take place.

Asymmetrical cell divisions. Ordinarily, cells that differentiate do so after arising from more or less symmetrical divisions of meristematic cells. However, in a number of cases differentiation starts with a parent cell which itself becomes structurally polarized preceding mitosis. Displacement of the nucleus to one end of the cell results in production of an eccentrically oriented mitotic spindle. The new cell plate then forms midway between the spindle, so that the genetically equivalent daughter nuclei resulting from the division are localized in different cytoplasmic environments. It has been assumed that some difference in cytoplasmic organization underlies the asymmetric division and the subsequent differences that develop between the two daughter cells, and that study of the phenomenon will aid in understanding the basis of differentiation. Examples of differentiated cells that are initiated by asymmetric divisions in this way include root hair cells, the stomatal complexes in many monocots, the siliceous-suberous cell pairs in barley leaves, the tube and generative cells of the pollen grain, leaf cells in *Sphagnum*, and sclereids in *Monstera*.

The first ultrastructural investigation of an asymmetrical mitosis was that by Avers on formation of the root hair initial in *Phleum* made before introduction of glutaraldehyde as a fixative. The subsequent discovery by Pickett-Heaps and Northcote that a dense band of microtubules appears at least in many plant cells just before mitosis has expanded considerable our knowledge of the fine structure of asymmetric divisions, although just how basic the presence of the band may be in determining subsequent events is in question.

In monocots the "stomatal complex" consists of a pair of guard cells and may also include one to several paired of surrounding subsidiary cells, depending on the genus. Ordinarily these complexes arise in rows, with the older stages in development being located farther from the meristem at the leaf base. Stebbins and colleagues have discussed the influence which apparently emanates from the guard mother cell and induces the epidermal cells on either side to divide asymmetrically to give the subsidiary cells.

Pickett-Heaps and Northcote studied the development of the stomatal complex in wheat, which at maturity consists of a pair of guard cells each flanked by a subsidiary cell. Three cell divisions are involved in development of the complex ; the first and second divisions are asymmetrical, while the third is symmetrical (Fig. 9.3). They found that the band of microtubules which appeared preceding mitosis in each asymmetric division was asymmetrically located, and that its position predicted the plane and position of the future cell plate. In Fig. 9.3. *a*, *c*, and *d*, each microtubular band is shown as two short heavy lines, representing a median section through the band normal to its plane.

Preceding the asymmetric first division, the nucleus migrates to one end of the cell and in preprophase the microtubules form a band next to the wall around the nucleus (Fig. 9.3). The asymmetric second division, which produces the two subsidiary cells, is particularly interesting because prior to mitosis the nuclei of the flanking epidermal cells come to rest against the wall shared with the guard mother cell, and the preprophase band appears adjacent to this wall and off to one side of the nucleus (Fig. 9.3). It also lies perpendicular to the plane of the band that preceded the first division. Subsequently the cell plate forms in an arc that joins the walls of the cells immediately above and below the guard mother cell, adjacent to the former position of the band (Fig. 9.3). The last division, in which the two guard cells are formed, is symmetrical (Fig 9.3), but the wall between the guard cells remains incomplete (Fig. 9.3).

Further investigations have raised some doubts about the relevance of the premitotic band to the subsequent plane and position of cell plate formation. Burgess and Northcote investigated cell division in *Phleum* roots and found preprophase microtubular bands, but surprisingly, they were symmetrically disposed in the cells, as were the preprophase nuclei, even when the subsequent metaphase plates and cell divisions were asymmetric. Furthermore, Pickett-Heaps has investigated asymmetric divisions in *Chara* and ascertained that although microtubules are present, they do not show the preprophase grouping equivalent to what is observed in *Triticum*. There is also now available a study by Heslop-Harrison of the asymmetric division giving the generative cell in orchid pollen. Although the spindle itself is asymmetrically located and the cell plate that is formed subsequently is hemispherical, no premitotic grouping of microtubules is observed in the expected regions.

In pursuing further the morphogenetic significance of the preprophase, and, Pickett-Heaps has treated dividing cells of wheat with caffeine, which inhibits cytokinesis. Examination of multinucleate cells produced in this way showed that the position of microtubular bands appearing in prophase was not noticeably affected by the size, shape, position, or number of nuclei in the cell. Thus, in Pickett-Heaps' view, the preprophase grouping may be a result rather than a cause of polarization. It may not be determinative in the preprophase positioning of the nucleus, but like the latter, simply an expression of a particular cytoplasmic environment. Quite possibly, formation of the microtubular band represents a preparatory step for mitosis, at the onset of which the microtubules or their subunits are incorporated into the spindle.

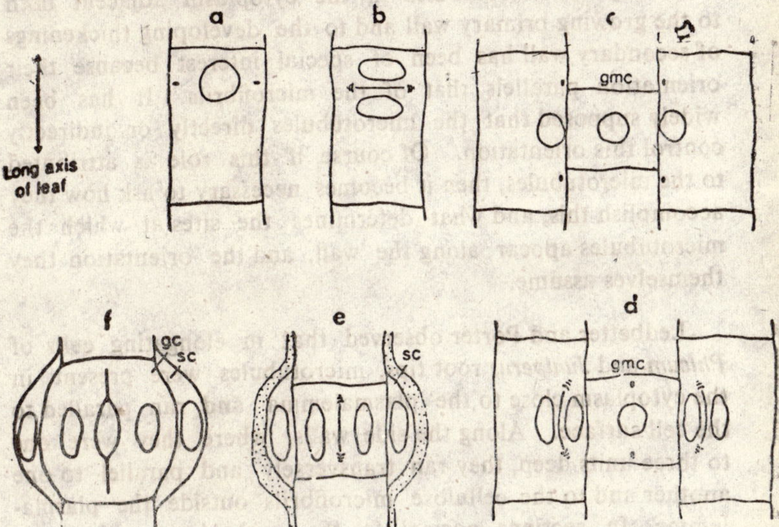

Fig. 9.2. Development of stomatal complex in *Triticum*.

a. Polarized primary stomatal mother cell. The two short, thick lines represent a group of microtubules in cross section and indicate the position and plane of the preprophase band.

b. The asymmetrical first division. The primary stomatal mother cell is dividing to give an epidermal cell and a smaller guard mother cell.

c. Guard mother cell (gmc) flanked by two epidermal cells. The two short thick lines in each cell represent the preprophase band of microtubules seen in section.

d. The asymmetrical second division. Lateral subsidiary cells are arising by asymmetric divisions in epidermal cells. Position and plane of preprophase band shown by short thick lines in the guard mother cell (gmc).

e. Symmetrical third division. The guard mother cell is producing two guard cells. S.c., subsidiary cell.

f. Two guard cells, each flanked by a subsidiary cell.

MICROTUBULES AND DEPOSITION OF CELL WALL MICROFIBRILS

Primary wall growth. For many years, students of wall structure and growth have recognized that the orientation of the cellulose microfibrils at the time of deposition determines the manner of cell enlargement and the shape of the cell, and that this orientation is under cytoplasmic control. Discovery of microtubules in the cytoplasm adjacent both to the growing primary wall and to the developing thickenings of secondary wall has been of special interest because their orientation parallels that of the microfibrils. It has been widely supposed that the microtubules directly or indirectly control this orientation. Of course, if this role is attributed to the microtubules, then it becomes necessary to ask how they accomplish this, and what determines the sites at which the microtubules appear along the wall, and the orientation they themselves assume.

Ledbetter and Porter observed that in elongating cells of *Phlcum* and *Juniperus* root tips, microtubules were present in the cytoplasm close to the plasmalemma and ran paralled to the cell surface. Along the side walls, where they were one to three units deep, they ran transversely and parallel to one another and to the cellulose microfibrils outside the plasmalemma. In sections normal to the protoplast surface the microtubules were never objected in contact with the plasmalemma, but were always separated from it by about 100 Å or more. At the end walls, both the microtubules and the wall microfibrils were randomly oriented.

Examples of the appearance of microtubules adjacent to the side walls of elongating cells are shown in Fig. 9.3.

Although microtubules near the primary wall in growing cells have undoubtedly been very widely observed since the publication of Ledbetter and Porter's original observations, they have generally received only passing notice, usually in connection with studies of secondary wall deposition modifferentiating cells. There are a few reports of microtubules adjacent to the primary walls of algae. Pickett-Heaps observed microtubules near the plasmalemma corresponding in alignment to the wall microfibrils in *Chara*. Nagai and Rebhun found microtubules near the wall in *Nitella*, but whether they were aligned parallel to the microfibrils is not clear from the micrographs. Kiermayer has reported microtubules running parallel to the cells surface and to one another beneath the plasmalemma in the desmid *Micrasterias*.

Although dividing cells have been the subject of study in a number of laboratories, these have yielded little or no information about microtubules in relation to be primary wall, since microtubules never occur next to the side or end walls in cells undergoing mitosis or cytokinesis. Pickett-Heaps and Northcote did, however, observe microtubules adjacent to the primary walls in premitotic cells of wheat.

With the evidence available, it seems likely that microtubules may be required next to primary walls in order for cellulose microfibrils oriented parallel to one another to be deposited. At any rate, whenever parallel microtubules are seen in the cytoplasm next to the primary wall, the most recently deposited wall microfibrils always appear to be similarly oriented. Whether microtubules also are associated typically with the random deposition of microfibrils is not clear ; as discussed later, they are not in some cases.

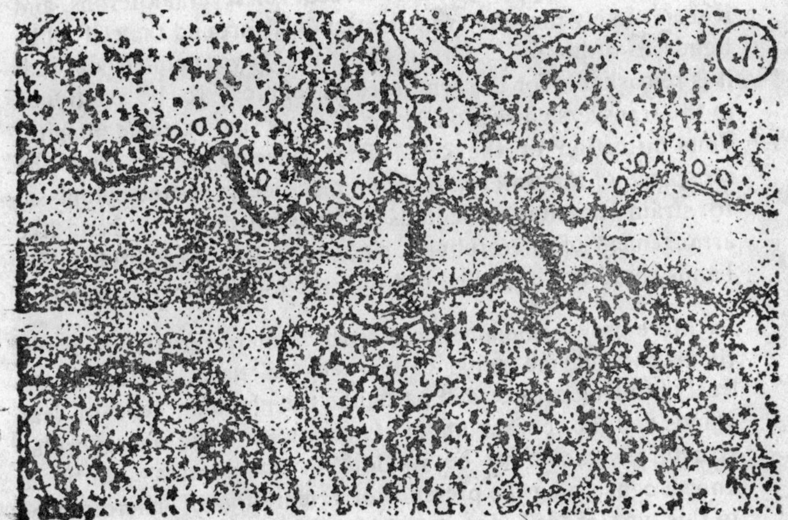

Fig. 9.3. Microtubules adjacent to side wall (primary wall) in a meristematic cell than a root up of white sweet clover (*Melilotus alba*). × 76,000.

Unfortunately, information on the occurrence of microtubules localized next to the wall is available for very few of the cell types classified according to the pattern of microfibril deposition and manner of wall growth. The observations of Ledbetter and Porter, and most of those since have been made on tissue cells which grow predominantly in length. Cells of this type include young vascular elements, fibers of various tissues, fusiform cambial initials, and parenchyma cells of apical meristems. The newly deposited microfibrils in these cells are usually oriented in a direction transverse to the root or shoot axis. Since the greatest coherence is along the axis of the transversely oriented microfibrils and the wall yields most readily to the turgor pressure perpendicular to this, the cells consequently elongate in the direction normal to that of the newly deposited microfibrils.

The observations available on this type of cell indicate that the microtubules next to the lateral walls are always oriented transversely like the newly deposited microfibrils. Individual microtubules frequently can be traced for several microns, and it seems likely that they run continuously around the cell "like hundreds of hoops". Green and King have pointed out that it is remarkable for the syntheses of cellulose to be continued in the transverse direction in the face of an extreme longitudinal distortion of the cell. They have suggested that the resistance of the wall synthesizing machinery of the cytoplasm to strain orientation could be explained by a band-like arrangement of microtubules, since until they were broken, the bands could not be pulled out into the longitudinal direction.

More or less isodiametric growth is characteristic of many tissue parenchyma cells, callus cells of tissue cultures, and freely growing cells or cell parts like the globular cells of leaf glands and the sporangium of *Phycomyces*. The microfibrillar structure is known to be random in some cases. Little is known concerning the presence or orientation of microtubules in this type of cell despite its frequency of occurrence and importance in physiological studies. Halperin and Jensen have reported that in carrot cell cultures, microtubules are not present when the cells are grown in the presence of auxin but appear when the auxin is removed. It would be interesting to know whether random and parallel microfibrils are deposited, respectively, under the two conditions.

Root hairs, pollen tubes, fungal hyphae, and the rhizoids of algae and fungi are examples of freely growing tubular cells or parts of cells with tip growth only. Both root hairs and pollen tubes have been studied ultrastructurally, permitting a comparison to be made of the relationship of microtubules to their patterns of microfibril deposition. In both, microfibrils are deposited in a random network at the tip, and then are pushed aside and become part of the side wall as the tip extends. Since the side wall does not extend, the random position of these microfibrils is not altered subsequently. Microtubules do not occur at the tip either in the root hair or the pollen tube, hence are not involved in the random deposition of the microfibrils which takes place at the growing tip.

In the pollen tube, no deposition of microfibrils occurs along the side wall, nor are microtubules known to occur in the adjacent cytoplasm. In the root hair, however, reinforcing microfibrils are deposited axially on the inner face of the wall, starting some distance from the tip (detectable at about 25μ from the tip in radish). Newcomb and Bonnett found that *axially* oriented microtubules occur in the cytoplasm next to the side wall where the axial microfibrils are being deposited. Although the microtubules extend forward into the region where only random microfibrils occur in the wall, they can have no influence on these microfibrils, which were deposited at the tip and remain immobilized thereafter. This point needs to be stressed, lest it be concluded that random deposition occurs in the presence of axially oriented microtubules. Probably the presence of microtubules in this region simply means that they are more easily detected than are the beginnings of axial microfibril deposition. The results furnish additional support for the hypothesis that, at least in higher plants, parallel deposition of microfibrils requires the presence of similarly oriented microtubules.

It is interesting to note that the *Phycomyces* sporangiophore resembles root hairs in depositing a reinforcing layer of axially oriented microfibrils (of chitin, not cellulose) on the inner surface of the side wall back of the growing tip. Although the fine structure of *Phycomyces* has been investigated, no microtubules have been reported.

Secondary wall deposition. Numerous studies have demonstrated that oriented microtubules occur in the cytoplasm adjacent to the developing secondary wall thickenings of differentiating xylem elements. The relationship was first shown by Hepler and Newcomb for parenchyma cells undergoing redifferentiation into tracheary elements in wounded *Coleus* stem, and shortly thereafter by Wooding and Northcote for secondary xylem elements differentiating from cambial initials in *Acer pseudoplatanus*. Other studies include those by Cronshaw and Bouck on differentiating xylem in *Avena* coleoptiles ; Cronshaw on xylem derived from cambium in *Acer rubrum* and on tracheilds differentiating from parenchyma cells in tobacco pith cultures : Esau, Cheadle and Gill on tracheary elements differentiating with helical thickenings in *Beta* and *Cucurbita* Pickett-Heaps and Northcote on differentiating xylem cells in the seylem and root of wheat ; and Srivastava and O'Brien on xylem differentiating from cambium in *Pinus*. In addition, Pickett-Heaps has used experimental approaches in contributing to the study of microtubules in xylem differentiation, namely by application of colchicine to wheat seedlings and by the use of radioisotopes to follow incorporation into the secondary wall thickenings of differentiating xylem in the same material.

Considered collectively, these studies present a remarkably consisted picture, and one which is essentially similar to that found for the primary wall. Except, apparently, for the earliest stages, microtubules are invariably present in the cytoplasm adjacent to the secondary wall thickenings or bands throughout their development. Moreover, the microfibrils within the bands run parallel to the microtubules, and since the microfibrils parallel the direction taken by the bands, the microtubules and bands likewise follow the same course. (It should be remembered that the microfibrils of a particular layer of the secondary wall retain their parallel orientation throughout development, in contrast to those of the primary wall which are gradually reoriented deeper in the wall as cell elongation occurs.)

In transections of a cell, the microtubules are seen to follow gently curving paths conforming to the curves of the wall thickenings. It seems likely that the microtubules associated with annular thickenings form complete circumferential rings in the cytoplasm. Whether in the case of a helical wall thickening the particular group of microtubules seen at any one level follows the course of the thickening for the entire length of the cell or only for a relatively short distance has not been determined.

As in the case of the primary wall, in suitable sections the microtubules usually appear to be separated from the plasmalemma by at least 100 A. They are not as concentrated as they sometimes are along the primary wall, but are rather sparsely and irregularly distributed around the thickenings, and most frequently only one unit deep. They are remarkably persistent, considering the lability of microtubules as a class, and can still be found around the secondary wall thickenings at late stages of differentiation when the cytoplasm is obviously undergoing degradation. Both smooth-surfaced and coated vesicles derived from the cytosomes are commonly observed among them. Other cytoplasmic bodies, however, do not appear to be associated with the microtubules or noticeably concentrated in the vicinity of the thickenings.

Pickett-Heaps and Northcote discovered that at an early stage in differentiation of a xylem element, when the thickenings of secondary wall first become evident, the microtubules occur along the primary wall between the thickenings, and do not appear over the thickenings until later. Possibly the presence of microtubules only along the primary wall at this stage indicates that secondary wall formation has not yet proceeded far enough to stop primary wall growth, and that initiation of secondary wall formation in a particular pattern over the primary wall surface involves a corresponding temporary disappearance of the microtubules.

The initial stages in the development of the secondary wall thickenings in the endothecial cells of the lily anther, studied by Heslop-Harrison differ from the above account for xylem. The thickenings in the endothecial cells consist of a basal plate of cellulose from which arise branching bars on the lateral walls. Before visible thickenings appear, oriented files of microtubules define zones that delimit the prospective sites of the bars. During the thickening process, the microtubules are restricted entirely to the vicinity of the bars. As expected, the cellulose microfibrils are oriented in the direction of the bars and microtubules.

Remarks on microtubules in relation to microfibril deposition. From the observed correlation of orientation between microtubules and the newly deposited cellulose microfibrils in both primary and secondary walls, it has been widely assumed that microfibrillar orientation at time of deposition is under microtubular control. Although Ledbetter and Porter suggested that this control might be achieved through the influence of the microtubules on cytoplasmic streaming, more recent evidence implicates much finer cytoplasmic filaments in streaming in a number of organisms. In the slime mold *Physarum*, which exhibits vigorous streaming, most investigators have been unable to find microtubules, but have observed numerous filaments about 50 Å in diameter. Since these closely resemble the muscle filaments 50 to 70 Å in diameter that are built from the globular protein actin, and since a protein has been isolated from slime mold that resembles actin closely and forms 50 Å filaments there is good reason to believe that actin or actin-like filaments may be of wide occurrence and may underlie the streaming phenomenon.

In the alga *Nitella*, whose growth has been analyzed intensively by Green, Nagai and Rebhun have also concluded that microfilaments are involved in developing the motive force for streaming. They found that bundles of microfilaments, each 50 Å in diameter and many microns long, were oriented parallel to the direction of streaming in the interface between the moving endoplasm and stationary ectoplasm where the site of motive force for streaming is believed to be located. The microtubules, however, were located near the wall in the stationary ectoplasm, where they could not be involved in streaming in any direct way.

Similar bundles of microfilaments were discovered in epidermal and parenchymatous cells of the oat coleoptile by O'Brien and Thimann and in cells of *Chara* by Pickett-Heaps.

The bundles observed by O'Brien and Thimann were composed of microfilaments 50 to 70 Å wide and of indefinite length. They paralleled the long axis of the cell and the direction of streaming, while the microtubules sometimes observed in the same cell were transversely oriented. The authors recognized the possibility that the microfilaments might be subunits of microtubules, and that if so, they would participate in streaming as microfilaments, but primarily in controlling shape when part of the microtubule.

Recently Hohl *et al.* have reported microfilaments of similar dimensions in the cytoplasm of *Acytostelium*, a cellular slime mold, where again they have been implicated in streaming. Microtubules are also encountered in the cytoplasm of *Acytostelium*, but are not numerous and are randomly oriented. On the other hand. Sabnis and jacobs have shown that in the coenocytic alga *Caulerpa*, the streaming endoplasm is filled with bundles of numerous, evenly spaced, oriented microtubules. They suggest that the microtubules may provide the framework, or may delimit areas of cytoplasmic substrate, upon which the motive force responsible for streaming is generated.

On balance, then, it appears that there is accumulating evidence for the presence of microfilaments in plant cells, but it is by no means clear that they will prove to be common cytoplasmic components. There is also the possibility, in view of the previously reviewed information, that microfilaments and microtubules are composed of chemically identical or closely related subunits.

Although there appears to be no evidence or support for the view that microtubules adjacent to the wall participate in the enzymatic synthesis of the cellulose microfibrils, it is generally agreed that they are involved "directly or indirectly" in orienting the microfibrils as they are deposited. This hypothesis has been greatly strengthened by experiments with colchicine, which has the same effect on microtubules located near the walls as it does on those of the spindle. In work on *Nitella*, where the microfibrils are deposited transversely, Green pointed out that in a simple mechanism to account for continued transverse synthesis, the orientation of new microfibrils could be based on that of pre-existing ones. When colchicine was applied, the cytoplasm was induced to put down a layer of random microfibrils, but then after removal of the drug, it was able again to deposit transverse microfibrils. Thus he showed that the order was in the synthetic machinery, not in the pre-existing microfibrils, and that this order could be re-established after removal of the drug.

The uniformity of spacing usually observed between the plasmalemma and the microtubules closest to it suggests that the two components may be attached, and several investigators have reported visual evidence for the presence of a connecting substance between them. Ledbetter and Porter suggested that the point of attachment might anchor a portion of the microtubules so that any undulations along it would induce a streaming motion. Ledbetter subsequently suggested that a stimulus arising from the microtubules might be transmitted in some way by the plasmalemma to the growing wall. Pickett-Heaps and Northcote suggested, as one of the possible roles of the microtubules, that they might exert specific directional control of particular enzyme systems especially concerned with formation of the organized wall microfibrils. A proposal along similar lines was made by Cronshaw, who reasoned that microtubules attached to the inner surface of the plasmalemma might, through movement, bring about the orientation of cellulose synthesizing enzymes located on the outer surface, resulting in the deposition of oriented microfibrils.

How and in what form the cellulosic component reaches the cell surface from the cytoplasm remains unknown, however. There is now considerable autoradiographic evidence that vesicles derived from the dictyosomes contribute materials to both primary and secondary walls. Also, dictyosome vesicles have been widely observed in abundance near both types of wall, and in some instances an apparent fusion of dictyosome vesicles with the plasmalemma has been reported. Nevertheless, it has not been claimed in the autoradiographic studies that cellulose precursors reach the wall via dictyosome vesicles. During secondary wall deposition the dictyosomes at least for a time are quite active in producing numerous large smooth-surfaced vesicles but whether these contain cellulose precursors is unknown.

If the vesicles were to contain cellulose precursors, then the microtubules could conceivably function by collecting these vesicles in particular regions, thus controlling the places in which new wall material were deposited. Although it is difficult to see how this could influence microfibril orientation, it might explain how the banded patterns of secondary wall arise. Experiments by Pickett-Heaps with colchicine suggest, in fact, that the microtubules may play this role. When colchicine was applied to wheat seedlings, virtually all microtubules disappeared from the differentiating xylem but deposition of secondary wall continued. However, the thickenings that appeared were malformed and lacked any regularity in size or distribution along the wall. It appeared also that within the thickenings the oriented microfibrillar texture was preserved.

This suggests that for the secondary wall the microtubules may not control the orientation of the microfibrils, but rather determine where the thickening will occur, perhaps by serving as tracts where vesicles and other cytoplasmic constituents are collected and deposited. Initially deposited microfibrils of the secondary wall might serve as templates in orienting those deposited subsequently. This does not rule out the influence of microtubules in determining the initial direction of orientation, however.

10

The Matter and Mechanics of Cells

The cell of a higher plant is a part of a living system, and within this system cells are dependent upon one another. The behaviour of the cell is influenced not only by its own chemical and physical state, but also by the nature of the surrounding cells and of the organism as a whole. Many materials—foods, mineral substances, water, gases—are continually moving, in the form of molecules or smaller particles, from one living cell to another.

The environment of cells includes not only other cells but also the external environment of water, soil, and air in which the plant lives. The outermost cells of the root are in direct contact with the soil and with the air and water of the soil. Many, perhaps most, of the living cells of the stem and leaf are exposed directly to the air through openings in the leaf and stem which connect with spaces extending through the plant.

In studying the cell, then, we are concerned with the physical laws that govern the movements of solids, liquids, and gases in so far as these affect plant life. If we are to understand the many processes and activities that go on in the cell and in the plant as a whole, we must know something about certain basic physical phenomena—among them the properties of colloids, diffusion, osmosis, evaporation, and capillarity.

MATERIALS OF THE CELL

The compounds in cells range from those as simple as carbon dioxide and water to the highly complex proteins pigments, hormones, and enzymes found in the cytoplasm, plastids, and vacuoles. But whatever their number and complexity, the chemical substances of the cell exist in three physical states : as solids,, liquids, and gases. Solids of the cell include the salts dissolved in the cell sap, proteins and fatty materials in suspension in the cytoplasm or vacuole, and the cellulose, lignin, and other components of the cell walls. Liquids are represented by the water of the cell sap. The gases, like many of the solids, are present in solution. The chief gases are carbon dioxide and oxygen. Another condition, not comparable to the first three, has already been mentioned. This is the colloidal state, outstandingly important in living organisms. It is different and less well known than the three physical states of matter.

As we have seen, protoplasm is a colloid, and all life is associated with the calloidal state. Consequently, an understanding of the nature of colloids will enable us to understand the behavior of protoplasm and hence some of the characteristics of life itself.

COLLOIDS

We shall understand colloids more easily if we first consider *suspensions* and *solutions*. A suspension can be prepared by stirring a spoonful of fine sand into a glass of water. The particles of sand in the water are so large that they may be seen with the naked eye or with a compound microscope. The sand will soon settle out of the water or may be removed by ordinary methods of filtration.

On the other hand, if salt or sugar is added to water, the resulting mixture is a solution. The solid (termed the *solute*) completely disappears in the water (the *solvent*). The particles of the solute are molecular in size or smaller. They cannot be seen, and the resulting mixture is physically so homogeneous that the solute cannot be separated from the solvent by settling out or by filtration. Gases also dissolve in solvents (oxygen and carbon dioxide both dissolve in water) and liquids dissolve in each other (mixtures of alcohol and water).

The particles of a colloid are intermediate in size between those of solutions and those of suspensions. Although salts and sugars are in solution in the protoplasmic fluids, protoplasm is predominantly composed of substances in the colloidal state. The colloidal state is characterized by the presence of insoluble particles so minute that they remain permanently suspended and cannot be removed by ordinary methods of filtration. These particles may be solid, liquid ; or gaseous. Although the colloid may appear homogeneous, like a solution, it is really nonhomogenous and consists of two phases, a *dispersion medium* and a *dispersed* phase. Milk is a good example of a colloidal suspension of particles of fat and protein dispersed in water. In protoplasm the dispersion medium is also water, whereas the dispersed phase consists of minute particles of protein, fatty materials, and other substances.

Colloidal particles range in size from those larger than most molecules to those just small enough to be invisible under the ordinary microscope. The unit of measurement of the dispersed particles in colloids is the millimicron (mμ), one millionth of a millimeter. Colloidal particles have diameters between 1 and 200 mμ. Even the smallest colloidal particle has a diameter about one hundred times that of the hydrogen atom. The largest particles approach in size the smallest bacteria (250 mμ). Systems in which the particles are smaller than 1 mμ are usually called solutions. Systems in which the particle is larger than 200 mμ are considered to be suspensions.

Because of the small size of colloidal particles the ratio of surface area to volume is very great. The extent of this surface can be appreciated by considering a single cubic millimeter of matter (about the size of a large cell). A cube this size has a surface of six square millimeters. But if the matter in it were divided into particles of colloidal size (1 mμ in diameter), the surface area of the particles would be increased to 6,000,000 square millimeters or 6 square meters. Thus, colloidal particles have great potentialities for surface activity.

Properties of Colloids

Colloidal systems have certain important and distinguishing features. Some of these properties are summarised below :

(1) *Filterability.* As stated earlier, colloids are very slow to diffuse. Colloids diffuse very slowly if at all through a semi-permeable membrane. This can be experimentally shown by placing starch paste (colloid) and sodium chloride solution (crystalloid) in a collodion bag and immersing it in water. The sodium chloride will diffuse out into the water and the starch will remain within the bag. In this manner one solution can be separated from another. This process is known as dialysis (Fig. 10.1).

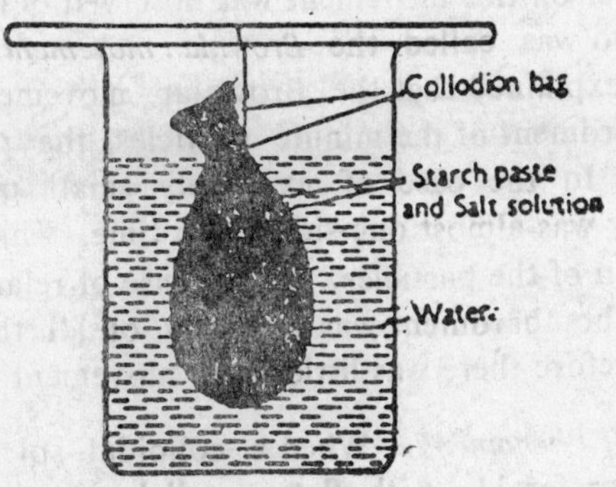

Fig. 10.1. Experiment to show dialysis.

The true solution can easily pass through ordinary filter papers without any separtation of dispersed phase and dispersion medium because the size of the pores of such filters ranges from 2 to 5 microns in diameter. The colloidal particles which range from 0.01 to 0.001 micron can also easily pass through such filters. But now ultrafilters have been designed which do not allow colloidal particles to pass through, whereas the particles (molecules and ions) of true solution can easily pass through.

(2) *Increased Surface Area and Adsorption.* It is well known that the total exposed surface area of a solid mass increases many times when it is subdivided into fine particles. The large surface is of vital importance in biological reactions. The interfaces of the minute colloidal particles are characteristically the seat of the phenomenon called *adsorption.* Adsorption consists of an interfacial concentrating of molecules or reactants. This is very important for chemical reactions. The larger the surface the greater is the process of absorption, facilitating a large number of complex chemical reactions in the protoplasm. This is why colloidal catalysts like platinum, etc., are used in many chemical reactions.

(3) *Brownian movement.* In 1829 *Robert Brown,* a botanist, while examining pollen grains in a drop of water under the microscope observed that they were in a rapid oscillatory motion. Later on this movement was observed in the particles of colloids and was called the *Brownian movement. Perrin,* a Frenchman, explained that the Brownian movement was due to the bombardment of the minute particles, the molecules of the solvent. In the case of extremely small particles the bombardment was almost one-sided at a time, which caused a zig-zag motion of the particles. In the case of relatively large particles, the bombardment was generally on all the sides at a time and therefore there was little or no movement in them.

(4) *Tyndall Phenomena.* When a colloidal sol is kept in a glass vessel, preferably with flat, parallel sides and a strong pencil of light is passed through it, the path of the light in the sol becomes luminous. This is called the *Tyndall Phenomena* and results from the scattering or diffraction of light. This phenomenon cannot be seen in the case of pure water or a true solution. The ultra-microscope is based on the principle of the Tyndall phenomenon. The limit of the resolving power of ordinary microscope is about 0.1μ. The ultramicroscope can be used for detecting the presence of particles in the size range of 0.001 μm to 0.1 μm, *i.e.*, in the colloidal range.

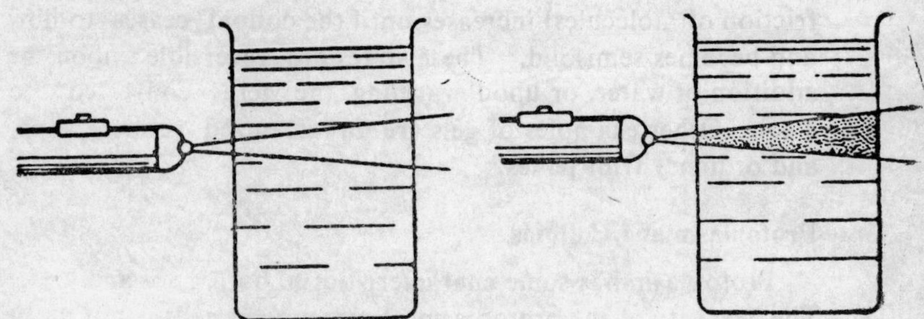

Fig. 10.2. Representation of the Tyndall effect. (*a*) True solution. (*b*) Collodial suspension.

(5) *Viscosity.* The viscosity of a fluid is its resistance to flow. While the viscosity of hydrophilic sols is almost similar to that of water, the viscosity of hydrophobic sols is usually greater than that of its dispersion medium.

(6) *Electrical Properties.* The dispersed particles of all hydrophobic solutions are electrically charged. A colloidal solution is electrically neutral because every charge carried on a micelle is neutralized by a charge of opposite value carried by ion in the dispersion medium. The movement of positively charged ions towards the negatively charged poles is termed *cataphoresis* or *electrophoresis*.

(7) *Flocculation.* The stability of a hydrophobic solution is maintained by the charge which micelles carry. When this charge is reduced to the point where there is no difference of electrical potential across the double layer it brings about *agglomeration* of the individual particles into flakes which settle down out of the system. The phenomenon is known as *flocculation*.

Forms of Colloids

If you have ever prepared a gelatin dessert, you have seen the two states in which some colloids exist. The gelatin does not dissolve in the hot water but forms a colloid which flows readily and possesses many of the physical properties of liquids. Such a colloid is called a *sol.* Upon cooling, the gelatin sol forms a jellylike colloid known as a *gel.* In a gel the viscosity (resistance to flow depending on the internal friction of molecules) increases until the colloid ceases to flow and becomes semisolid. These states are reversible ; upon the addition of water, or upon warming, the gel is converted into a sol. Other examples of gels are unvulcanized rubber, soap, and ordinary fruit jellies.

Protoplasm and Colloids

Protoplasm has some characteristics of both a sol and a gel. The viscosity of the protoplasm of many plant cells is not more than eight times that of water—much less than that of glycerine or castor oil. Protoplasm in an active state flows freely, a property of a sol, yet it is elastic, which is characteristic only of a gel. Thus it does not conform to the picture of a colloidal sol, in which particles suspended in water are independent of each other and move about freely and at random.

It is difficult to conceive that protoplasm could carry on such complicated activities as the transmission of stimuli and assimilation unless some degree of structure were present. Life seems to be associated with some fundamental kind of organization. As a result of studies with such modern tools as X-rays and the polarizing microscope, various theories on the structure of protoplasm have been advanced. It has been postulated, for example, that, in addition to the presence of larger granules, droplets, and other particles, there exists in the basic ground substance of protoplasm an ultramicroscopic pattern due to the presence of characteristically complex protein molecules. These protein molecules may be aggregated into long chains, rods, or fibers, which are connected at intervals by side chains, forming a latticelike framework. The cavities of this spongy framework are filled with water, and carbohydrates, fatty materials, salts, and other materials are loosely associated with the lattice. This structure may change from time to time ; for example, the fibers may contract or fold, thus producing globular aggregates, still connected by side chains. Although such theories are suggestive, the complexity of protoplasm is such that no theory of its organization has yet met with general acceptance.

THE KINETIC-MOLECULAR THEORY

An understanding of the behaviour of liquids and gases in the plant cell depends upon a knowledge of the kinetic-molecular theory. This theory conceives of gases and liquids as composed of small particles called molecules. An essential aspect of the theory is that, under ordinary conditions, molecules of all substances are in motion at all times and therefore possess *kinetic energy*, or energy of motion.

In samples of a liquid or gas, the motion of the molecules is random, in all directions. At any one instant a molecule is moving in a straight line with a uniform velocity. But the distance traveled will be very short, because of collisions with other molecules, and such collisions may cause a change both in velocity and in the direction of movement. The average velocity of all the molecules will, however, remain the same if the temperature does not change. Particles of colloidal size or larger possess little movement of their own.

The erratic, zig-zag movement of these particles, known as Brownian movement, results from bombardment by the particles of the dispersion medium. This movement was first observed by a botanist, Robert Brown, in 1828, while he was studying a suspension of granular particles set free by crushing pollen grains. When observed through a microscope, the tiny particles, suspended in water, were seen to be in a state of continual zig-zag motion. Later studies showed that particles of colloidal size also exhibit this motion in similar circumstances. The cause is the constant bombardment of the colloidal particles by the molecules of the dispersion medium. Because colloidal particles are so small, they are moved by this molecular activity, and this movement can be observed in the ultramicroscope. Brownian movement has been accepted as evidence to support the validity of the kinetic-molecular theory.

The effect of molecular bombardment of particles larger than colloidal size is easily seen under the microscope. The zig-zag motion occurs because the forces of impact from all directions are not always equal, and it frequently happens that the number of collisions on one side is momentarily greater than on another, so that the particles are pushed in one direction, only to be pushed in a new direction by another set of collisions. As a result, the particles appear to vibrate or follow an irregular path. This movement may be observed in protoplasm and in bacteria and other small bodies. It should be distinguished from motility brought about by organs of locomotion.

DIFFUSION

Should the instructor spill a teaspoonful of a concentrated solution of ammonia on his demonstration table, the students sitting near by would become acutely aware of the fact before very long. Sooner or later even those in the far corners of the room would notice the biting odor of the ammonia. Even if air currents in the room were reduced to a minimum, the molecules of ammonia would spread by *Diffusion*. This process involves the movement of molecules from a region where they are in high concentration to a region where they are less concentrated. The molecules of ammonia gas, freed from the spilled liquid, move off in random directions through the air. The inevitable collisions with one another and with air molecules send them in every direction. As more and more ammonia molecules spread through the room, they reach a concentration sufficient to produce a detectable odor. While the ammonia molecules are moving from the region of their greater concentration to areas of lower concentration, the air molecules are doing the same thing. They are moving from the region of greater concentration (of air molecules) to the region of less—that is, to the region of the spilled ammonia, where the presence of the ammonia molecules means there are fewer of the molecules making up the air. This process continues until the ammonia molecules are distributed equally in the room. As long as there are differences in concentration, the movement away from the area of concentration will continue. Air currents may aid in the distribution of the ammonia, but the process of diffusion goes on independently of this action.

Diffusion in liquids is similar to diffusion in gases, for the molecules of water and the dissolved particles of the solute are also in a state of constant motion. The rate of diffusion in a solution is, however, much slower. Diffusion in a solution may be illustrated by placing a crystal of potassium permanganate at the bottom of a jar of water. As the salt dissolves, the purple colour gradually spreads upward until the water is uniformly coloured. While the particles of the solute are diffusing upward, moving from a region of high concentration to one of lower, water molecules are diffusing downward, where the concentration of water is lower because of the presence of the potassium permanganate.

Diffusion Pressure

Take a porous pot fitted with a delivery tube, the other end of which is dipping in a beaker of water. On inverting a jar of H_2 gas over the pot, bubbles of gas will be seen coming out in the beaker of water. This is due to the fact that on inverting a jar of H_2 gas over porous pot molecules of H_2 gas diffuse through the pores of porous pot and increase the pressure inside and some of the air is pushed out which is seen in the form of bubbles. This simple experiment shows that diffusing particles have pressure. This pressure is known as *diffusion pressure* and is proportional to the concentration or number of diffusing particles. It may be remembered that the diffusion is due to diffusion pressure and not that diffusion pressure is the result of diffusion.

Like gases, liquids and solids also possess a diffusion pressure, although the existence of such a pressure in their case becomes evident only under specific conditions. For example, if a piece of copper sulphate is put in a tumbler of water the molecules of $CuSO_4$ start diffusing into the water (Fig. 10.3). Near the crystal of copper sulphate concentration of diffusing molecules will be more so they have higher diffusion pressure than in some other part of the water where concentration of $CuSO_4$ molecules is less.

Just as gases have a pressure, whether or not any diffusion s occurring, so do liquids and solids have a diffusion pressure

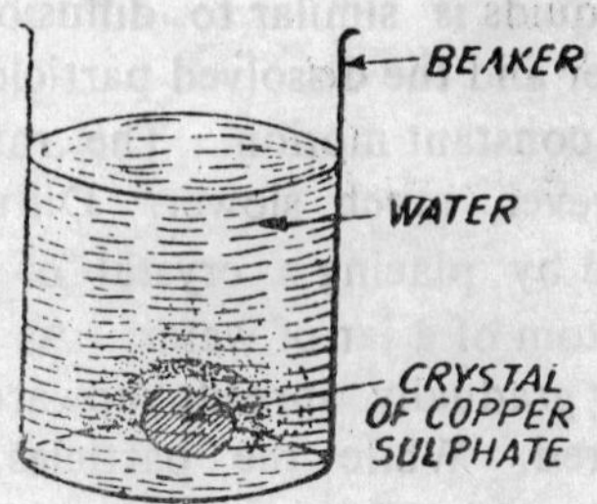

Fig. 10.3. Diffusion of Copper Sulphate.

whether or not any diffusion is taking place. Diffusion pressure is cause of diffusion, not its result. It is a property of diffusing particles because of their kinetic energy.

Principle of Independent Diffusion

If in a two-chamber box in which chambers are in contact through a small opening, 80% CO_2 and 20% O_2 are filled in

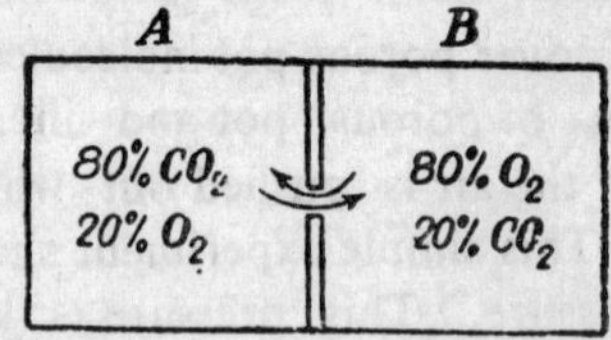

Fig. 10.4. Principle of Independent Diffusion.

chamber 'A' Fig. 10.4) and 80% O_2 and 20% CO_2 are filled in chamber 'B'; on noting the movement of CO_2 and oxygen we find that CO_2 moves from chamber A to chamber B and O_2 moves from chamber B to chamber A. Thus the movements of CO_2 and O_2 are from their higher concentration to their lower concentration independent of the presence of other gases. Principle of independent diffusion states that the movement of a substance into other medium is independent of the presence of the other substances and takes place from the place of its higher concentration to its lower concentration.

Factors affecting Rate of Diffusion

From the rate of diffusion, we understand the speed with which diffusion occurs. Various factors affecting the rate of diffusion are :

(*i*) *Temperature*. With an increase in temperature the speed of diffusion increases. It is because of the fact that the kinetic activity of the molecules of the diffusing substance increases on heating.

(*ii*) *Density of the Diffusing Substance*. Rate of diffusion is inversely proportional to the square root of the density of the diffusing substance. From density we understand the size and weight of the particles of the diffusing substance. Lighter the particles greater is their speed and therefore greater is the rate of diffusion. Oxygen and hydrogen have atomic weight 16 and 1 respectively. Therefore, hydrogen diffuses four times faster than oxygen.

(*iii*) *Density of the Medium through which Diffusion occurs*. Take a bell jar. Place a tube of bromine gas in it. The gas will start diffusing into the atmosphere of bell jar. Note the time taken by brown vapours of bromine for completely filling the jar. Now completely exhaust the air from the bell jar and repeat the experiment. Note again the time taken for complete diffusion of Br_2. It will be found to be less than the previous time. Thus denser the medium into which diffusion occurs lesser is the rate of diffusion. This simple experiment explains the fact that why rate of diffusion of water (transpiration) from leaves at higher attitudes is more than at lower altitudes.

(*iv*) *Diffusion Gradient*. Take two equal sized, two-chamber boxes. Fill 100% CO_2 in chamber A, 20% CO_2 and 80% O_2 in chamber B, 60% CO_2 and 40% O_2 in chamber A′ and 40% CO_2 and 60% O_2 in chamber B′. If we note the direction of movement of gases in the first set CO_2 (100%) from chamber. A will move towards the chamber B, and in second set CO_2 (60%) will move from chamber A′ to chamber B′. Thus the direction of movement of CO_2 in both the sets is the same but the rate of diffusion in the two cases will be different.

In the first set for every 100 molecules of CO_2 moving from A to B 20 are moving from B to A so that there will be a net increase of 80 molecules in chamber B.

In the second set for every 60 molecules of CO_2 moving from A′ to B′ 40 are moving from B′ to A′. There is thus a net increase of only 20 molecules in B′. Comparing the two cases it is clear that for diffusion of the *same number of molecules*, CO_2 in chamber A will take comparatively less time for its diffusion into B than the time taken for diffusion of CO_2 from chamber A′ to B′. Thus higher the difference in concentration between two regions higher is the speed of diffusion.

It is clear from the above experiment that rate of diffusion is directly proportional to the concentration of gas at the two ends. We already know that diffusion pressure is directly proportional to the concentration. Therefore, the rate of diffusion is directly proportional to difference in diffusion pressure between the two ends.

Now if we increase the length of the tube (distance) between chambers of boxes with the same concentration of gases in respective chambers, the rate of diffusion is inversely proportional to the distance ; greater the distance between the two chambers lesser is the speed of diffusion. Therefore,

$$\text{Rate of diffusion} = \frac{\text{Difference in DP at two ends}}{\text{Distance through which diffusion occurs}}$$

These two factors, *i.e.*, difference in DP at two ends of the gas and distance through which diffusion takes place constitute the *Diffusion gradient*. Higher the diffusion gradient, higher is the rate of diffusion. Therefore,

$$\text{Diffusion gradient} = \frac{\text{Difference in DP at two ends}}{\text{Distance through which diffusion occurs}}$$

$$\text{D.G.} = \text{K}$$

where K is a constant factor.

Importance of Diffusion

Diffusion is important in plants. Transpiration is a diffusion process and involves the movement of water from higher to lower concentration. The principle of independent diffusion is important for photosynthesis and respiration. The air present in the mesophyll has greater concentration of CO_2 than outside air and because of the principle of independent diffusion entrance of O_2 from outside to inside and exit of CO_2 from inside to outside goes on at the same time.

OSMOSIS

Tea diffuses through a teabag rapidly because the teabag is *permeable* to the tea—it lets the tea through. But a teabag made of aluminum foil would not work at all, because aluminum is *impermeable* to tea. Cell membranes cannot be described as either permeable or impermeable. Most membranes are *differentially* permeable—that is, they are more permeable to some substances than to others.

Plant membranes are usually permeable to water, but they may or may not be permeable to substances dissolved in the water. When water moves through a differentially permeable membrane, from a region of high concentration of water to one of lower concentration of water, the process is termed *osmosis*. Osmosis is thus a special case of diffusion. It is the diffusion of water through a membrane. If a solute is moving through a membrane, the movement is not osmosis, but diffusion. Both osmosis and diffusion are important in living cells and may occur simultaneously.

Osmosis may be demonstrated by means of a membrane separating two regions differing in their concentration of water (Fig. 10.5). A membrane, such as parchment paper, is tied firmly over the base of a thistle tube. The base of the tube is then filled with a sugar solution and the apparatus placed in a jar of water. The parchment-paper membrane is somewhat permeable to the molecules of the sugar, but more permeable to water. Because of the presence of the sugar molecules, the concentration of water within the thistle tube is lower than the concentration outside, in the jar of water. Water molecules pass through the membrane in both directions, since it is permeable to water. But during any given period of time more water molecules will move inward, because of the greater concentration of water outside, than will move outward. Hence the net movement of water will be into the sugar solution.

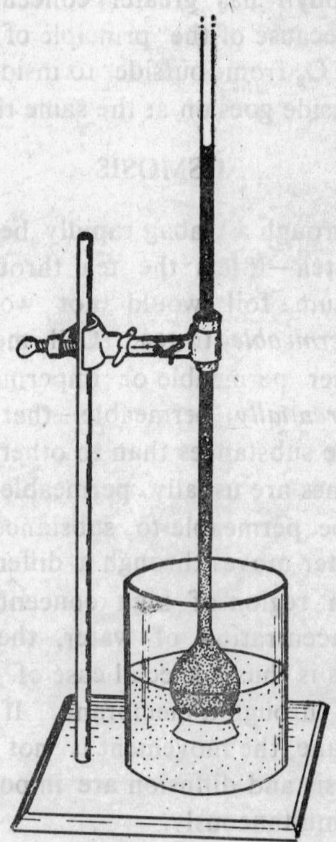

Fig. 10.5. Demonstration of osmosis. The thistle tube is closed by a differentially permeable membrane fastened over the lower end. The enlarged base of the tube was filled with a sugar solution and placed in a jar of water. Water has moved through the membrane, and the resulting pressure has pushed the sugar solution up the tube.

Because of this inward flow, the volume of the liquid inside increases, and hence the liquid rises in the stem of the thistle tube.

The thistle tube may be compared roughly to the vacuolated plant cell. In the plant cell, however, two membranes are interposed between the cell sap and the environment. The outermost of these membranes, the cell wall, is for all practical purposes a permeable membrane that does not interfere with the movement of water and solutes into or out of the cell. It is the cytoplasm lining the cell wall which acts as the differentially permeable membrane.

OSMOTIC PRESSURE AND TURGOR PRESSURE

Osmotic pressure is generally defined as the actual pressure which develops in a solution when it is separated from pure water by means of a strictly semi-permeable membrane.

The qualitative concept of osmotic pressure as the hydrostatic pressure produced by osmosis is inadequate. In the strict quantitative sense, the term osmotic pressure is not used for non-equilibrium (*i.e.*, temporary or even steady-state) pressures but only for the final equilibrium pressure (above atmospheric) attained as a result of osmosis under specific standard conditions. Since forces occur in pairs, osmotic pressure is measured by applying an increasing pressure until the applied pressure stops osmosis. The osmotic pressure is, therefore, also defined as the pressure required to stop osmosis when a solution is separated from pure water by a semipermeable membrane at 20°C. The applied pressure counteracts the effect of the solute on the free energy of the solvent. This change in free energy of water produced by the solute is called *osmotic potential* (π) of the solution.

Only the standard equilibrium pressure, which is called the osmotic pressure is numerically equal to the osmotic potential (though opposite in sign). It is given a negative sign since it is not a real pressure. Addition of solute makes the osmotic potential increasingly negative. Addition of solute also decreases the free energy of the solvent. Dilution of the solution with solvent will increase the value of osmotic potential.

Osmotic quantities are usually measured in terms of atmospheric pressures. The recent trend, however, is to measure them in *bars* (a unit of the metric system : 1 atm. = 1.01 bars).

It should be noted that any reference to osmotic pressure must indicate the temperature at which the pressure is measured *viz.*, 0°C, 20°C, or ambient temperature. Again it is not proper to assume that the osmotic pressures of two isotonic solutions are always equal. Equimolar solutions of a small molecule like urea and large one like sucrose have very different osmotic pressure. These points are clearly illustrated by Table 10.1.

Van't Hoff's equation relating the osmotic pressure to the concentration of the solution is expressed,

$$\pi V = n_s RT$$

where π is osmatic pressure in bars or atmospheres, V is the volume of the solvent, n_s the moles of solute, R is the gas constant (0.0820 litre atm/degree mole), and T is the absolute temperature. 1 mole of solute in 1 litre of solvent at 273°θ gives a value for π of 22.4 atm, or 22.7 bars. This equation is, however, not applicable to electrolytes and to solutes which become hydrated or bind water molecules.

Table 10.1. Osmotic potentials (bars) of solutions of given molal[1] and molar[2] concentrations at 20°C.

Conc.	*Molal Sucrose* 0°C	*Molal Sucrose* 20°C	*Molar Sucrose at 20°C*	*Molar KCl at 20°C*	*Molar NaCl at 20°C*
0.1	—2.4	—2.6	—2.7	—4.7	—4.4
0.2	—4.7	—5.2	—5.4	—8.9	—8.5
0.3	—7.0	—7.7	—8.2	—12.8	—12.6
0.4	—9.4	—10.2	—11.2	—16.9	—16.8
0.5	—11.9	—13.0	—14.5	—20.8	—21.2
0.6	—14.3	—15.6	—18.0	—24.4	—26.0
0.7	—16.8	—18.3	—21.7	—28.7	—30.1
0.8	—19.4	—21.1	—25.8	—32.6	—34.7
0.9	—22.1	—24.0	—30.1	—36.4	—39.1
1.0	—24.8	—26.9	—35.1	—39.7	—43.8

1. A molal solution is 1 gm. molecular weight+1000 grams of water.
2. A molar solution contains 1 gram molecular weight in 1 litre of solution.

The osmotic pressure of a solution is therefore, a rating of the potential maximum pressure which can be developed in a solution as a result of osmosis. Osmotic pressure of a solution is also said to be equal to the gas pressure, which the solute would exert if it were present as a gas at that temperature, in a volume equal to the volume of the solution.

The pressure which develops in a cell from time to time due to the osmotic diffusion of water is called the *turgor pressure*. The turgor pressure is variable unlike the osmotic pressure, which is always constant. During endosmosis the turgor pressure increases gradually in a solution and when it has reached its maximum limit it is equal to the osmotic pressure, provided the other liquid with which it is in a state of equilibrium is pure water or pure solvent.

WATER POTENTIAL (DPD)

Liquids, like gases and solutes, have a diffusion pressure. A pure solvent is supposed to have maximum diffusion pressure. When certain solute particles are added to the pure solvent, the diffusion pressure of the resulting solution will be lowered. The amount, by which the diffusion pressure of a solution is lower than that of its solvent, is called the *water potential* or *diffusion pressure deficit* (DPD). It can also be interpreted as follows :

The diffusion of water depends upon its chemical potential (free energy per mole). The wall pressure (Pw), increases the free energy of the cell's water. The sum of osmotic potential and wall pressure denotes a net change in chemical potential of the water relative to that of pure water at atmospheric pressure and at the same temperature. This net change in the chemical potential is called *water potential* (ψ). It is always negative (the maximum value being zero for pure water). The water potential controls the rate and direction of diffusion. Osmosis occurs from a region of higher water potential (lower DPD) to a region of lower water potential (higher DPD).

In case of a deficit in its diffusion pressure a solution will always tend to wipe it out by absorbing solvent particles. In other words water potential or D.P.D. is an index of the absorbing power of a solution. It is also sometimes called the *suction pressure* (S.P.).

Matric Potential

The hydrating colloid or gel-like material *e.g.*, the cell wall is called matrix, and its water potential is called the *matric potential.* Whereas osmotic potential is controlled by the presence of solutes, the matric potential is influenced by hydrophilic colloids. The matric potential will be lowest (most negative) for a dry material. When the matrix starts absorbing water, the matric potential increase toward a maximum value of zero when it is in equilibrium with pure water at atmospheric pressure. The relationship between degree of hydration and matric potential is called the *moisture release curve.*

Electro-Osmosis

Unlike *electrophoresis* where the charged solute particles migrate through the solvent to the anode or cathode, in electro-osmosis the solvent moves through fixed and charged particles under the influence of an electrical potential gradient. Negative charges in the wall of a pore induce positive charges in the surface layers of water molecules within the pore and the charged water then moves in the direction of the negative pole in response to an electro-chemical potential gradient. Conversely, positively charged pore surfaces induce negative charges in the water layer and cause movement in the reverse direction.

RELATIONSHIP OF O.P. T.P., AND D.P.D.

The D.P.D. of a solution is initially equivalent to its osmotic pressure. Thus if a particular amount of solute in a solution causes an O.P. of 10 atm., the diffusion pressure deficit of that solution will be equal to 10 atm. Thus, D.P.D.=O.P.

In a cell, during osmosis, the increasing *turgor pressure* forces the cytoplasm out against the cell wall. Since the cell wall is rather rigid, it will exert an equal and opposite pressure called the *wall pressure* (W.P) on the protoplasm and the water content of the cell. The W.P. will add to the diffusion pressure of the solution and hence a corresponding decrease in the D.P.D. will take place.

The relationship can be conveniently expressed in the equation

D.P.D. (S.P.) = O.P. − T.P. (W.P.)

The decreasing effect of increasing tugor pressure on the water potential or diffusion pressure deficit of cell is expressed.

Methods for Measuring Osmotic Pressure of Cell

(1) *Plasmolytic Method.* Plant cells, *e.g.*, epidermal peelings of Roheo leaf are placed in slightly rising grades of sucrose solution under similar conditions of temperature and pressure and for the same length of time.

The cells from different grades of solutions are then examined under microscope and that concentration of solution is found out in which cells are at incipient plasmolysis stage. The temperature of solution is recorded and the osmotic pressure of solution showing incipient plasmolysis is calculated using Van't Hoff's formula OP=CRT where C is molar concentration of the solution, R=0.082, T=(273+Temperature of solution in centigrade).

The average osmotic pressure of the cells under investigation is considered to be equal to the osmotic pressure of the solution in which incipient plasmolysis stage appears. The value obtained by this method is usually greater than the actual osmotic pressure of the cells since plasmolysis of cells is preceded by a shrinkage in their total volume which makes the cell sap slightly concentrated. For calculating the actual OP of cell sap the following formula is used

$$OP_1 = \frac{OP_2 \times V_2}{V_1}$$

where OP_1 is the real osmotic pressure, OP_2 is the calculated osmotic pressure, V_1 is the original volume of the cell at the time of starting the experiment and V_2 is the volume at the stage of incipient plasmolysis.

At the stage of incipient plasmolysis turgor pressure is zero, therefore, DPD = OP. Thus, in fact, we measure the diffusion pressure deficit or suction pressure of the cell which is equal to the osmotic pressure.

(2) *Cryoscopic Method.* There is a direct relationship between osmotic pressure and lowering of vapour pressure elevation of the boiling point and depression of freezing point of solutions. Hence osmotic pressure of cell sap can be calculated from measurement of any of these physical values.

The cell sap is extracted from the cell whose osmotic pressure is to be determined. The depression in freezing point is determined for the cell sap. The depression in freezing point means the decrease in freezing temperature of sap from freezing temperature of pure water which is the solvent of cell sap. If the depression in freezing point is known osmotic pressure of the cell sap can be calculated from the formula $OP=\frac{22.4\times\Delta}{1\ 86}$ where Δ represents depression in freezing point. Similar calculation for determining OP of cell sap can be done by noting elevation of boiling point and lowering of vapour pressure.

(3) *Manometric Method.* It is a direct method for measuring the osmotic pressure of solutions. A manometric tube is fitted in the solution in which osmosis is occurring. The maximum height to which water rises in osmometer is a measure of the osmotic pressure of that solution.

Factors affecting Osmotic Pressure of Plant Cells

Osmotic pressure of a solution is determined by the relative concentration of solvent and solute. More the concentration of solute higher is the osmotic pressure of solution and higher the concentration of solvent smaller is the osmotic pressure of the solution. The various factors affecting the concentration of solvent and solute and hence osmotic pressure of cells are discussed below :

A. Factors affecting the Concentration of Solvent

(*i*) *Transpiration.* If the loss of water (transpiration) is high the relative concentration of solute in the cells will increase and thus the OP of cell sap will also increase. Thus higher the rate of transpiration higher is the osmotic pressure of plant cells.

(*ii*) *Absorption.* More the absorption of water, higher is its concentration in cells and, therefore, lesser is the OP.

Since under normal conditions transpiration is equal to absorption therefore two factors cancel the effect of one another.

B. Factors affecting Concentration of Solute

(*iii*) *Photosynthesis.* During day photosynthesis takes place and sugars formed are accumulated in the cell sap thus increasing the concentration of solute. Therefore OP rises during the day. At night when no photosynthesis takes place sugars formed during day are oxidised and osmotic pressure decreases. If a graph is plotted at regular interval OP shows diurnal variations, *i.e.*, up to noon there is a regular rise in OP. Then OP falls down and is minimum at midnight after which it again shows a regular rise. This regular rise and fall in osmotic pressure during day and night is known as diurnal variation.

(*iv*) *Mineral Nutrients.* Higher intake of mineral nutrients increases the osmotic pressure of the cell sap. That is why plants growing in halophytic conditions show higher OP.

Negative Osmosis

We have seen that when water is separated from a solution by a semipermeable membrane, water moves into the solution. Sometimes a reverse process takes place. Water is withdrawn from the solution and moves towards the water, making the solution stronger. This is called *negative osmosis* or *anasmosis*. Bartell found that osmosis depended upon the size of the pores of the membrane. When the membrane has pores whose diameter was 0.9μ, no osmosis took place. With pore size between 0.4 and 0.1μ negative osmosis occurred. When the pore size was less than 0.1μ normal osmosis was exhibited. Water oozing out of certain cells at night is due to this phenomenon of negative osmosis, the water coming from the cell sap.

Magnitude of Osmotic Pressure in Plant Cells

The magnitude of osmotic pressure in the plant cells shows a considerable range of variation. In land plants it varies from 5 to 30 atmospheres and in aquatic plants from 1 to 3 atmospheres. Plants of arid regions possess high osmotic pressures, and in plants of saline soils the value of osmotic pressure may reach very high magnitudes. The highest osmotic pressure on record is of 202.5 atomspheres for the marsh xerophyte *Atriplex confertifolia.*

The osmotic pressure varies not only in different plants and in different habitats but also varies considerably in different tissues and organs of the same plant. Thus, the cells of the inner cortex of the root possess greater osmotic pressure than those of the outer cortex ; leaves usually have higher osmotic pressure than the roots ; and the higher leaves on a plant usually have higher osmotic pressures than the lower leaves.

The osmotic pressure of plant cells shows pronounced diurnal and seasonal fluctuations. In the leaves the osmotic pressure increases during daylight hours, reaching its maximum a little after noon and begins to fall in the evening reaching its minimum a little after midnight. This is due to the accumulation of organic compounds as a result of photo-synthesis in daylight and their translocation during the night. In the perennial plants of the semidesert regions of the Punjab the osmotic pressures are likely to be at their highest in the hot dry months (May-June) and fall rapidly with the advent of monsoon rains.

The Roles of Osmosis

The absorption of water by the root hairs from the soil and the movement of water from one living cell to another within the plant are partly brought about by osmosis. Osmosis is also mainly responsible for the movement of water from the non-living xylem elements into living cells. On osmosis depends the stretched and rigid condition of cells known as turgor which is responsible for keeping the young stems erect and leaves extended. It is also responsible for much of the force developed in growing tissues. The force developed in young roots growing through the soil and the force which enables the young seedling to break through the soil are chiefly osmotic in nature. Many of the turgor movements, for example those concerned in the opening and closing of flowers and the sleep movements of leaves described are probably dependent in part on osmosis. Another important function of osmosis is the control that it exercises over the distribution of water in the different parts of the plant after it has been absorbed from the soil. When, for example, the tissues of the leaves or shoots of a land plant become deficient in water as a result of transpiration, their osmotic concentration will increase and they will as a consequence withdraw water from other tissue with weaker concentration. Chandler (1914) found that in the case of cherries, apples and cucumbers when the branches bearing leaves and green fruits were allowed to wilt, the leaves continued fresh for a longer time than the fruits. The reason is that unripe fruits are poor in soluble sugars and, therefore, have low osmotic pressure. The leaves lose water and the osmotic concentration of their cells increases. As a consequence, the leaves withdraw water from the latter and remain fresh. When the fruits are ripe their sugar content increases and their osmotic concentration almost equals that of the leaves with the result that the latter cannot withdraw water from the fruits and consequently wilt early. The osmotic concentration of cells plays an important role in determining their *resistance to drought and frost*.

PERMEABILITY

The passage of substances into and out of the living cell depends upon the property of permeability of the membranes of the cell. A membrane which allows both water as well as solute particles to pass through it is called *permeable*. Some membranes which are impermeable to some substances and are permeable to other substances are known as *semi-permeable*. Many biologically important membranes allow some substances to pass through them much more readily than others and are hence known as *selectively* or *differentially permeable*. A plant cell normally has a differentially permeable membrane. The permeability of the complex cell membrane depends on the nature of the surrounding particles as well as on the changing conditions inside and outside the cell.

Of all the substances, water has the most rapid rate of diffusion through cell membranes. The membrane is also very much permeable to gases such as carbon dioxide, oxygen, and nitrogen and fat-solvent compounds such as alcohol, ether, and chloroform. Many organic substances such as mono and disaccharides, fatty acids, glycerol, amino acids as well as strong electrolytes *i.e.*, salts, and bases can penetrate cell membranes at a much slower rate. The cell membranes are almost impermeable to polysaccharides, phospholipids and proteins.

THEORIES OF CELL PERMEABILITY

A number of theories have been proposed to explain the permeability of the cell membrane towards a variety of substances. Only some important theories are described here.

1. "Retention Pressure" Theory

Traube (1867) proposed his "*Haftdruck*" or "retention pressure" theory. According to him the permeation of solutes through a membrane depended on their relative greatness of affinity towards the membrane substances as compared with that of water. Though not very wrong, his theory lacked conviction with the workers.

2. Solution Theory

According to *Overton* (1895), cell membranes are made up of lipids or fats and therefore, they are permeable to substances which are readily soluble in them. This theory gets support from the fact that permeability of the membrane towards different substances corresponds to their solubility in lipid.

The theory explains the accumulation of acids in the vacuole. Those commonly found in the vacuole (oxalic, malic, citric acids) are lipid insoluble and, therefore, cannot leak back into the cytoplasm. Lipid-soluble acids (acetic, lactic, pyruvic acids) are never found in the vacuole.

There are several objections against the theory :

(*a*) According to some workers hydrophobic proteins present in the membrane can also favour the permeation of lipid-soluble *i.e.*, the hydrophobic substances and, therefore, it is not essential to assume the presence of lipids in the membranes for the purpose of permeability as was done by Overton.

(*b*) Since lipid-soluble substances are often surface-active some workers assume that the easy permeation of these substances is not primarily due to their solubility in the lipids of the plasma membrane, but rather due to their absorption at lipid-water interfaces.

(*c*) The theory also fails to explain the entry of many substances (sugars, amino acids, minerals) which are not soluble in lipids.

3. Mosaic-Hypothesis

Nathansohn (1904) proposed that the plasma membrane was composed of a mosaic of both lipids and proteins, and that the lipid-soluble substances entered through the lipoidal areas and the lipid-insoluble substances entered through the proteineous parts. According to the mosaic hypothesis while lipid-soluble substances could enter easily through the membrane (an observed fact) the entry of other substances would vary from time to time depending upon variation in the protein part of the membrane. The theory, however, lacks experimental support.

4. Sieve or Ultrafiltration Theory

Ruhland and *Kuster* studied the behaviour of cells with respect to over 80 different acidic and basic dyes.

According to *W. Ruhland* (1912), *Kuster* (1911, 1918) and *Hoffmann* (1925) the cell membranes have very fine pores which act as molecular sieves or as ultra filter. These sieves allow smaller molecules to pass through more rapidly than the larger ones. Glucose molecules, therefore, diffuse more readily than the larger sucrose molecules through a cell membrane. *Seifriz* (1936) believes that the pores change in diameter according to surrounding conditions. The sieve is thus not a fixed one but is capable of dynamic change and adjustment. A change in the size of the pores can be caused by aggregation or coagulation, by peptization or dispersion, by reorientation of surface molecules and by a change in sign and magnitude of electric charge.

The theory, however, does not apply equally to different substances having similar size of molecules. It has also been shown that certain membranes are permeable to the large molecules of alkaloids while they are impermeable to the relatively small molecules of amino acids.

5. Colloidal Theory

The protoplasm and its membrane constitute a polyphase colloidal system. The colloidal theories assume changes in viscosity, phase inversion, or electrical reactions to be responsible for changes in permeability of the cell membrane. While *Spaeth* (1916) throught that the changes in viscosity or the concentration of the dispersed phase controls the permeability of a membrane, *Clowes* (1916) was of the view that inversion of the phase *i.e.*, change in the places of continuous and discontinuous phase brings about changes in the permeability.

The assumed phase reversal in the cell membrane has, however, been not substantiated.

6. Electrocapillary Theory

According to *Michaelis*, *Scarth* and *Lloyd* the membranes have pores lined with charged groups of membrane proteins. In the case of an electrolyte this facilitates the entry of positively charged particles through a negatively charged pore and *vice versa*. The non-electrolytes can, however, enter the membrane only if their molecules happen to be smaller than the size of the pore.

Plasmolysis

When a cell is surrounded by a relatively concentrated solution of table salt, the solution penetrates the cell wall and comes into contact with the cytoplasmic membrane, which is less permeable to the solute. Under these conditions, the concentration of water within the vacuole is relatively greater than the concentration of water in the external solution. As a consequence, water leaves the cell by osmosis, and both the cell wall and the cytoplasm tend to contract. The cell wall, although elastic, is relatively rigid, and soon reaches its limits of contraction. The cytoplasm, on the other hand, has been held in contact with the cell wall only by turgor pressure. This pressure drops as water moves out of the cell. As a result, the cytoplasm shrinks away from the cell wall, as can be observed under the microscope (Fig. 10.6).

This shrinkage of the protoplasm from the cell wall as a result of lack of turgor pressure to hold it in place is known as *plasmolysis*. The protoplasm forms one or several masses in the center of the cell, sometimes retaining fine cytoplasmic connections with the cell wall. Thin slices of beet or potato placed in a salt solution quickly give evidence of the effects of plasmolysis. They lose their crispness and become limp and flaccid.

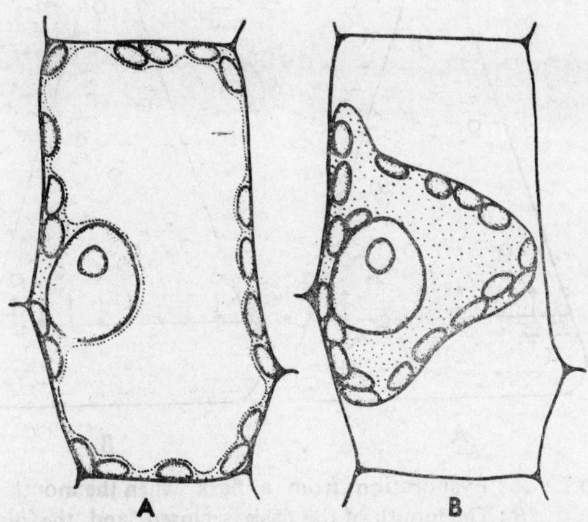

Fig. 10.6. Plasmolysis. A : Normal cell. B : Plasmolyzed cell.

Dead cells, because they have lost their differentially permeable properties, cannot be plasmolyzed. Weeds are sometimes destroyed by sprinkling salt around them. Rain forms a strong salt solution which plasmolyzes and kills the active root cells.

EVAPORATION

The transfer of water between one mature living plant cell and another is explained on the basis of osmosis. But under certain conditions water is transferred directly to the air. Since this process occurs in the leaf, its operation is worth noting.

Here again we are concerned with the kinetic energy of the molecules of liquids and gases. In a narrow-mouthed flask partly filled with water (Fig. 10.7A), the molecules of the liquid are in a state of active motion, moving at random in very short paths. The velocity of the movement increases as the temperature rises.

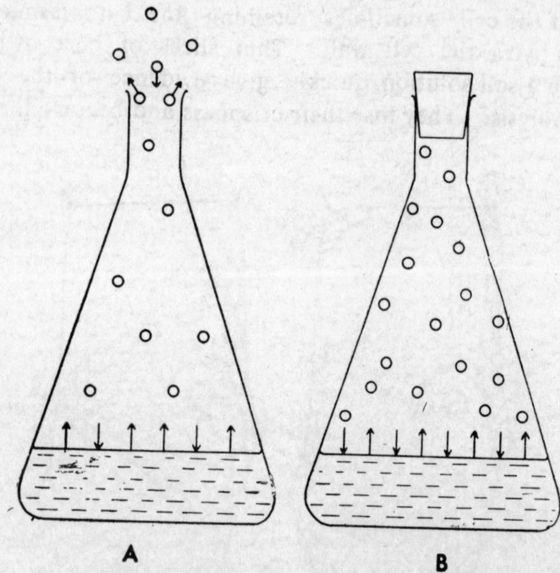

Fig. 10.7. *A* : Evaporation from a flask when the mouth is open. *B* : The mouth of the flask is closed, and the air within the flask becomes saturated with water-vapour molecules.

This motion is counteracted, in some degree, by intermolecular forces, or forces of cohesion, between the water molecules. However, some molecules, moving with greater velocity than the average, possess sufficient kinetic energy to overcome the attraction between molecules and escape, or *evaporate*, into the air of the flask. At the same time, water molecules in the air, likewise in a state of active motion, strike the water surface and become a part of the liquid water. This is condensation, the reverse of evaporation. If the rate at which molecules leave the water is greater than the rate at which they enter it, the volume of the liquid is reduced.

If the mouth of the flask remains closed (Fig. 10.7B), the concentration of molecules of water in the air shortly becomes so great that they will strike the water surface as frequently as other molecules escape from the liquid. The air in the flask is then saturated with water vapour. Since the rate of entering and leaving has reached a state of equilibrium, the volume of the liquid undergoes no further change.

If the mouth of the flask remains open, however (Fig. 10.7A), an interchange of molecules will go on between the interior of the flask and the outer air. A greater number of water molecules will diffuse out of the flask than enter during any period of time, since such molecules are in greater concentration within the flask than in the air outside. This diffusion reduces the concentration of water molecules inside the flask and hence reduces the rate of condensation. Since evaporation continues at a constant rate, the volume of liquid in the flask will decrease until no more remains. On the other hand, if the air outside the flask is saturated with water vapour, the number of molecules entering the flask will equal the number leaving, and the volume of the liquid in the flask will not decrease.

We can now understand why clothes dry very slowly on a humid day, when the air is close to its saturation point. The rate at which water vapour molecules escape from the surface of the clothes is the same as though the air were dry at the same temperature. But since the air is close to saturation, with many water molecules in the air, the rate of condensation is greater than on a dry day, and so the clothes dry slowly.

The loss of water from plant cells into the air is determined by the same factors. This loss occurs chiefly from the wet surfaces of cells within the leaf. The rate at which it proceeds depends upon the ratio between the rate at which molecules of water strike this water surface and the rate at which they leave such a surface. Environmental factors which affect this ratio are discussed later.

CAPILLARITY

Capillarity, or capillary movement, accounts for the upward movement of water in a tube of small diameter. The height to which water will rise depends upon the diameter of the tube. The smaller the diameter, the higher the thread of water rises against the pull of gravity.

Capillarity has frequently been suggested as the mechanism by which water rises in trees, but this explanation may be discarded, because there is no evidence to support it. Nevertheless, certain of the forces involved in capillary movement are involved in the accepted explanation of the ascent of water. Chief among these is the attraction of water molecules for one another. Such intermolecular attraction is so great that water molecules possess a high degree of *cohesion*. This cohesion resists stresses tending to separate the molecules. Cohesion is particularly evident when water is confined in thread-like columns. In addition to the attractive forces of water, another important factor in capillarity is *adhesion*, the attraction of water molecules to substances that are wetted by water.

The surface of the water partly filling a capillary tube (Fig. 10.8) is concave. It is termed a *meniscus*. The sides of the tube above the meniscus are coated with a thin film of water, the result of the condensation from the water vapour of the air. Because the water on the margin of the meniscus is continuous with the water film lining the tube, the total surface of the meniscus is very much greater than is apparent to the eye. This surface, like any other water surface, is under the influence of forces which cause it to contract to the smallest possible area. There is sufficient attraction (adhesion) between the film of water molecules and the sides of the tube to offset the pull of gravity. At the same time the contraction of the meniscus results in a pull at the top of the water column. Because of cohesive forces between the molecules within the liquid, this pull, or stress, is transmitted downward. As a result, the water column moves upward against the pull of gravity.

Capillarity is responsible for the movement of water for short distances through the soil. It may also be involved in the process of imbibition, which is of importance in plants. Aside from these, capillarity is not of great significance in plant processes.

IMBIBITION

In addition to osmosis, the process of *imbibition* is involved in the movement of water into the plant cell. This term is applied to the absorption of water by solid colloidal materials. This inward movement of water is accompanied by a swelling of the solid and an increase in weight. Imbibition results in the swelling of starch and gelatin when water is added. The increase in volume of seeds in water or moist soid, and the swelling of dry fruits in water, are both due to imbibition. The swelling and warping of damp wood is caused by the imbibition of water by the cellulose. This explains why doors and windows stick in damp weather. When a wooden boat is put into water there is usually some leakage at the seams. But

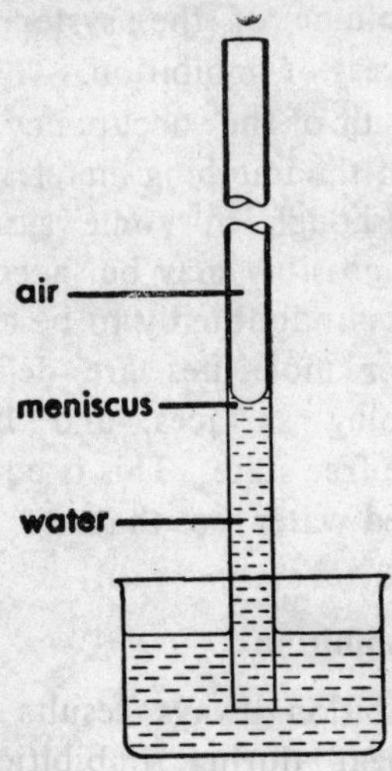

Fig. 10.8. Rise of water in a capillary tube.

after soaking for a few days, the swelling of the planks, due to imbibition, is usually sufficient to make the hull watertight.

Mature, living cells function as an osmotic system in which water moves from the vacuole of one cell to the vacuole of the next through cell walls and cytoplasmic membranes. This movement can occur, however, only if the cell wall and cytoplasmic membranes are already wet with imbibed water. Imbibition in living cells thus precedes osmosis, and water is continuous from one cell to another even though water movement is not proceeding. Imbibition is especially significant in meristematic cells, which characteristically have small, numerous vacuoles. The concentration of colloidal materials in such vacuoles usually permits them to imbibe water from the cytoplasm, which, in turn, imbibes water from the cell wall. Increase in size of young cells is, therefore, more the result of imbibition than of osmosis.

Volume Changes in Imbibition

The volume of the *imbibant* always increases during imbibition. The final volume of the entire system (liquid+imbibant) is always less, however, than the sum of the initial volumes of the liquid and the imbibing substance. In other words, a contraction in the volume of the system, liquid+imbibant, occurs during the process of imbibition. This volume shrinkage is not principally a result of the occupancy of minute spaces within the substance of the imbibing material by the liquid, as might be surmised, although in some systems a part of the contraction in volume possibly may be accounted for in this way. The explanation is undoubtedly to be ascribed to the fact that the adsorbed water molecules are definitely oriented in relation to the adsorbing surfaces, and hence occupy less space than when in the free state. This is equivalent to a compression of the adsorbed water so that its density is greater than that of free water.

Energy Relations of Imbibition

The process of imbibition always results in the release of heat. Liberation of heat during imbibition may be easily detected by allowing dry starch or some other material with a high imbibitional capacity to imbibe water while contained in a calorimeter, and noting the change in temperature (Table 10.2).

Table 10.2. Heat of Imbibition of Dry Starch (Data of Rodewald, 1897).

Percent water imbibed	*Heat evolved, g-cal-per g. starch*
0.23	28.11
2.39	22.60
6.27	15.17
11.65	8.43
15.68	5.21
19.52	2.91

The adsorption of water molecules which occurs when they are imbibed results in a loss of a large part of their kinetic energy, which reappears in the system as heat energy. The essential energy change in the process of imbibition is a loss of kinetic energy by the adsorbed molecules, and its transference to the other molecules of the system. Their increased kinetic activity is the cause of the observed rise in temperature. As shown in Table 10.2, the greatest evolution of heat accompanies the initial stages of imbibition. This is to be expected, as a larger proportion of the first molecules of water imbibed would be firmly adsorbed than of those which enter the imbibant later.

The Effect of Temperature on Imbibition

As shown in Fig. 10.9 the rate of imbibition increases with rise in temperature.

Osmotic Effects on Imbibition

Water moves by imbibition into a substance only when its diffusion pressure exceeds the diffusion pressure of the water in the imbibant. The introduction of a solute into water invariably has the effect, as already discussed, of reducing the diffusion pressure of the water in the resulting solution as compared with pure water. The osmotic pressure of a solution is a measure of its diffusion-pressure deficit, insofar as it results from the presence of solutes. Hence the magnitude of the osmotic pressure influences both the rate of imbibition and the equilibrium water content of an imbibant.

Shull (1916) used the seeds of cocklebur (*Xanthium pennsylvanicum*) as the imbibant in a comprehensive study of osmotic effects on imbibition. Samples of seeds of this species were allowed to come to equilibrium with solutions of sodium chloride and lithium chloride of different osmotic pressures, after which the amount of water which had been imbibed was determined. (Table 10.3). The coats of these seeds are permeable to water, but virtually impermeable to these solutes.

The general principle shown by the results of these experiments is that, with increase in the osmotic pressure of the solution in which the seeds are immersed, the amount of water

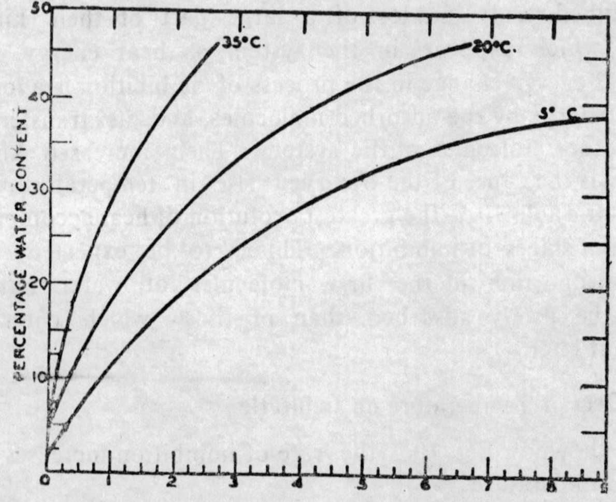

Fig. 10.9. Relation between temperature and imbibition of water by *Xanthium* seeds. Data of Shull (1920).

held by imbibition per unit of dry weight at the equilibrium point decreases. Since at equilibrium the diffusion pressures of the water in the imbibing substance and in the surrounding liquid must be equal, the basis for this osmotic effect upon imbibition is evident. If the diffusion pressure in the solution is relatively low (high osmotic pressure), the diffusion of less water into the imbibing substance is necessary to raise the diffusion pressure of the water in it to a value equal to the diffusion pressure of the water in the circumambient liquid than if the diffusion pressure in the surrounding liquid is hight. The relation between the osmotic pressure and the amount of water imbibed is not, however, a strictly proportional one. At the lower end of the range of concentrations employed, an increase of a few atmospheres in osmotic pressure causes a decrease in the amount of water imbibed, in terms of the air dry weight of the seeds, of nearly 15 per cent. Near the upper end of this range of concentrations an increase in osmotic pressure of several hundred atmospheres is required for an equivalent change in the volume of water imbibed. This is further evidence that the first increments of water passing into any imbibant are

Table 10.3. Imbibition of water by Cocklebur seeds immersed in solutions of different osmotic pressures.

Volume molar concentration of solutions	*Osmotic pressure of solutions, atmos.*	*Water imbibed by seeds at equilibrium (48 hours). Per cent of air dry weight*
H_2O	0.0	51.58
0.1 MNaCl	3.8	46.33
0.2	7.6	45.52
0.3	11.4	42.05
0.4	15.2	40.27
0.5	19.0	38.98
0.6	22.8	35.18
0.7	26.6	32.85
0.8	30.4	31.12
0.9	34.2	29.79
1.0	38.0	26.73
2.0	72.0	18.55
4.0	130.0	11.76
Sat. NaCl	375.0	6 35
Sat. LiCl	965.0	—0.29

held by tremendously greater forces than those which are imbibed subsequently.

Imbibition Pressure

Pressures, sometimes of an enormous magnitude, are often exerted by a swelling imbibant if it is confined in some manner. Blocks of rock, for example, have been successfully quarried by drilling holes along the desired cleavage planes, driving tight-fitting wooden stakes into the holes, and pouring water on the stakes. The pressure developed upon imbibition of the water by the wood is sufficient to split the rock.

The term "imbibition pressure" is best employed, however, not as a designation for the actual pressures which develop during imbibition, but with a meaning analogous to that of "osmotic pressure," as an index of certain properties of the imbibant. Used in this sense imbibition pressure is an index of (1) the potential maximum pressure which can develop in an imbibant as a result of imbibition, and (2) the diffusion-pressure deficit of the water in an imbibant as long as its free expansion is not impeded in any way. In a solution, the greater amount of water present in proportion to a given amount of solute, the less the osmotic pressure. Likewise, in an imbibant, the greater the amount of water present in proportion to a given amount of imbibing substance, the less the imbibition pressure. Air-dry pea seeds, for example, have a greater imbibition pressure than half-swollen pea seeds ; pea seeds which have reached a condition of maximum swelling have an imbibition pressure of zero.

The *actual* pressures which develop as a result of imbibition, like those which develop as a result of osmosis, can logically be termed *turgor pressures*. No turgor pressure develops in an unconfined imbibant any more than it does in an unconfined solution. Only when complete swelling is prevented, as by enclosure within an inelastic wall, does a turgor pressure develop within an imbibant.

Quantitative Aspects of Imbibition

The diffusion-pressure deficit of the water in an imbibant is equal to the imbibition pressure less the turgor pressure just as in an osmotic system the diffusion-pressure deficit of the water is equal to the osmotic pressure less the turgor pressure. This relation is expressed in the following equation :

$$D.P.D. = I.P. - T.P.$$

Since no turgor pressure develops in an unconfined imbibant, for such systems this equation simplifies to :

$$D.P.D. = I.P.$$

If an imbibant with an imbibition pressure of 100 atm. is immersed in pure water, the initial diffusion-pressure deficit of the water in the imbibant is 100 atm. ; when equilibrium has been attained, the imbibition pressure and diffusion-pressure deficit of the water in the imbibant have both become zero. If such an imbibant is immersed in a solution with an osmotic pressure of 20 atm., when equilibrium has been attained both the imbibition pressure and the diffusion-pressure deficit of the water in the imbibant will be 20 atm. In unconfined imbibants equilibrium of the diffusion-pressure deficits of the water in them with the water in outside solutions occurs solely by adjustments in imbibition pressures.

If an imbibant with an initial imbibition pressure of 100 atm. is enclosed in an inclastic. water permeable wall, and immersed in pure water the diffusion-pressure deficit of the water in the imbibant at equilibrium will be zero, not because of a decrease in imbibition pressure but because of the development of a turgor pressure of 100 atm. If a solution of 20 atm. osmotic pressure is substituted for the pure water outside of the imbibing system, then the diffusion-pressure deficit of the water in the imbibant at equilibrium will be 20 atm., again not because of a decrease in imbibition pressure, but because of the development of a turgor pressure of 80 atm.

Measurement of Imbibition Pressures

The magnitudes of imbibition pressures have been measured in two different ways. One method has been to counterbalance the pressure developed during imbibition by means of a mechanical pressure. Such a method was used by Reinke as long ago as 1879. He stacked disks of dried fronds of *Laminaria* (a sea weed) in a hollow metal cylinder, and inserted above the disks a metal piston bearing a platform at its upper end. Water was then brought into contact with the dry disks. By placing on the platform weights of just sufficient mass to prevent

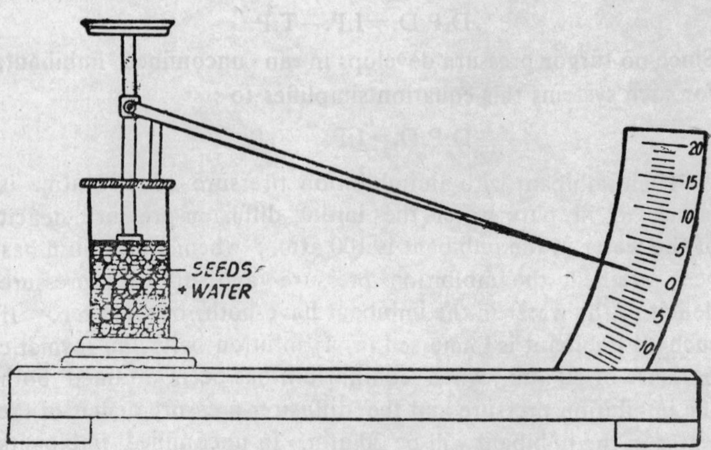

Fig. 10.10. Apparatus to measure the imbibitional pressure.

swelling of the kelp, the magnitude of the pressure developed was measured. By this procedure there is measured, in effect, the magnitude of the turgor pressure which is equal to the imbibition pressure, *i.e.*, which keeps the diffusion-pressure deficit of the water in the imbibant at zero and thus prevents any movement of water into the imbibant.

Imbibition pressure can also be measured by an osmotic method. The necessary procedure is to find a solution in which the diffusion-pressure deficit (osmotic pressure) is just great enough to prevent any movement of water into the imbibant. The osmotic pressure of such a solution is equal to the imbibition pressure of the imbibant. This method, in effect, results in the determination of the diffusion-pressure deficit of the water in the imbibant under such conditions that it is equal to the imbibition pressure, *i.e.*, when there is no turgor pressure.

Imbibition pressures of air-dry seeds may be as high as approximately 1000 atm (Table 10.3).